Geometry of Pseudo-Finsler Submanifolds

Geometry of Pseudo-Finsler Submanifolds

by

Aurel Bejancu

and

Hani Reda Farran

Department of Mathematics and Computer Science,
Kuwait University, Kuwait

KLUWER ACADEMIC PUBLISHERS
DORDRECHT / BOSTON / LONDON

A C.I.P. Catalogue record for this book is available from the Library of Congress.

ISBN 978-90-481-5601-6

Published by Kluwer Academic Publishers,
P.O. Box 17, 3300 AA Dordrecht, The Netherlands.

Sold and distributed in North, Central and South America
by Kluwer Academic Publishers,
101 Philip Drive, Norwell, MA 02061, U.S.A.

In all other countries, sold and distributed
by Kluwer Academic Publishers,
P.O. Box 322, 3300 AH Dordrecht, The Netherlands.

Printed on acid-free paper

CONTENTS

PREFACE

Finsler geometry is the most natural generalization of Riemannian geometry. It started in 1918 when P. Finsler [1] wrote his thesis on curves and surfaces in what he called generalized metric spaces. Studying the geometry of those spaces (which where named Finsler spaces or Finsler manifolds) became an area of active research. Many important results on the subject have been brought together in several monographs (cf., H. Rund [3], G. Asanov [1], M. Matsumoto [6], A. Bejancu [8], P.L. Antonelli, R.S. Ingarden and M. Matsumoto [1], M. Abate and G. Patrizio [1] and R. Miron [3]).

However, the present book is the first in the literature that is entirely devoted to studying the geometry of submanifolds of a Finsler manifold. Our exposition is also different in many other respects. For example, we work on pseudo–Finsler manifolds where in general the Finsler metric is only non-degenerate (rather than on the particular case of Finsler manifolds where the metric is positive definite). This is absolutely necessary for physical and biological applications of the subject. Secondly, we combine in our study both the classical coordinate approach and the modern coordinate–free approach. Thirdly, our pseudo–Finsler manifolds $\mathbb{F}^m = (M, M', F^*)$ are such that the geometric objects under study are defined on an open submanifold M' of the tangent bundle TM, where M' need not be equal to the entire $TM^0 = TM \backslash \theta(M)$. Finally, our techniques rely heavily on the concept of vectorial Finsler connection (cf., A. Bejancu [1], [4]). The vertical vector bundle VM' of M' is used in this respect. One important strength of this technique is that it enables us to handle simultaneously and easily the geometrical objects induced by the classical Cartan, Berwald, and Rund connections.

We have been very careful to give all the necessary definitions and background to make the book as self–contained as possible. Rigorous proofs are given for almost all the results. Only lengthy calculations have been omitted to avoid possible distractions from the main ideas. We cover a wide area of theorems on the subject ranging from basic propositions to the latest results, including some results of the authors that appear here for the first time. This makes the book suitable for all researchers working on pseudo–Finsler geometry in general and on pseudo–Finsler submanifolds in particular. It is our hope that students and researchers wishing to be introduced to the subject will find in this book a great deal of help.

In the first chapter we present our approach to the geometry of pseudo–Finsler manifolds using the vertical vector bundle. Thus the whole Finslerian tensor calculus on $\mathbb{F}^m$ is developed by using cross–sections of VM'

(Finsler vector fields) and of V^*M' (Finsler covector fields). This enables us to give the main concepts and results we need in developing the theory of pseudo–Finsler submanifolds. In this chapter we introduce (for the first time) the notion of generalized Landsberg manifolds. We also give a new setting for pseudo–Finsler manifolds of constant curvature (cf., A. Bejancu and H.R. Farran [3], [4]).

A general theory of pseudo–Finsler submanifolds is developed in Chapter 2. We consider an arbitrary Finsler connection on the ambient pseudo–Finsler manifold and by using some vectorial Finsler connections and the induced Finsler connection we obtain all the Gauss–Codazzi–Ricci equations of a pseudo–Finsler submanifold. We pay special attention to the Cartan, Berwald, and Rund connections on the ambient manifold. The main difference between the theory of pseudo–Finsler submanifolds and theory of pseudo–Riemannian submanifolds is that in the former the induced Finsler connection, in general, does not coincide with the intrinsic Finsler connection. For this reason we present a comparison between the induced and intrinsic geometric objects on a pseudo–Finsler submanifold with respect to the above classical Finsler connections.

Some special pseudo–Finsler immersions are studied in Chapter 3. We present characterizations of totally geodesic pseudo–Finlser submanifolds and show that, in general, they inherit the main properties of the ambient manifold. Also, we study pseudo–Finsler immersions with parallel vertical bundle and parallel pseudo–Finsler normal bundle. Finally, we find a large class of proper Finsler submanifolds of a C–reducible Finsler manifold on which the induced and intrinsic Finsler connections coincide.

Chapters 4 and 5 are devoted to the study of some particular pseudo–Finsler submanifolds with respect to their codimension. Thus in Chapter 4 we study the geometry of curves in a Finsler manifold. Here we construct a Frenet frame and obtain the corresponding Frenet equations. This enables us to state a Fundamental Theorem for curves in Finsler manifolds and to give theorems on the reduction of the codimension of a Finsler immersion. The general theory developed in Chapters 2 and 3 is applied in Chapter 5 to the study of pseudo–Finsler hypersurfaces. When the ambient manifold is a pseudo–Minkowski space we stress the major role of the Rund connection in our study. In particular, we show that totally umbilical pseudo–Finsler hypersurfaces are either totally geodesic or pseudo–Riemannian. The last section of the chapter is devoted to a Minkowskian approach to the theory of pseudo–Finsler hypersurfaces. In this case all the induced geometric objects and structure equations live on the base manifold, since the Minkowskian unit normal depends on position only. At this moment it is not clear whether this approach (completely different from the one we developed throughout the book) can be extended to the study of pseudo–Finsler submanifolds of arbitrary codimension.

In the last chapter, using the geometry of Finsler manifolds of dimensions 2 and 3, we study Finsler surfaces. When the ambient space is a Minkowski space we state a *Theorema Egregium*, provided that the induced and intrinsic connections coincide on the surface. Finally, we show that proper Finsler surfaces with vanishing mean curvature must be open subsets of a Minkowski plane.

In concluding this preface, we would like to thank Kluwer Academic Publishers for their great patience and kind cooperation. Our thanks are also due to Kuwait University for inviting Professor A. Bejancu as a visiting professor to the Department of Mathematics and Computer Science during the academic year 1998/1999. This was absolutely necessary for the completion of the book. Finally, we would like to thank Mr. Muhammad Shahid Rafiqui for the excellent job of typing the entire manuscript.

Kuwait, June 30, 2000 A. Bejancu and H.R. Farran

CHAPTER 1

PSEUDO–FINSLER MANIFOLDS

This chapter reviews the main topics of the geometry of pseudo–Finsler manifolds. It gives the basic background needed in the rest of the book. Here, we use the vertical vector bundle to develop a Finsler tensorial calculus. We define the main geometric objects necessary for the study of Finsler geometry. In particular, we describe in detail the Berwald, Cartan, and Rund connections. Finally, we give a new setting for pseudo–Finsler manifolds of constant curvature and present some special pseudo–Finsler manifolds.

1. Definitions and Examples

Let M be a real m–dimensional smooth manifold and TM be the tangent bundle of M (cf., O. Neill [1], p. 26). A coordinate system in M is denoted by $\{(\mathcal{U},\varphi); x^1,\cdots,x^m\}$ or briefly $\{(\mathcal{U},\varphi); x^i\}$, where $\mathcal{U}$ is an open subset of M, $\varphi : \mathcal{U} \longrightarrow \mathbb{R}^m$ is a diffeomorphism of $\mathcal{U}$ onto $\varphi(\mathcal{U})$, and $(x^1,\cdots,x^m) = \varphi(x)$ for any $x \in \mathcal{U}$. Denote by π the canonical projection of TM on M and by T_xM the fibre at $x \in M$, i.e., $T_xM = \pi^{-1}(x)$. The coordinate system $\{(\mathcal{U},\varphi); x^i\}$ in M defines a coordinate system $\{(\mathcal{U}^*,\Phi); x^1,\cdots,x^m, y^1,\cdots,y^m\} = \{(\mathcal{U}^*,\Phi); x^i, y^i\}$ in TM, where $\mathcal{U}^* = \pi^{-1}(\mathcal{U})$ and $\Phi : \mathcal{U}^* \longrightarrow \mathbb{R}^{2m}$ is a diffeomorphism of $\mathcal{U}^*$ on $\varphi(\mathcal{U}) \times \mathbb{R}^m$, and $(x^1,\cdots,x^m,y^1,\cdots,y^m) = \Phi(y_x)$, for any $x \in \mathcal{U}$ and $y_x \in T_xM$. For short we denote by (x,y) the coordinates of y_x.

Next, consider another coordinate system $\{(\tilde{\mathcal{U}},\tilde{\varphi}); \tilde{x}^i\}$ in M such that $\mathcal{U} \cap \tilde{\mathcal{U}} \neq \phi$. Then the local coordinates (x,y) and $(\tilde{x},\tilde{y})$ on TM are related by

$$\begin{cases} \tilde{x}^i = \tilde{x}^i(x^1,\cdots,x^m), \\ \tilde{y}^i = B^i_j(x) y^j, \end{cases} \tag{1.1}$$

where we set $B^i_j(x) = \partial\tilde{x}^i/\partial x^j$. As a consequence of (1.1) the local frame fields $\{\partial/\partial x^i, \partial/\partial y^i\}$ and $\{\partial/\partial\tilde{x}^i, \partial/\partial\tilde{y}^i\}$ satisfy

$$\frac{\partial}{\partial x^i} = B^j_i(x)\frac{\partial}{\partial\tilde{x}^j} + B^j_{ik}(x)y^k\frac{\partial}{\partial\tilde{y}^j}; \qquad B^j_{ik}(x) = \frac{\partial^2\tilde{x}^j}{\partial x^i\partial x^k}, \tag{1.2}$$

and

$$\frac{\partial}{\partial y^i} = B^j_i(x)\frac{\partial}{\partial\tilde{y}^j}. \tag{1.3}$$

Analogously we deduce that the local coframe fields $\{dx^i, dy^i\}$ and $\{d\tilde{x}^i, d\tilde{y}^i\}$ on the cotangent bundle T^*M of M are related by

$$d\tilde{x}^i = B^i_j(x)dx^j, \tag{1.4}$$

and

$$d\tilde{y}^i = B^i_{jk}(x)y^j dx^k + B^i_j(x)dy^j. \tag{1.5}$$

Let M' be a non–empty open submanifold of TM such that $\pi(M') = M$ and $\theta(M) \cap M' = \phi$, where θ is the zero section of TM. Suppose that $M'_x = T_xM \cap M'$ is a **positive conic set**, i.e., for any $k > 0$ and $y \in M'_x$ we have $ky \in M'_x$. Clearly, the largest M' satisfying the above conditions is $TM^0 = TM\backslash\theta(M)$, which is usually taken for the definition of a Finsler manifold. *In the present book we do not assume that M' is necessarily TM^0.*

We now consider a smooth function $F : M' \longrightarrow (0,\infty)$ and take $F^* = F^2$. Then suppose that for any coordinate system $\{(\mathcal{U}', \Phi'); x^i, y^i\}$ in M', the following conditions are fulfilled:

($\boldsymbol{F_1}$) *F is positively homogeneous of degree one with respect to $(y^1, \cdots, y^m)$, i.e., we have*

$$F(x^1, \cdots, x^m, ky^1, \cdots, ky^m) = kF(x^1, \cdots, x^m,\ y^1, \cdots, y^m), \tag{1.6}$$

for any $(x,y) \in \Phi'(\mathcal{U}')$ and any $k > 0$.

($\boldsymbol{F_2}$) *At any point $(x,y) \in \Phi'(\mathcal{U}')$*

$$g_{ij}(x,y) = \frac{1}{2}\frac{\partial^2 F^*}{\partial y^i\partial y^j}(x,y), \quad i,j \in \{1,\cdots,m\}, \tag{1.7}$$

are the components of a positive definite quadratic form on $\mathbb{R}^m$.

We say that the triple $\mathbb{F}^m = (M, M', F)$, with F satisfying ($\boldsymbol{F_1}$) and ($\boldsymbol{F_2}$), is a **Finsler manifold**, and F is the **fundamental function** of $\mathbb{F}^m$.

Certainly, condition ($\boldsymbol{F_2}$) is not appropriate for some applications of Finsler geometry. To remove this inconvenience we consider a positive integer $q < m$ and a smooth function $F^* : M' \longrightarrow \mathbb{R}$, where M' is as above.

Moreover, suppose that for any coordinate system $\{(\mathcal{U}', \Phi') : x^i, y^i\}$ in M', the following conditions are fulfilled:

(F_1^*) *F^* is positively homogeneous of degree two with respect to* $(y^1, \cdots, y^m)$, i.e., *we have*

$$F^*(x^1, \cdots, x^m, \; ky^1, \cdots, ky^m) = k^2 F^*(x^1, \cdots, x^m, \; y^1, \cdots, y^m), \qquad (1.8)$$

for any point $(x, y) \in \Phi'(\mathcal{U}')$ *and* $k > 0$.

(F_2^*) *At any point* $(x, y) \in \Phi'(\mathcal{U}')$, $g_{ij}(x, y)$ *defined by the same formula* (1.7) *are the components of a quadratic form on* $\mathbb{R}^m$ *with* q *negative eigenvalues and* $m - q$ *positive eigenvalues,* $0 < q < m$.

In this case we say that $\mathbb{F}^m = (M, M', F^*)$ is an **indefinite Finsler manifold** of index q. If, in particular, $q = 1$ we say that $\mathbb{F}^m$ is a **Finsler manifold of Lorentzian signature.** The **fundamental function** for an indefinite Finsler manifold $\mathbb{F}^m = (M, M', F^*)$ is the function $F : M' \longrightarrow [0, \infty)$ locally given by

$$F(x, y) = |F^*(x, y)|^{\frac{1}{2}}, \; \forall (x, y) \in \Phi'(\mathcal{U}'). \qquad (1.9)$$

As is well known, since 1918, when Finsler [1] published his thesis on curves and surfaces in the so called generalized spaces, many papers have been devoted to Finsler manifolds (see references in Rund [3], Matsumoto [6] and Asanov [1]). In contrast, the indefinite Finsler manifolds have been studied by few people (see Beem [1]–[4], Bejancu [9], Bejancu - Deshmukh [1], Bejancu–Farran [1], Miron [1], Teodorescu [1]). As most of the results we present in the book are true for both Finsler manifolds and indefinite Finsler manifolds, by a **pseudo–Finsler manifold** $\mathbb{F}^m = (M, M', F^*)$ we shall understand that $\mathbb{F}^m$ is either a Finsler manifold or an indefinite Finsler manifold. Thus by combining the above definitions we may think of a pseudo–Finsler manifold as a triple $\mathbb{F}^m = (M, M', F^*)$ satisfying (F_1^*) and (F_2^*) with $0 \leq q < m$.

Using mainly the homogeneity properties of both F and F^*, we shall derive some useful identities relating these functions and their partial derivatives. First, taking into account (F_1^*) and (1.9) we see that the fundamental function F of a pseudo–Finsler manifold $\mathbb{F}^m$ is positively homogeneous of degree one, i.e., F satisfies (1.6). Then we differentiate (1.6) with respect to k and by setting $k = 1$ we obtain

$$y^i \frac{\partial F}{\partial y^i} = F, \qquad (1.10)$$

which implies

$$y^i \frac{\partial^2 F}{\partial y^i \partial y^j} = 0. \qquad (1.11)$$

In case $\mathbb{F}^m$ is a Finsler manifold, replace F^* by F^2 in (1.7) and obtain

$$g_{ij} = F\frac{\partial^2 F}{\partial y^i \partial y^j} + \frac{\partial F}{\partial y^i}\frac{\partial F}{\partial y^j}. \tag{1.12}$$

Using (1.10)–(1.12) we deduce that

$$g_{ij}y^j = F\frac{\partial F}{\partial y^i}, \tag{1.13}$$

and

$$g_{ij}y^i y^j = F^*. \tag{1.14}$$

Note that (1.14) holds also for an indefinite Finsler manifold. Indeed, we first differentiate (1.8) with respect to k and, by setting $k = 1$, we infer that

$$y^i\frac{\partial F^*}{\partial y^i} = 2F^*. \tag{1.15}$$

Then using (1.7) and differentiating (1.15) with respect to y^j we derive

$$2y^i g_{ij} = y^i\frac{\partial^2 F^*}{\partial y^i \partial y^j} = \frac{\partial F^*}{\partial y^j}, \tag{1.16}$$

which together with (1.15) implies (1.14). Finally, differentiating the last equality in (1.16) with respect to y^k we obtain

$$y^i\frac{\partial^3 F^*}{\partial y^i \partial y^j \partial y^k} = 0,$$

which yields

$$\frac{\partial g_{ij}}{\partial y^k}(x,y)y^i = 0; \qquad \frac{\partial g_{ij}}{\partial y^k}(x,y)y^j = 0\ ; \qquad \frac{\partial g_{ij}}{\partial y^k}(x,y)y^k = 0, \tag{1.17}$$

on any pseudo–Finsler manifold $\mathbb{F}^m$.

As a direct consequence of (F_2) and (F_2^*) we deduce that

$$\det[g_{ij}(x,y)] \neq 0, \quad \forall (x,y) \in \Phi'(\mathcal{U}'), \tag{1.18}$$

that is, $[g_{ij}(x,y)]$ is an invertible $m \times m$ matrix. On the contrary, from (1.11) we infer that

$$\det\left[\frac{\partial^2 F}{\partial y^i\, \partial y^j}\right] = 0,$$

which, together with (1.12), imply that

$$\det\left[g_{ij} - \frac{\partial F}{\partial y^i}\frac{\partial F}{\partial y^j}\right] = 0. \tag{1.19}$$

Now we consider an open positive conic subset D of $\mathbb{R}^m$ and a smooth real function f on D. Since most of the local components of the tensor fields with which we are dealing throughout the book are positively homogeneous functions, we recall the following general definition. We say that f is positively homogeneous of degree r if we have

$$f(ky^1, \cdots, ky^m) = k^r f(y^1, \cdots, y^m), \tag{1.20}$$

for any $k > 0$ and $(y^1, \cdots, y^m) \in D$. In this case it follows that $\partial f/\partial y^i$, $i \in \{1, \cdots m\}$ are positively homogeneous of degree $r - 1$. Also, we recall the following well known theorem.

THEOREM 1.1 (Euler's Theorem) *A smooth function f on D is positively homogeneous of degree r if and only if it satisfies the condition*

$$y^i \frac{\partial f}{\partial y^i} = rf. \tag{1.21}$$

Based on the above assertions about positively homogeneous functions, we may state the following remark.

Remark 1.1. (i) $\partial F^*/\partial y^i$ are positively homogeneous of degree one with respect to $(y^1, \cdots y^m)$.
(ii) $\partial F/\partial y^i$ and g_{ij} are positively homogeneous of degree zero with respect to $(y^1, \cdots y^m)$.
(iii) The equations (1.10), (1.11), (1.15) and (1.17) can also be deduced by using (1.21) and the homogeneity of functions involved therein. ■

Next, we shall introduce three hypersurfaces in each tangent space of a pseudo–Finsler manifold $\mathbb{F}^m = (M, M', F^*)$. More precisely, at any $x \in M$ consider $M'_x = T_xM \cap M'$ and define

$$IM_x^+ = \{y \in M'_x;\ F^*(x,y) = 1\}\,, \qquad IM_x^- = \{y \in M'_x\,;\ F^*(x,y) = -1\}\,,$$

and

$$\Lambda M_x = \{y \in M'_x\,;\ F^*(x,y) = 0\}\,.$$

Thus IM_x^+, IM_x^- and ΛM_x are hypersurfaces in T_xM which we call the **positive indicatrix**, the **negative indicatrix**, and the **null (lightlike) Finsler cone** at x, respectively. Clearly, IM_x^- and Λ_x are defined only for indefinite Finsler manifolds. If, in particular, $\mathbb{F}^m$ is a Riemannian manifold then there exists only IM_x^+, which is a unit sphere. When $\mathbb{F}^m$ is a pseudo–Riemannian manifold of index $0 < q < m$, IM_x^+, IM_x^-, and ΛM_x are the unit pseudo–sphere, the unit pseudo–hyperbolic space and the null (lightlike) cone, respectively (cf., O'Neill [1], p. 110).

As far as we know, the geometry of the bundles

$$IM^+ = \bigcup_{x\in M} IM_x^+;\ IM^- = \bigcup_{x\in M} IM_x^-;\ \Lambda M = \bigcup_{x\in M} \Lambda M_x,$$

has not been investigated in the case of a pseudo–Finsler manifold. When $\mathbb{F}^m$ is a Riemannian manifold IM^+ is just the sphere bundle over M, and basic results about its geometry can be found in Blair [1], pp. 131–138. Also some interesting differential geometric properties of the indicatrix bundle over a Finsler manifold can be found in Matsumoto [3].

We close the section with some examples of pseudo–Finsler manifolds.

Example 1.1. Let M be a real m–dimensional pseudo–Riemannian manifold endowed with a pseudo–Riemannian metric $g = (g_{ij})$ of index $0 \leq q < m$, (cf., O'Neill [1], p. 54). It is easy to see that $\mathbb{F}^m = (M, M', F^*)$, where $M' = TM^0$ and $F^*(x, y) = g_{ij}(x)y^iy^j$, is a pseudo–Finsler manifold. It is important to note that this example is the only one with the property that F^* is smooth on the whole tangent bundle TM (see Lemma 1.4.1 in Abate–Patrizio [1]). ■

Any pseudo–Finsler manifold that is not pseudo–Riamanian is called a **proper pseudo–Finsler manifold**.

Next, we consider an open set M'' of $\mathbb{R}^m \backslash \{0\}$ and construct the open set $M' = \mathbb{R}^m \times M''$ of $T\mathbb{R}^m \cong \mathbb{R}^{2m}$. Suppose there exists a smooth function $F : M' \longrightarrow (0, \infty)$ which satisfies $(\boldsymbol{F}_1)$ and $(\boldsymbol{F}_2)$ and depends on $(y^1, \cdots, y^m) \in M''$ alone. Then the Finsler manifold $\mathbb{F}^m = (\mathbb{R}^m, M', F)$ is called a **Minkowski space**. In case there exists a smooth function $F^* : M' \longrightarrow \mathbb{R}$ satisfying $(\boldsymbol{F}_1^*)$ and $(\boldsymbol{F}_2^*)$ and depending on $(y^1, \cdots, y^m) \in M''$ alone, we say that $\mathbb{F}^m = (\mathbb{R}^m, M', F^*)$ is an **indefinite Minkowski space** of index $0 < q < m$. Combining these definitions we may say that a **pseudo–Minkowski space** is a triple $\mathbb{F}^m = (\mathbb{R}^m, M', F^*)$ satisfying $(\boldsymbol{F}_1^*)$ and $(\boldsymbol{F}_2^*)$ with $0 \leq q < m$.

We recall that the Euclidean structure and the pseudo–Euclidean structure of index $0 < q < m$ on $\mathbb{R}^m$ are defined by

$$F(x, y) = \left(\sum_{i=1}^{m} (y^i)^2\right)^{\frac{1}{2}}, \tag{1.22}$$

and

$$F^*(x, y) = -\sum_{i=1}^{q} (y^i)^2 + \sum_{a=q+1}^{m} (y^a)^2, \tag{1.23}$$

respectively. Then it is easy to see that $(\mathbb{R}^m, \mathbb{R}^m \backslash \{0\}, F)$ is a Minkowski space, while $(\mathbb{R}^m, \mathbb{R}^m \backslash \{0\}, F^*)$ is an indefinite Minkowski space. Next, we present some examples of pseudo–Minkowski spaces whose functions F and F^* do not come necessarily from a scalar product. Any pseudo–Minkowski space that is not pseudo–Euclidean (Euclidean) is said to be a **proper pseudo–Minkowski space.**

Example 1.2. Let $M'' = \{(y^1, y^2, y^3) \in \mathbb{R}^3;\ y^1 > 0,\ y^2 > 0,\ y^3 > 0\}$ and $M' = \mathbb{R}^3 \times M''$. Define $F : M' \longrightarrow (0, \infty)$ by

$$F(x, y) = \frac{(y^1)^2}{\sqrt{y^2 y^3}}. \tag{1.24}$$

Then by straightforward calculations using (1.7) and (1.24) we obtian

$$[g_{ij}(x, y)] = \frac{(y^1)^2}{y^2 y^3} \begin{bmatrix} 6 & -2\dfrac{y^1}{y^2} & -2\dfrac{y^1}{y^3} \\ -2\dfrac{y^1}{y^2} & \left(\dfrac{y^1}{y^2}\right)^2 & \dfrac{1}{2}\dfrac{(y^1)^2}{y^2 y^3} \\ -2\dfrac{y^1}{y^3} & \dfrac{1}{2}\dfrac{(y^1)^2}{y^2 y^3} & \left(\dfrac{y^1}{y^3}\right)^2 \end{bmatrix}, \tag{1.25}$$

which is positive definite on M'. Hence $\mathbb{F}^3 = (\mathbb{R}^3, M', F)$ is a Minkowski space. ∎

Example 1.3. Let $M'' = \{(y^1, y^2, y^3, y^4) \in \mathbb{R}^4;\ y^1 > 0,\ y^2 > 0,\ y^3 > 0\}$ and $M' = \mathbb{R}^4 \times M''$. Then define the function $F^* : M' \longrightarrow (0, \infty)$ by

$$F^*(x, y) = (y^4)^2 + \left(\frac{y^2 y^3}{y^1}\right)^2. \tag{1.26}$$

In this case we have

$$[g_{ij}(x, y)] = \begin{bmatrix} 3\dfrac{(y^2 y^3)^2}{(y^1)^4} & -2\dfrac{y^2 (y^3)^2}{(y^1)^3} & -2\dfrac{(y^2)^2 y^3}{(y^1)^3} & 0 \\ -2\dfrac{y^2 (y^3)^2}{(y^1)^3} & \left(\dfrac{y^3}{y^1}\right)^2 & 2\dfrac{y^2 y^3}{(y^1)^2} & 0 \\ -2\dfrac{(y^2)^2 y^3}{(y^1)^3} & 2\dfrac{y^2 y^3}{(y^1)^2} & \left(\dfrac{y^2}{y^1}\right)^2 & 0 \\ 0 & 0 & 0 & 1 \end{bmatrix}. \tag{1.27}$$

It is easy to check that $\mathbb{F}^4 = (\mathbb{R}^4, M', F^*)$ is an indefinite Minkowski space of index $q = 1$, that is, $\mathbb{F}^4$ is a pseudo–Finsler manifold of Lorentzian signature. ■

The next example shows that the same F^* can define Finsler structures of different indices (and hence different pseudo–Finsler manifolds), when different open submanifolds of TM are taken.

Example 1.4. Let $M'' = \{(y^1, y^2) \in \mathbb{R}^2\ ;\ y^1 \neq 0,\ y^2 \neq 0,\ \dfrac{y^1}{y^2} + \sqrt[3]{4} > 0\}$ and $M' = \mathbb{R}^2 \times M''$. Then define the function $F^* : M' \longrightarrow \mathbb{R}$ by

$$F^*(x, y) = (y^2)^2 + \frac{(y^1)^3}{y^2},$$

and obtain

$$[g_{ij}(x, y)] = \begin{bmatrix} 3\dfrac{y^1}{y^2} & -\dfrac{3}{2}\left(\dfrac{y^1}{y^2}\right)^2 \\ -\dfrac{3}{2}\left(\dfrac{y^1}{y^2}\right)^2 & 1 + \left(\dfrac{y^1}{y^2}\right)^3 \end{bmatrix}.$$

Consider $M^* = \left\{(x^1, x^2, y^1, y^2) \in M';\ \dfrac{y^1}{y^2} > 0\right\}$ and conclude that $\mathbb{F}^2 = (\mathbb{R}^2, M^*, F^*)$ is a Minkowski space, while $\tilde{\mathbb{F}}^2 = (\mathbb{R}^2, M' \backslash M^*, F^*)$ is an indefinite Minkowski space of index $q = 1$. ■

We now present some Finsler manifolds which will provide us with a large class of Berwald manifolds (see Section 8).

Example 1.5. Let D be an open set of $\mathbb{R}^2$ and $f : D \to \mathbb{R}$ be a smooth function. Consider $(m, n) \in \mathbb{R}^2 \backslash \{(2, 0), (0, 2)\}$ such that $m + n = 2$, and define the function

$$F^*(x, y) = 2e^{f(x^1, x^2)}(y^1)^m (y^2)^n. \tag{1.28}$$

Clearly, F^* is a positively homogeneous function of degree two with respect to (y^1, y^2). Moreover, F^* is a smooth function on $M' = D \times M''$, where M'' is an open set in $\mathbb{R}^2$. Finally, by direct calculations we obtain

$$g_{ij}(x, y) = e^{f(x^1, x^2)}(y^1)^{m-2}(y^2)^{n-2} \begin{bmatrix} m(m-1)(y^2)^2 & mny^1y^2 \\ mny^1y^2 & n(n-1)(y^1)^2 \end{bmatrix}. \tag{1.29}$$

Certainly, for $m = 1$ and $n = 1$ we obtain a Lorentzian structure on D. It is easy to check that for $m \in (0, \infty) \backslash \{1\}$ and $n \in (0, \infty) \backslash \{1\}$, $\mathbb{F}^2 =$

(D, M', F^*) is a proper pseudo–Finsler manifold of Lorentzian signature, while for the rest of pairs $(m, n), \mathbb{F}^2$ is a proper Finsler manifold. ■

The next example illustrates that Finsler structures can be constructed by using some other geometrical structures on manifolds.

Example 1.6. Let $(M, \alpha = (\alpha_{ij}(x)))$ be an m–dimensional Riemannian manifold and $\beta = (\beta_i(x))$ be a 1-form on M. Consider a coordinate system $\{(\mathcal{U}, \varphi); x^i\}$ in M and on $\varphi(\mathcal{U}) \times \mathbb{R}^m$ define

$$F(x, y) = \left(\alpha_{ij}(x)y^iy^j\right)^{\frac{1}{2}} + |\beta_i(x)y^i|. \tag{1.30}$$

Using (1.1), it is easy to check that we obtain a global positive function on TM^0 which we denote by the same symbol F. Clearly F is a smooth function on an open submanifold M' of TM such that $\pi(M') = M$, $\theta(M) \cap M' = \phi$ and M'_x is a positive conic set for each $x \in M$. Moreover we prove the following important result.

THEOREM 1.2 *Let $(M, \alpha = (\alpha_{ij}(x)))$ be an m–dimensional Riemannian manifold and $\beta = (\beta_i(x))$ a 1–form on M. Then $\mathbb{F}^m = (M, M', F)$ with F given by (1.30) is a Finsler manifold.*

PROOF. It is easy to see that F satisfies (F_1). For (F_2), we consider a point $(x_0, y_0) \in \varphi(\mathcal{U}) \times \mathbb{R}^m$ such that $\beta_i(x_0)y_0^i > 0$. Then there exists a neighborhood $\vartheta \subset \varphi(\mathcal{U}) \times \mathbb{R}^m$ of (x_0, y_0) such that $\beta_i(x)y^i > 0$ for any $(x, y) \in \vartheta$. Thus (1.30) becomes

$$F(x, y) = \|y\| + \beta_i(x)y^i, \ \forall (x, y) \in \vartheta, \tag{1.31}$$

where we set $\|y\| = (\alpha_{ij}(x)y^iy^j)^{\frac{1}{2}}$. By straightforward calculations using (1.12) and (1.31) we deduce that

$$\begin{aligned} g_{ij}(x, y) &= \frac{F(x, y)}{\|y\|^3}\left(\alpha_{ij}(x)\|y\|^2 - y_iy_j\right) \\ &\quad + \left(\frac{y_i}{\|y\|} + \beta_i(x)\right)\left(\frac{y_j}{\|y\|} + \beta_j(x)\right), \end{aligned} \tag{1.32}$$

where we set $y_i = \alpha_{ik}(x)y^k$. Consider a non–zero vector $v = (v^i) \in \mathbb{R}^m$ and fix a point $(x, y) \in \vartheta$. Then, using the Cauchy–Schwarz inequality with respect to $\alpha_{ij}(x)$, from (1.32) we obtain

$$\begin{aligned} g_{ij}(x, y)v^iv^j &= \frac{F(x, y)}{\|y\|^3}\left(\|v\|^2\|y\|^2 - (\alpha(v, y))^2\right) \\ &\quad + \left(\left(\frac{y_i}{\|y\|} + \beta_i(x)\right)v^i\right)^2 \\ &\geq 0. \end{aligned}$$

We should prove that the above inequality becomes equality only for $v = 0$. If we suppose $\|v\|^2\|y\|^2 - (\alpha(v,y))^2 = 0$ for a $v \neq 0$, then there exists $\lambda \neq 0$ such that $v = \lambda y$. Thus we have

$$g_{ij}(x,y)v^i v^j = \lambda^2 F^2(x,y) > 0.$$

Hence $g_{ij}(x,y)$ are the components of a positive definite quadratic form on $\mathbb{R}^m$. In case $\beta_i(x_0)y_0^i < 0$ the only change is that everywhere in the proof $\beta_i(x)$ is replaced by $-\beta_i(x)$. ■

In 1941 Randers [1] was the first to introduce a fundamental function given by

$$F(x,y) = \left(\alpha_{ij}(x)y^i y^j\right)^{\frac{1}{2}} + \beta_i(x)y^i. \tag{1.33}$$

Certainly, in this case β and α should satisfy some conditions such that F is a non–negative function (see Antonelli et al. [1], p. 43). We shall call an $\mathbb{F}^m = (M, M', F)$ with F given by either (1.30) or (1.33) a **Randers manifold (space)**. Applications of Randers manifolds in physics can be found in Asanov [1], Eliopoulos [2], Horváth [1], [2], Ingarden [1], Ingarden–Tamássy [1], Stephenson [1], [2] and Stephenson–Kilmister [1]. Also, the geometry of Randers manifolds was intensively studied by the Japanese School on Finsler geometry (cf., Matsumoto [2], Hashiguchi et al. [1], Shibata–Shimada–Azuma–Yasuda [1], Yasuda–Shimada [1]).

More generally, suppose that α is a pseudo–Riemannian metric on M and β is, as above, a 1–form on M. Then we use the notations:

$$\alpha(x,y) = \left(|\alpha_{ij}(x)y^i y^j|\right)^{\frac{1}{2}} \quad \text{and} \quad \beta(x,y) = \beta_i(x)y^i,$$

and consider a non–negative smooth function $F(x,y) = f(\alpha(x,y), \beta(x,y))$, where f is supposed to be positively homogeneous of degree one in (α, β). Finsler manifolds with fundamental function F as above have been introduced by Matsumoto [1] and their geometry was investigated in several papers (see Matsumoto [7]). The function F is called an (α,β)**–metric**. Thus $\mathbb{F}^m = (M, M', F)$ is said to be a pseudo–Finsler manifold with an (α,β)–metric. Interesting applications of this class of pseudo–Finsler manifolds in physics and biology can be found in Antonelli et al. [1]. Clearly, when $F = \alpha + |\beta|$, we obtain a Randers manifold. Also, **Kropina manifolds** (cf., Kropina [1]) with fundamental function $F = \alpha^2/|\beta|$ fall in this same general category of pseudo–Finsler manifolds. ■

2. Vertical Vector Bundle and Finsler Tensor Fields

The purpose of this section is to define the vertical vector bundle of a pseudo–Finsler manifold, and to show its major role in Finsler geometry. More precisely, we shall see that any Finsler tensor field is a section of a certain vector bundle constructed by means of the vertical vector bundle.

Let $F^m = (M, M', F^*)$ be a pseudo–Finsler manifold of index $0 \leq q < m$. Consider the tangent mapping $\pi_* : TM' \longrightarrow TM$ of the submersion $\pi : M' \longrightarrow M$ and define the vector bundle $VM' = Ker\ \pi_*$. As locally $\pi^i(x, y) = x^i$, we obtain

$$\pi^i_*\left(\frac{\partial}{\partial x^j}\right) = \delta^i_j \quad \text{and} \quad \pi^i_*\left(\frac{\partial}{\partial y^j}\right) = 0,$$

on a coordinate neighborhood $\mathcal{U}' \subset M'$. Hence $\{\partial/\partial y^i\}$ is a basis of $\Gamma(VM'_{|\mathcal{U}'})$, which implies that VM' is an integrable distribution of rank m on M'. We call VM' the **vertical vector bundle** of $\mathbb{F}^m$.

Next, denote by π' the canonical projection of TM' on M' and consider its restriction on VM' which we denote by the same symbol. A smooth section of VM' is a smooth mapping $X : M' \longrightarrow VM'$, such that $\pi' \circ X = 1_{M'}$ (see Section 3). Locally, on a coordinate neighborhood $\mathcal{U}' \subset M'$ we have

$$X = X^i(x, y)\frac{\partial}{\partial y^i}, \tag{2.1}$$

where X^i are smooth functions on $\mathcal{U}'$. Then using (1.3) and (2.1) we deduce that

$$\tilde{X}^i(\tilde{x}, \tilde{y}) = X^j(x, y)B^i_j(x), \tag{2.2}$$

on the overlapping domain of two coordinate neighborhoods $\mathcal{U}'$ and $\tilde{\mathcal{U}}'$ in M'. Hence the local components of X are transformed with respect to (1.1) exactly as the components of a vector field on M are transformed with respect to the coordinate transformations on M. Taking into account that Finsler geometry is dealing with vector fields whose components satisfy (2.2) (cf., Rund [3], p. 52), we call any smooth section of VM' a **Finsler vector field** on $\mathbb{F}^m$.

In particular, on each coordinate neighborhood $\mathcal{U}' \subset M'$ we define the vector field

$$L = y^i\frac{\partial}{\partial y^i}. \tag{2.3}$$

Then using (2.3), (1.1) and (1.3) we obtain a global section of VM' that we still denote by L and call it the **Liouville vector field** on M'. In classical Finsler geometry, L is known as the **supporting element** in $\mathbb{F}^m$ (cf., Cartan [1], p. 4, Rund [3], p. 66).

Since Finsler vector fields are sections of the vertical vector bundle, we can construct the whole Finsler tensor algebra in exactly the same way as the usual tensor algebra on the manifold M, but using VM' instead of TM. To this end we denote by V^*M' the dual vector bundle of VM'. Then a **Finsler 1–form** is a smooth section of V^*M'. Suppose $\{\theta^1, \cdots, \theta^m\}$ is a dual basis to $\{\partial/\partial y^1, \cdots, \partial/\partial y^m\}$, i.e., $\theta^i(\partial/\partial y^j) = \delta^i_j$. Then each $\omega \in \Gamma(V^*M')$ is locally written as

$$\omega = \omega_i(x, y)\theta^i, \tag{2.4}$$

where $\omega_i(x, y) = \omega(\partial/\partial y^i)$. Using (1.3) and (2.4) we derive

$$\omega_i(x, y) = \tilde{\omega}_j\,(\tilde{x}, \tilde{y})\, B^j_i\,(x), \tag{2.5}$$

and

$$\tilde{\theta}^i\,(\tilde{x}, \tilde{y}) = \theta^j\,(x, y) B^i_j(x), \tag{2.6}$$

with respect to (1.1)

More generally, a **Finsler tensor field** of type (r, s) on $\mathbb{F}^m$ is a $(r+s)$–multilinear mapping

$$T : (\Gamma(V^*M'))^r \times (\Gamma(VM'))^s \longrightarrow \mathcal{F}(M').$$

Locally, T is given by m^{r+s} smooth functions

$$T^{i_1 \cdots i_r}_{j_1 \cdots j_s}(x, y) = T\left(\theta^{i_1}, \cdots, \theta^{i_r}, \frac{\partial}{\partial y^{j_1}}, \cdots, \frac{\partial}{\partial y^{j_s}}\right)(x, y), \tag{2.7}$$

which satisfy

$$\tilde{T}^{h_1 \cdots h_r}_{k_1 \cdots k_s}\,(\tilde{x}, \tilde{y})\, B^{k_1}_{j_1}\,(x) \cdots B^{k_s}_{j_s}\,(x) = T^{i_1 \cdots i_r}_{j_1 \cdots j_s}(x, y) B^{h_1}_{i_1}\,(x) \cdots B^{h_r}_{i_r}\,(x) \tag{2.8}$$

with respect to (1.1). In this way a **Finsler tensor** of type (r, s) at the point $y_x \in M'_x$ is an element of the vector space

$$T^r_s\,(VM')_{y_x} = L_{r+s}\left((V^*M')^r_{y_x} \times (VM')^s_{y_x}, \mathbb{R}\right).$$

The vector bundle

$$T^r_s(VM') = \bigcup_{y_x \in M'} T^r_s(VM')_{y_x},$$

over M' is called the **Finsler tensor bundle** of type (r, s) of $\mathbb{F}^m$. In particular, $T^1_0(VM') = VM'$ and $T^0_1(VM') = V^*M'$ are called the **Finsler vector bundle** and the **Finsler covector bundle** of $\mathbb{F}^m$ respectively.

In what follows we define an isomorphism that will be needed in coming sections (Section 4 in particular). This isomorphism H is defined by

$$\left\{\begin{array}{l} H:\Gamma\left(L((VM')^r, VM')\right)\longrightarrow\Gamma(T_r^1(VM')) \\ H(T)\left(\omega, X_1,\cdots,X_r\right)=\omega\left(T(X_1,\cdots,X_r)\right), \end{array}\right. \tag{2.9}$$

for any $T\in\Gamma\left(L((VM')^r, VM')\right)$, $\omega\in\Gamma(V^*M'), X_a\in\Gamma(VM')$, $a\in\{1,\cdots r\}$.

A Finsler tensor field T of type (r,s) is **contravariant symmetric (covariant symmetric)** if it is invariant with respect to transposing of any two of its arguments in $(V^*M')^r$ $((VM')^s)$. When each such transposition produces a change in sign of the original tensor we say that T is **contravariant skew–symmetric (covariant skew–symmetric)**. In particular, for Finsler tensor fields of type $(r,0)$ and $(0,r), r\geq 2$, we simply call them **symmetric** or **skew–symmetric Finsler tensor fields**. A skew–symmetric Finsler tensor field of type $(0,r)$ is called a **Finsler r–form**.

Now, besides the Liouville vector field L on M' we shall define some other geometric objects on $\mathbb{F}^m$ and give their fundamental properties. First, on each coordinate neighborhood $\mathcal{U}'$ of M' we define

$$\left\{\begin{array}{l} g_{|\mathcal{U}'}:\Gamma\left((VM')_{|\mathcal{U}'}\right)^2\longrightarrow\mathcal{F}(M') \\ g_{|\mathcal{U}'}(V,W)(x,y)=g_{ij}(x,y)V^i(x,y)W^j(x,y), \end{array}\right. \tag{2.10}$$

for any Finsler vector fields V and W with local components (V^i) and (W^i) respectively, where g_{ij} are functions given by (1.7). From (2.10) we derive

$$g_{ij}(x,y)=g_{|\mathcal{U}'}\left(\frac{\partial}{\partial y^i},\frac{\partial}{\partial y^j}\right)(x,y). \tag{2.11}$$

By straightforward calculations using (1.7) and (1.3) we infer that

$$\tilde{g}_{kh}(\tilde{x},\tilde{y})B_i^k(x)B_j^h(x)=g_{ij}(x,y), \tag{2.12}$$

with respect to the transformation of coordinates (1.1) on M'. Due to (2.12) and (2.2) we obtain a global Finsler tensor field g of type $(0,2)$ on $\mathbb{F}^m$ which is defined on each $\mathcal{U}'\subset M'$ by (2.11) and thus it has (g_{ij}) as local components. Clearly g is a symmetric Finsler tensor field. We call g a **Finsler metric** or an **indefinite Finsler metric**, according as the condition (F_2) or (F_2^*) is satisfied, respectively. In general, when we refer to a pseudo–Finsler manifold $\mathbb{F}^m$, we call g the **pseudo–Finsler metric** of $\mathbb{F}^m$. *Finally we note that g can be thought of as a Riemannian (pseudo–Riemannian) metric on the Finsler vector bundle VM'.*

3. Linear Connections on Vector Bundles

A linear connection is one of the main components of a vectorial Finsler connection and a Finsler connection which are treated in Sections 4 and 5 respectively. Here we introduce the concept of a linear connection on a vector bundle by using both the coordinate free and the coordinates method. The Levi–Civita connection on a semi–Riemannian manifold M is presented as a particular linear connection on the tangent bundle TM.

Let $\xi = (E, \pi, M)$ be a vector bundle, where M, π and E are the real m-dimensional manifold, the projection mapping of E on M and the real $(m+r)$–dimensional total space of ξ respectively. As each fibre $E_x = \pi^{-1}(x)$ of ξ is of dimension r we say that ξ is a vector bundle of rank r. For simplicity we often say that E is a vector bundle over M. For any submanifold N of M denote by $E_{|N}$ the vector bundle whose fibres are E_x for all $x \in N$.

Next, we consider a vector bundle chart $\{(\mathcal{U}^*, \Phi);\ x^i, y^a\}, i \in \{1, \cdots, m\}$, $a \in \{1, \cdots, r\}$ of E induced by the coordinate system $\{(\mathcal{U}, \varphi);\ x^i\}$ on M. Then the transformation of coordinates on E is given by

$$\begin{cases} \tilde{x}^i = \tilde{x}^i\,(x^1, \cdots, x^m) \\ \tilde{y}^a = L^a_b\,(x^1, \cdots, x^m)\, y^b, \end{cases} \tag{3.1}$$

where L^a_b are smooth functions locally defined on M.A **smooth section** of E on an open set $\mathcal{U}$ of M is a smooth mapping $X : \mathcal{U} \longrightarrow E$ which satisfies $\pi \circ X = 1_{\mathcal{U}}$, where $1_{\mathcal{U}}$ is the identity mapping on $\mathcal{U}$.

In previous sections we have seen some examples of vector bundles like: TM, T^*M, VM', V^*M' and $T^r_s(VM')$. Later in the book we shall present new examples of vector bundles.

A subset F of E is said to be a **vector subbundle** of E if for each $x \in M$ there exists a vector bundle chart $\{(\mathcal{U}^*, \Phi)\}$ of E and a natural number $1 \leq p < r$ such that $\Phi\,(\pi^{-1}(\mathcal{U}) \cap F) = \mathcal{U} \times \mathbb{R}^p$. It follows that F is also a vector bundle over M and any fibre F_x is isomorphic to $\mathbb{R}^p$. In particular, a vector subbundle of the tangent bundle TM of M is called a **distribution** on M.

For any $X \in \Gamma(TM)$ with local components $\{X^i\}$ and $f \in \mathcal{F}(M)$ define

$$X(f) = X^i(x)\frac{\partial f}{\partial x^i}. \tag{3.2}$$

The Lie bracket of two vector fields X and Y on M is the vector field $[X, Y]$ defined by

$$[X, Y]f = X(Y(f)) - Y(X(f)), \tag{3.3}$$

for any $f \in \mathcal{F}(M)$. If D is a distribution of rank p on M, we say that it is **involutive** if, for any $X, Y \in \Gamma(D)$, we have $[X, Y] \in \Gamma(D)$. An **integral**

manifold of D is a submanifold N of M such that $T_xN \subset D_x$, for any $x \in N$. We say that D is **integrable** if, for any point $x \in M$, there exists a coordinate system $\{(\mathcal{U}, \varphi);\ x^i\}$ such that all the submanifolds of $\mathcal{U}$ given by $x^\alpha = c^\alpha,\quad \alpha \in \{p+1, \cdots, m\},\ c^\alpha \in \mathbb{R}$, are integral manifolds of D. A maximal connected integral manifold of D is called a **leaf** of D. The following theorem of Frobenius (cf., Chevalley [1], p. 94 and Sternberg [1], p. 132) is crucial in studying the differential geometry of D.

THEOREM 3.1 *A distribution D on M is integrable, if and only if it is involutive. Moreover, through every point $x \in M$ there passes a unique leaf N_x of D and any other connected integral manifold containing x is a submanifold of N_x.*

By using the $\mathcal{F}(M)$–module $\Gamma(E)$ of sections of a vector bundle E over M we may introduce, as in the case of the tangent bundle, the differential forms and tensor fields on E. More precisely, a **differential 1–form** on the vector bundle E is an $\mathcal{F}(M)$–linear mapping $\omega : \Gamma(E) \longrightarrow \mathcal{F}(M)$. Denote by E^* the dual vector bundle of E and say that a $(p+q)$–multilinear mapping

$$S : (\Gamma(E^*))^p \times (\Gamma(E))^q \longrightarrow \mathcal{F}(M),$$

is a **tensor field** of type (p, q) on vector bundle E. Finally, consider two vector bundles E and F over M and define an **F–valued differential q–form** on E as a q–multilinear mapping $\omega : (\Gamma(E))^q \longrightarrow \Gamma(F)$, such that

$$\omega(X_{\sigma(1)}, \cdots, X_{\sigma(q)}) = sgn\ \sigma\ \omega\,(X_1, \cdots, X_q)\,,$$

for any permutation σ of $\{1, \cdots, q\}$.

Next, we say that the mapping ∇ which assigns to any vector field X on M the differential operator

$$\nabla_X : \Gamma(E) \longrightarrow \Gamma(E);\ Y \longrightarrow \nabla_X Y,\ \forall Y \in \Gamma(E),$$

is a **linear connection** on the vector bundle ξ if it satisfies:

$$\nabla_{fX+Y} Z = f\nabla_X Z + \nabla_Y Z, \tag{3.4}$$

and

$$\nabla_X(fZ + U) = f\nabla_X Z + X(f)Z + \nabla_X U, \tag{3.5}$$

for any $X, Y \in \Gamma(TM),\ Z, U \in \Gamma(E)$ and $f \in \mathcal{F}(M)$. The differential operator ∇_X is called the **covariant derivative** with respect to X. For a differential 1–form ω on E we define its covariant derivative by

$$(\nabla_X \omega)(Z) = X(\omega(Z)) - \omega(\nabla_X Z), \tag{3.6}$$

for any $X \in \Gamma(TM)$ and $Z \in \Gamma(E)$. More generally, the covariant derivative of a tensor field S of type (p, q) on E is defined by

$$
\begin{aligned}
(\nabla_X S)(\ \omega^1, \cdots, \omega^p, Z_1, \cdots, Z_q) \\
&= X(S(\omega^1, \cdots, \omega^p, Z_1, \cdots, Z_q)) \\
&- \sum_{s=1}^{p} S(\omega^1, \cdots, \nabla_X \omega^s, \cdots, \omega^p,\ Z_1, \cdots, Z_q) \\
&- \sum_{t=1}^{q} S(\omega^1, \cdots, \omega^p,\ Z_1, \cdots, \nabla_X Z_t, \cdots, Z_q), \qquad (3.7)
\end{aligned}
$$

for any $\omega^s \in \Gamma(E^*)$ and $Z_t \in \Gamma(E)$, $s \in \{1, \cdots, p\}$, $t \in \{1, \cdots, q\}$. By using an isomorphism similar to the one given by (2.9), we may think of a tensor field of type $(1, q)$ on E as a q–linear mapping $S : (\Gamma(E))^q \longrightarrow \Gamma(E)$. Then its covariant derivative is given by

$$
\begin{aligned}
(\nabla_X S)(Z_1, \cdots, Z_q) &= \nabla_X (S(Z_1, \cdots, Z_q)) \qquad (3.8) \\
&- \sum_{t=1}^{q} S(Z_1, \cdots, \nabla_X Z_t, \cdots, Z_q),
\end{aligned}
$$

for any $X \in \Gamma(TM)$ and $Z_t \in \Gamma(E)$. An arbitrary tensor field S is said to be **parallel** with respect to ∇ if for any $X \in \Gamma(TM)$, we have $\nabla_X S = 0$. If, in particular, there exists a parallel pseudo–Riemannian metric g with respect to ∇ on the vector bundle E, we say that ∇ is a **metric linear connection** on E.

We now consider three vector bundles E, E', and E'' over M endowed with the linear connections ∇, ∇', and ∇'', respectively. Then the covariant derivative ∇^* of an $\mathcal{F}(M)$–bilinear mapping $S : \Gamma(E) \times \Gamma(E') \longrightarrow \Gamma(E'')$ is defined by

$$
(\nabla^*_X S)(Y, Z) = \nabla''_X(S(Y, Z)) - S(\nabla_X Y, Z) - S(Y, \nabla'_X Z), \qquad (3.9)
$$

for any $X \in \Gamma(TM)$, $Y \in \Gamma(E)$ and $Z \in \Gamma(E')$.

The **curvature form** of a linear connection ∇ on E is an $L(E, E)$–valued differential 2–form Ω on M given by

$$
\Omega(X, Y) = \nabla_X \circ \nabla_Y - \nabla_Y \circ \nabla_X - \nabla_{[X,Y]}, \qquad (3.10)
$$

for any $X, Y \in \Gamma(TM)$. In particular, when $E = TM$, we say that ∇ is a **linear connection** on M. In this case, the curvature form Ω defines the **curvature tensor field** R by

$$
R(X, Y)Z = \Omega(X, Y)Z = \nabla_X(\nabla_Y Z) - \nabla_Y(\nabla_X Z) - \nabla_{[X,Y]}Z, \qquad (3.11)
$$

for any $X, Y, Z \in \Gamma(TM)$. Moreover, for a linear connection ∇ on M there exists a **torsion tensor field** T given by

$$T(X,Y) = \nabla_X Y - \nabla_Y X - [X,Y], \tag{3.12}$$

for any $X, Y \in \Gamma(TM)$.

We now suppose that there exists a pseudo–Riemannian metric g on M (see Example 1.1). Then there exists a unique linear connection ∇ on M satisfying the following conditions:
(i) g is parallel with respect to ∇, i.e., we have

$$(\nabla_X g)(Y,Z) = X(g(Y,Z)) - g(\nabla_X Y, Z) - g(Y, \nabla_X Z) = 0, \tag{3.13}$$

for any $X, Y, Z \in \Gamma(TM)$.
(ii) ∇ is a **torsion–free linear connection**, i.e.,

$$\nabla_X Y - \nabla_Y X = [X,Y], \tag{3.14}$$

for any $X, Y \in \Gamma(TM)$.
This connection is called the **Levi–Civita connection** and it is given by

$$\begin{aligned} 2g(\nabla_X Y, Z) &= X(g(Y,Z)) + Y(g(X,Z)) - Z(g(X,Y)) \\ &\quad + g([X,Y],Z) + g([Z,X],Y) - g([Y,Z],X), \end{aligned} \tag{3.15}$$

for any $X, Y, Z \in \Gamma(TM)$. As is well known (see Yano–Kon [1], p. 41) a **Killing vector field** X is characterized by $\mathcal{L}_X g = 0$, where $\mathcal{L}_X$ is the Lie derivative with respect to X, given by

$$\mathcal{L}_X g(Y,Z) = g(\nabla_Y X, Z) + g(\nabla_Z X, Y) \tag{3.16}$$

for any $Y, Z \in \Gamma(TM)$.

We close this section with the local expressions for the geometric objects we defined on the vector bundle ξ. First, let $\{\partial/\partial x^i\}$, $i \in \{1, \cdots, m\}$ and $\{S_a\}$, $a \in \{1, \cdots, r\}$ be a local frame field on $\mathcal{U} \subset M$ and a basis for $\Gamma(E_{|\mathcal{U}})$ respectively. Then we define the local components $\Gamma_a{}^b{}_i(x)$ of ∇ by

$$\nabla_{\partial/\partial x^i} S_a = \Gamma_a{}^b{}_i(x) S_b. \tag{3.17}$$

Consider another coordinate neighborhood $\tilde{\mathcal{U}} \subset M$ such that $\mathcal{U} \cap \tilde{\mathcal{U}} \neq \phi$. Then we have

$$\frac{\partial}{\partial x^i} = B_i^j(x) \frac{\partial}{\partial \tilde{x}^j}, \tag{3.18}$$

and

$$S_a = G_a^b(x) \tilde{S}_b, \tag{3.19}$$

where G_a^b are smooth functions on $\mathcal{U}\cap\tilde{\mathcal{U}}$ such that $\left[G_a^b(x)\right]$ is an invertible $r\times r$ matrix for any $x\in\mathcal{U}\cap\tilde{\mathcal{U}}$.

Finally, using (3.4), (3.5) and (3.17)–(3.19) we deduce that

$$\Gamma_{a\ i}^{\ b}(x)G_b^d(x) = B_i^j(x)G_a^c(x)\tilde{\Gamma}_{c\ j}^{\ d}(\tilde{x}) + \frac{\partial G_a^d(x)}{\partial x^i}. \tag{3.20}$$

Conversely, suppose that there exists mr^2 smooth functions $\Gamma_{a\ i}^{\ b}(x)$ on the domain of each local chart of M satisfying (3.20). Then we obtain a linear connection on E locally given by (3.17).

Hence we may state the following theorem:

THEOREM 3.2 *There exists a linear connection ∇ on the vector bundle $\xi = (E,\pi,M)$ if and only if, on the domain of each local chart of M, there exist mr^2 smooth functions $\Gamma_{a\ i}^{\ b}(x)$ satisfying (3.20) with respect to the transformations (3.18) and (3.19).*

Next, we take $X = X^i(\partial/\partial x^i)$ and $Z = Z^aS_a$ and using (3.4), (3.5) and (3.17) we derive

$$\nabla_X Z = Z^a_{\ |i}X^iS_a, \tag{3.21}$$

where

$$Z^a_{\ |i} = \frac{\partial Z^a}{\partial x^i} + Z^b\Gamma_{b\ i}^{\ a}. \tag{3.22}$$

Similarly, (3.6) is locally expressed by

$$(\nabla_X\omega)\, Z = \omega_{a|i}X^iZ^a, \tag{3.23}$$

where we set

$$\omega_{a|i} = \frac{\partial\omega_a}{\partial x^i} - \omega_b\Gamma_{a\ i}^{\ b}, \tag{3.24}$$

and ω_a are the local components of ω given by

$$\omega_a = \omega(S_a).$$

We now consider $\{\theta^a\}$ as the dual basis of $\{S_a\}$ and define the local components of a tensor field S of type (p,q) on E by

$$S^{b_1\cdots b_p}_{a_1\cdots a_q} = S\left(\theta^{b_1},\cdots,\theta^{b_p},\ S_{a_1},\cdots,S_{a_q}\right).$$

Then, using (3.7) and (3.21)–(3.24) we deduce that

$$(\nabla_X S)(\omega^1,\cdots,\omega^p,\ Z_1,\cdots,Z_q) = S^{b_1\cdots b_p}_{a_1\cdots a_q|i}X^i\omega^1_{b_1}\cdots\omega^p_{b_p}Z_1^{a_1}\cdots Z_q^{a_q},$$

where

$$\omega^h = \omega^h_{b_h}\theta^{b_h}, \quad Z_k = Z^{a_k}_k S_{a_k}, \; h \in \{1, \cdots, p\}, \; k \in \{1, \cdots, q\},$$

and

$$\begin{aligned} S^{b_1\cdots b_p}_{a_1\cdots a_q|i} &= \frac{\partial S^{b_1\cdots b_p}_{a_1\cdots a_q}}{\partial x^i} + \sum_{h=1}^{p} S^{b_1\cdots b_{h-1} c b_{h+1}\cdots b_p}_{a_1\cdots a_q}\Gamma_c{}^{b_h}{}_i \\ &\quad - \sum_{k=1}^{q} S^{b_1\cdots b_p}_{a_1\cdots a_{k-1} c\, a_{k+1}\cdots a_q}\Gamma_{a_k}{}^c{}_i. \end{aligned} \tag{3.25}$$

We now derive local expressions for the curvature form Ω and curvature and torsion tensor fields defined by (3.10)–(3.12). First we see that the Lie bracket (3.3) has the local expression

$$[X, Y] = \left(X^j \frac{\partial Y^i}{\partial x^j} - Y^j \frac{\partial X^i}{\partial x^j}\right)\frac{\partial}{\partial x^i}. \tag{3.26}$$

Then, using (3.10), (3.21), (3.22), and (3.26) we infer that

$$\Omega(X, Y)Z = R_b{}^a{}_{ij} Z^b Y^i X^j S_a,$$

where we have put

$$R_b{}^a{}_{ij} = \frac{\partial \Gamma_b{}^a{}_i}{\partial x^j} - \frac{\partial \Gamma_b{}^a{}_j}{\partial x^i} + \Gamma_b{}^c{}_i\Gamma_c{}^a{}_j - \Gamma_b{}^c{}_j\Gamma_c{}^a{}_i. \tag{3.27}$$

In case of a linear connection ∇ of M we put

$$\nabla_{\partial/\partial x^j}\frac{\partial}{\partial x^i} = \Gamma_i{}^k{}_j\frac{\partial}{\partial x^k},$$

and by using (3.11) and (3.12) we obtain that

$$R(X, Y)Z = R_h{}^k{}_{ij} Z^h Y^i X^j \frac{\partial}{\partial x^k} \quad \text{and} \quad T(X, Y) = T_i{}^k{}_j Y^i X^j \frac{\partial}{\partial x^k},$$

where

$$R_h{}^k{}_{ij} = \frac{\partial \Gamma_h{}^k{}_i}{\partial x^j} - \frac{\partial \Gamma_h{}^k{}_j}{\partial x^i} + \Gamma_h{}^\ell{}_i\Gamma_\ell{}^k{}_j - \Gamma_h{}^\ell{}_j\Gamma_\ell{}^k{}_i, \tag{3.28}$$

and

$$T_i{}^k{}_j = \Gamma_i{}^k{}_j - \Gamma_j{}^k{}_i. \tag{3.29}$$

Finally, the local coefficients of the Levi–Civita connection are called the **Christoffel symbols** and they are given by

$$\Gamma_i{}^k{}_j = \frac{1}{2}g^{kh}\left(\frac{\partial g_{hi}}{\partial x^j} + \frac{\partial g_{hj}}{\partial x^i} - \frac{\partial g_{ij}}{\partial x^h}\right). \tag{3.30}$$

4. Vectorial Finsler Connections

In the present section we shall discuss vectorial Finsler connections which, as we shall see in the book, play an important role in studying the geometry of Finsler submanifolds. As a non–linear connection is one of the main geometric objects involved in the definition of a vectorial Finsler connection, we start the section with this concept and present its main properties.

Let $\mathbb{F}^m = (M, M', F^*)$ be a pseudo–Finsler manifold. A complementary distribution HM' to VM' in TM' is called a **non–linear connection** or a **horizontal distribution** on M'. This connection plays a fundamental role and is a powerful tool in studying Finsler geometry (see Abate–Patrizio [1], Barthel [2], Kawaguchi [1],[2], Matsumoto [6] and Miron–Anastasiei [1]).

In order to obtain a theorem on the existence of non–linear connections we start with a decomposition

$$TM' = HM' \oplus VM'. \tag{4.1}$$

Then we take a local frame field $\{X_i, \partial/\partial y^i\}$ on $\mathcal{U}' \subset M'$ adapted to (4.1), i.e., $X_i \in \Gamma(HM')$ and $\partial/\partial y^i \in \Gamma(VM')$. Thus we have

$$\frac{\partial}{\partial x^i} = A_i^j(x,y)X_j + N_i^j(x,y)\frac{\partial}{\partial y^j}, \tag{4.2}$$

where A_i^j and N_i^j are smooth functions which are locally defined on M'. Hence the transition matrix from the local frame field $\{\partial/\partial x^i, \partial/\partial y^i\}$ to $\{X_j, \partial/\partial y^j\}$ is

$$\Lambda = \begin{bmatrix} A_i^j(x,y) & 0 \\ N_i^k(x,y) & \delta_h^k \end{bmatrix}.$$

As Λ is a nonsingular matrix it follows that the $m \times m$ matrix $[A_i^j(x,y)]$ is also a nonsingular matrix. Thus the set of local vector fields $\{\delta/\delta x^1, \cdots, \delta/\delta x^m\}$ given by

$$\frac{\delta}{\delta x^i} = A_i^j(x,y)X_j$$

is a basis in $\Gamma(HM'_{|\mathcal{U}'})$. In this way (4.2) becomes

$$\frac{\delta}{\delta x^i} = \frac{\partial}{\partial x^i} - N_i^j(x,y)\frac{\partial}{\partial y^j}. \tag{4.3}$$

Denote by $\widetilde{N}_h^k$ the functions in (4.3) given with respect to another coordinate system $\{(\widetilde{\mathcal{U}}', \widetilde{\Phi}'); \tilde{x}^i, \tilde{y}^i\}$ on M' such that $\mathcal{U}' \cap \widetilde{\mathcal{U}}' \neq \phi$. Then by using (1.2), (1.3) and (4.3) we obtain

$$N_i^j(x,y)B_j^k(x) = \widetilde{N}_h^k(\tilde{x},\tilde{y})B_i^h(x) + B_{ih}^k(x)y^h. \tag{4.4}$$

Conversely, suppose that on the domain $\mathcal{U}'$ of each local chart of M' there exist m^2 smooth functions N_i^j satisfying (4.4) with respect to the transformation of coordinates on M'. Then we define by (4.3) m linear independent vector fields on $\mathcal{U}'$. Moreover, (4.3) and (4.4) yield

$$\frac{\delta}{\delta x^i} = B_i^j(x)\frac{\delta}{\delta \tilde{x}^j}. \tag{4.5}$$

Hence on M' we obtain a globally defined distribution HM' which is locally spanned by $\{\delta/\delta x^i\}$. Since HM' is a distribution of rank m on M' and $\{\delta/\delta x^i\}$ do not belong to $\Gamma(VM')$ we conclude that HM' is a complementary distribution to VM' in TM'. Therefore we can state the following result.

THEOREM 4.1 *There exists a non–linear connection HM' on M' if and only if, on the domain of each local chart on M', there exist m^2 smooth functions N_i^j satisfying (4.4) with respect to the coordinate transformation (1.1) on M'.*

A non–linear connection HM' enables us to define an almost product structure on M' as follows. Consider a vector field X on M'. Then locally we have

$$X = X^i\frac{\delta}{\delta x^i} + \dot{X}^i\frac{\partial}{\partial y^i}, \tag{4.6}$$

where X^i and $\dot{X}^i$ satisfy

$$\widetilde{X}^j = B_i^j X^i \;\; ; \;\; \dot{\widetilde{X}}^j = B_i^j \dot{X}^i, \tag{4.7}$$

with respect to (1.1). Then we define

$$\begin{cases} Q : \Gamma(TM') \longrightarrow \Gamma(TM') \\ QX = \dot{X}^i\dfrac{\delta}{\delta x^i} + X^i\dfrac{\partial}{\partial y^i}. \end{cases} \tag{4.8}$$

By (1.3), (4.5) and (4.7) it follows from (4.8) that QX does not depend on the local chart on M'. Moreover, $Q^2 = I$ and therefore Q is an almost product structure on M'. We call Q the **associate almost product structure** to the non–linear connection HM'.

Further, denote by h and v the projection morphisms of TM' to HM' and VM' respectively. Then we have

$$(a) \quad Q \circ h = v \circ Q \qquad \text{and} \qquad (b) \quad Q \circ v = h \circ Q. \tag{4.9}$$

By means of Q and v we define the $\mathcal{F}(M')$–bilinear mapping

$$\begin{cases} T_1 : \Gamma(VM') \times \Gamma(VM') \longrightarrow \Gamma(VM') \\ T_1(X, Y) = -v\,[QX, QY],\ \forall X, Y \in \Gamma(VM'). \end{cases} \tag{4.10}$$

According to the isomorphism defined by (2.9) we see that T_1 is a Finsler tensor field of type (1,2). By direct calculations using (4.10), (4.8), (4.3) and (3.26) we deduce that

$$T_1\left(\frac{\partial}{\partial y^j}, \frac{\partial}{\partial y^i}\right) = \left[\frac{\delta}{\delta x^i}, \frac{\delta}{\delta x^j}\right] = R^k{}_{ij}\frac{\partial}{\partial y^k}, \tag{4.11}$$

where we have set

$$R^k{}_{ij} = \frac{\delta N^k_i}{\delta x^j} - \frac{\delta N^k_j}{\delta x^i}. \tag{4.12}$$

From (4.10) it follows that $T_1 = (R^k{}_{ij})$ is the obstruction to the integrability of the horizontal distribution. More precisely, we have the following result:

THEOREM 4.2 *The horizontal distribution HM' is involutive if and only if $T_1 = 0$ on M', or equivalently, on each coordinate neighborhood $\mathcal{U}' \subset M'$ we have $R^k{}_{ij} = 0$.*

Now, combining the two concepts, namely, the linear connection on a vector bundle and the non–linear connection on M', we introduce the concept of vectorial Finsler connection (cf., Bejancu [1]).

Let $\mathbb{F}^m = (M, M', F^*)$ be a pseudo–Finsler manifold and $\xi = (E, \rho, M')$ be a vector bundle of rank r over M'. Suppose that there exist a linear connection ∇ on E and a non–linear connection HM' on M'. Then we say that the pair $VFC = (HM', \nabla)$ is a **vectorial Finsler connection** on the vector bundle E. Throughout the book we shall see that vectorial Finsler connections actually play the same role in Finsler geometry as linear connections on vector bundles in Riemannian geometry.

We close the section by a local approach of a vectorial Finsler connection and of its corresponding differential operators. First, we note that due to Theorems 3.2 and 4.1, a vectorial Finsler connection is locally determined by some smooth functions satisfying (3.20) and (4.4). However, for our purpose we consider the local frame field $\{\delta/\delta x^i,\ \partial/\partial y^i\}$ adapted to the decompoorition (4.1) and set

$$\nabla_{\delta/\delta x^i} S_a = F_a{}^b{}_i S_b, \tag{4.13}$$

and

$$\nabla_{\partial/\partial y^i} S_a = C_a{}^b{}_i S_b. \tag{4.14}$$

Now suppose that we change both the local coordinates on M' and the local basis in $\Gamma(E)$. Then using (1.3), (3.4), (3.5), (3.19), (4.5), (4.13) and (4.14) we infer that

$$F_a{}^b{}_i(x,y)G^c_b(x,y) = \widetilde{F}_d{}^c{}_j(\tilde{x},\tilde{y})G^d_a(x,y)B^j_i(x) + \frac{\delta G^c_a}{\delta x^i}(x,y), \tag{4.15}$$

and

$$C_a{}^b{}_i(x,y)G^c_b(x,y) = \widetilde{C}_d{}^c{}_j(\tilde{x},\tilde{y})G^d_a(x,y)B^j_i(x) + \frac{\partial G^c_a}{\partial y^i}(x,y). \tag{4.16}$$

Therefore we may state the following general result:

THEOREM 4.3 *There exists a vectorial Finsler connection $VFC = (HM'$ $, \nabla)$ on a vector bundle E over M' if and only if on the domain of each local chart on M' there exists a triple $(N^j_i, F_a{}^b{}_i, C_a{}^b{}_i)$ satisfying (4.4), (4.15) and (4.16).*

Next, we consider a section $Z = Z^a S_a$ of E and using (3.4), (3.5), (4.13), (4.14), and (4.6) we easily deduce that

$$\nabla_X Z = (X^i Z^a{}_{|i} + \dot{X}^i Z^a{}_{||i})S_a, \tag{4.17}$$

where we set

$$Z^a{}_{|i} = \frac{\delta Z^a}{\delta x^i} + Z^b F_b{}^a{}_i, \tag{4.18}$$

and

$$Z^a{}_{||i} = \frac{\partial Z^a}{\partial y^i} + Z^b C_b{}^a{}_i. \tag{4.19}$$

We call (4.18) and (4.19) the **horizontal covariant derivative** and the **vertical covariant derivative** of Z respectively. Thus a vectorial Finsler connection on E induces two covariant derivatives for sections of E. Certainly, we should have the same situation when we apply ∇ to an arbitrary tensor field on E.

We first consider a differential 1–form $\omega = \omega_a \theta^a$ on E and by using (3.4)–(3.6), (4.6), (4.13) and (4.14) we obtain

$$(\nabla_X \omega)\, Z = (\omega_{a|i} X^i + \omega_{a||i}\dot{X}^i)Z^a, \tag{4.20}$$

where we have set

$$\omega_{a|i} = \frac{\delta \omega_a}{\delta x^i} - \omega_b F_a{}^b{}_i, \tag{4.21}$$

and

$$\omega_{a||i} = \frac{\partial \omega_a}{\partial y^i} - \omega_b C_a{}^b{}_i. \tag{4.22}$$

Analogously, we call (4.21) and (4.22) the **horizontal covariant derivative** and the **vertical covariant derivative** of ω with respect to the vectorial Finsler connection VFC.

More generally, we consider a tensor field S of type (p, q) on E. Then by using (3.7), (4.17), and (4.20) we derive

$$\begin{aligned}(\nabla_X S)(\omega^1, \cdots, \omega^p, Z_1, \cdots, Z_q) \qquad (4.23)\\ = (S_{a_1\cdots a_q|i}^{b_1\cdots b_p} X^i + S_{a_1\cdots a_q||i}^{b_1\cdots b_p} \dot{X}^i)\omega_{b_1}^1 \cdots \omega_{b_p}^p Z_1^{a_1} \cdots Z_q^{a_q},\end{aligned}$$

where

$$\begin{aligned} S_{a_1\cdots a_q|i}^{b_1\cdots b_p} &= \frac{\delta S_{a_1\cdots a_q}^{b_1\cdots b_p}}{\delta x^i} + \sum_{s=1}^{p} S_{a_1\cdots a_q}^{b_1\cdots c b_{s+1}\cdots b_p} F_c{}^{b_s}{}_i \\ &\quad - \sum_{t=1}^{q} S_{a_1\cdots c a_{t+1}\cdots a_q}^{b_1\cdots b_p} F_{a_t}{}^c{}_i, \end{aligned} \tag{4.24}$$

and

$$\begin{aligned} S_{a_1\cdots a_q||i}^{b_1\cdots b_p} &= \frac{\partial S_{a_1\cdots a_q}^{b_1\cdots b_p}}{\partial y^i} + \sum_{s=1}^{p} S_{a_1\cdots a_q}^{b_1\cdots c b_{s+1}\cdots b_p} C_c{}^{b_s}{}_i \\ &\quad - \sum_{t=1}^{q} S_{a_1\cdots c a_{t+1}\cdots a_q}^{b_1\cdots b_p} C_{a_t}{}^c{}_i. \end{aligned} \tag{4.25}$$

Thus (4.24) and (4.25) define the **horizontal covariant derivative** and the **vertical covariant derivative** of the tensor field S with respect to VFC.

In particular, let g be a pseudo–Riemannian metric on E with local components $g_{ab} = g(S_a, S_b)$. Then we say that VFC is an **h–metric vectorial Finsler connection** and a **v–metric vectorial Finsler connection** if $\nabla_{hX} g = 0$ and $\nabla_{vX} g = 0$ respectively for any $X \in \Gamma(TM')$. In case $\nabla_X g = 0$ for any $X \in \Gamma(TM')$ we say that VFC is a **metric vectorial Finsler connection.**

In order to determine the local expression for the curvature form of ∇ of VFC we use (4.3) and by direct calculations obtain

$$\left[\frac{\delta}{\delta x^i}, \frac{\partial}{\partial y^j}\right] = \frac{\partial N_i^k}{\partial y^j}\frac{\partial}{\partial y^k}. \tag{4.26}$$

The local components of Ω are denoted by $R_{a\ ij}^{\ b}, P_{a\ ij}^{\ b}, S_{a\ ij}^{\ b}$, that is, we have

$$\Omega\left(\frac{\delta}{\delta x^j}, \frac{\delta}{\delta x^i}\right) S_a = R_{a\ ij}^{\ b} S_b; \qquad \Omega\left(\frac{\partial}{\partial y^j}, \frac{\delta}{\delta x^i}\right) S_a = P_{a\ ij}^{\ b} S_b$$

$$\Omega\left(\frac{\partial}{\partial y^j}, \frac{\partial}{\partial y^i}\right) S_a = S_{a\ ij}^{\ b} S_b.$$

Then by using (3.10), (4.6), (4.17), (4.11) and (4.26) we infer that

$$\Omega(X,Y)Z = (R_{a\ ij}^{\ b} Y^i X^j + P_{a\ ij}^{\ b} Y^i \dot{X}^j + S_{a\ ij}^{\ b} \dot{Y}^i \dot{X}^j) Z^a S_b,$$

and the local components of Ω are given by

$$R_{a\ ij}^{\ b} = \frac{\delta F_{a\ i}^{\ b}}{\delta x^j} - \frac{\delta F_{a\ j}^{\ b}}{\delta x^i} + F_{a\ i}^{\ c} F_{c\ j}^{\ b} - F_{a\ j}^{\ c} F_{c\ i}^{\ b} + R^k_{\ ij} C_{a\ k}^{\ b}, \tag{4.27}$$

$$P_{a\ ij}^{\ b} = \frac{\partial F_{a\ i}^{\ b}}{\partial y^j} - \frac{\delta C_{a\ j}^{\ b}}{\delta x^i} + F_{a\ i}^{\ c} C_{c\ j}^{\ b} - C_{a\ j}^{\ c} F_{c\ i}^{\ b} + \frac{\partial N_i^k}{\partial y^j} C_{a\ k}^{\ b}, \tag{4.28}$$

$$S_{a\ ij}^{\ b} = \frac{\partial C_{a\ i}^{\ b}}{\partial y^j} - \frac{\partial C_{a\ j}^{\ b}}{\partial y^i} + C_{a\ i}^{\ c} C_{c\ j}^{\ b} - C_{a\ j}^{\ c} C_{c\ i}^{\ b}. \tag{4.29}$$

We call $R_{a\ ij}^{\ b}, P_{a\ ij}^{\ b}$ and $S_{a\ ij}^{\ b}$ the **horizontal curvature tensor**, the **mixed curvature tensor**, and the **vertical curvature tensor** of VFC, respectively.

5. Finsler Connections

Let $\mathbb{F}^m = (M, M', F^*)$ be a pseudo–Finsler manifold. Then a **Finsler connection** on $\mathbb{F}^m$ is a pair $FC = (HM', \nabla)$ where HM' is a non–linear connection on M' and ∇ is a linear connection on the vertical vector bundle VM'. In other words, a Finsler connection is nothing but a vectorial Finsler connection on VM'.

Certainly, for a reader who is concerned with Finsler connections in their classical form, it is important to show that, starting with the pair $FC = (HM', \nabla)$ we obtain the whole calculation apparatus of Finsler geometry. To this end we consider a coordinate system $\{(\mathcal{U}', \Phi'); x^i, y^i\}$ on M' and take a frame field $\{\delta/\delta x^i, \partial/\partial y^i\}$ on M' adapted to the decomposition (4.1). Then, locally we set

$$\nabla_{\delta/\delta x^j} \frac{\partial}{\partial y^i} = F_{i\ j}^{\ k}(x,y) \frac{\partial}{\partial y^k}, \tag{5.1}$$

and

$$\nabla_{\partial/\partial y^j}\frac{\partial}{\partial y^i} = C_i{}^k{}_j(x,y)\frac{\partial}{\partial y^k}, \tag{5.2}$$

where $F_i{}^k{}_j$ and $C_i{}^k{}_j$ are smooth functions on M'. As in the case of vectorial Finsler connections (see (4.15) and (4.16)) by using (1.3), (4.5), (5.1) and (5.2) we find

$$F_i{}^k{}_j(x,y)B_k^h(x) = \widetilde{F}_s{}^h{}_t(\tilde{x},\tilde{y})B_i^s(x)B_j^t(x) + B_{ij}^h(x), \tag{5.3}$$

and

$$C_i{}^k{}_j(x,y)B_k^h(x) = \widetilde{C}_s{}^h{}_t(\tilde{x},\tilde{y})B_i^s(x)B_j^t(x), \tag{5.4}$$

where $\widetilde{F}_s{}^h{}_t$ and $\widetilde{C}_s{}^h{}_t$ are the functions in (5.1) and (5.2) with respect to a coordinate system $\{(\widetilde{\mathcal{U}}',\widetilde{\Phi}');\tilde{x}^i,\tilde{y}^i\}$ such that $\mathcal{U}'\cap\widetilde{\mathcal{U}}'\neq\phi$.

Therefore, based on Theorem 4.3, we state the following theorem:

THEOREM 5.1 *There exists a Finsler connection $FC = (HM',\nabla)$ on $\mathbb{F}^m$ if and only if, on the domain of each local chart on M', there exists a triple $(N_i^k(x,y), F_i{}^k{}_j(x,y), C_i{}^k{}_j(x,y))$ satisfying (4.4), (5.3) and (5.4) with respect to a transformation of coordinates on M'.*

As in the case of vectorial Finsler connections, we have the horizontal and vertical covariant derivatives defined by a Finsler connection. First, for any Finsler vector field $Z = Z^i\partial/\partial y^i$ and any vector field X on M' locally given by (4.6), by using (4.17)–(4.19) we obtain

$$\nabla_X Z = (X^j Z^k{}_{|j} + \dot{X}^j Z^k{}_{||j})\frac{\partial}{\partial y^k}, \tag{5.5}$$

where we set

$$Z^k{}_{|j} = \frac{\delta Z^k}{\delta x^j} + Z^i F_i{}^k{}_j, \tag{5.6}$$

and

$$Z^k{}_{||j} = \frac{\partial Z^k}{\partial y^j} + Z^i C_i{}^k{}_j. \tag{5.7}$$

We call $Z^k{}_{|j}$ and $Z^k{}_{||j}$ the **horizontal covariant derivative** and the **vertical covariant derivative** of Z with respect to the Finser connection FC. As an example, the horizontal and vertical covariant derivatives of the Liouville vector field (see (2.3)) are given by

$$y^k{}_{|j} = y^i F_i{}^k{}_j - N_j^k, \tag{5.8}$$

and

$$y^k{}_{||j} = \delta_j^k + y^i C_i{}^k{}_j, \tag{5.9}$$

respectively,

Next, by using the local coefficients $N^i_j(x,y)$ of the non–linear connection HM' we define

$$\delta y^i = dy^i + N^i_j dx^j. \tag{5.10}$$

Then (1.4), (1.5), (4.4) and (4.10) imply

$$\delta \tilde{y}^i = B^i_j(x)\delta y^j,$$

with respect to a transformation of coordinates on M'. Thus $\{\delta y^i\}$ satisfy (2.6) and hence they are Finsler 1–forms on $\mathbb{F}^m$. Moreover, since $\delta y^i\,(\partial/\partial y^j) = \delta^i_j$, we conclude that $\{\delta y^1, \cdots, \delta y^m\}$ is a basis in $\Gamma(V^*M'_{|\mathcal{U}'})$ which is dual to the basis $\{\partial/\partial y^1, \cdots, \partial/\partial y^m\}$ in $\Gamma(VM')$. Therefore a Finsler 1–form ω on $\mathbb{F}^m$ is locally expressed as follows:

$$\omega = \omega_i(x,y)\delta y^i,$$

where ω_i are smooth functions on $\mathcal{U}'$ satisfying (2.5). In this case (4.20), (4.21) and (4.22) become

$$\nabla_X \omega Z = (\omega_{j|i} X^i + \omega_{j||i} \dot{X}^i) Z^j, \tag{5.11}$$

$$\omega_{j|i} = \frac{\delta \omega_j}{\delta x^i} - \omega_k F_j{}^k{}_i, \tag{5.12}$$

and

$$\omega_{j||i} = \frac{\partial \omega_j}{\partial y^i} - \omega_k C_j{}^k{}_i, \tag{5.13}$$

respectively. Thus (5.12) and (5.13) represent the **horizontal covariant derivative** and the **vertical covariant derivative** of ω with respect to the Finsler connection FC.

Finally, we consider a Finsler tensor field S of type (p,q) whose local components are defined by (cf., (2.7))

$$S^{k_1\cdots k_p}_{j_1\cdots j_q} = S\left(\delta y^{k_1}, \cdots, \delta y^{k_p}, \ \frac{\partial}{\partial y^{j_1}}, \cdots, \frac{\partial}{\partial y^{j_q}}\right).$$

Then (4.23), (4.24), and (4.25) become

$$\begin{aligned}(\nabla_X S)&(\omega^1, \cdots, \omega^p, Z_1, \cdots, Z_q)\\ &= \left(S^{k_1\cdots k_p}_{j_1\cdots j_q|i} X^i + S^{k_1\cdots k_p}_{j_1\cdots j_q||i} \dot{X}^i\right) \omega^1_{k_1} \cdots \omega^p_{k_p} Z^{j_1}_1 \cdots Z^{j_q}_q,\end{aligned} \tag{5.14}$$

$$S^{k_1\cdots k_p}_{j_1\cdots j_q|i} = \frac{\delta S^{k_1\cdots k_p}_{j_1\cdots j_q}}{\delta x^i} + \sum_{s=1}^{p} S^{k_1\cdots r k_{s+1}\cdots k_p}_{j_1\cdots j_q} F_r{}^{k_s}{}_i$$
$$- \sum_{t=1}^{q} S^{k_1\cdots k_p}_{j_1\cdots r j_{t+1}\cdots j_q} F_{j_t}{}^r{}_i, \tag{5.15}$$

and

$$S^{k_1\cdots k_p}_{j_1\cdots j_q||i} = \frac{\partial S^{k_1\cdots k_p}_{j_1\cdots j_q}}{\partial y^i} + \sum_{s=1}^{p} S^{k_1\cdots r k_{s+1}\cdots k_p}_{j_1\cdots j_q} C_r{}^{k_s}{}_i$$
$$- \sum_{t=1}^{q} S^{k_1\cdots k_p}_{j_1\cdots r j_{t+1}\cdots j_q} C_{j_t}{}^r{}_i, \tag{5.16}$$

respectively, where $\{\omega^1, \cdots, \omega^p\}$ and $\{Z_1, \cdots, Z_q\}$ are Finsler 1–forms and Finsler vector fields on $\mathbb{F}^m$. Thus (5.15) and (5.16) give the **horizontal covariant derivative** and the **vertical covariant derivative** of S with respect to FC.

We say that S is h–**parallel** (v–**parallel**) with respect to FC if $\nabla_{hX} S = 0$ ($\nabla_{vX} S = 0$) for any $X \in \Gamma(TM')$ respectively. In particular, the pseudo–Finsler metric $g = (g_{ij})$ (see (2.11)) is h–parallel and v–parallel with respect to FC if

$$g_{ij|k} = \frac{\delta g_{ij}}{\delta x^k} - g_{rj} F_i{}^r{}_k - g_{ir} F_j{}^r{}_k = 0, \tag{5.17}$$

and

$$g_{ij||k} = \frac{\partial g_{ij}}{\partial y^k} - g_{rj} C_i{}^r{}_k - g_{ir} C_j{}^r{}_k = 0, \tag{5.18}$$

respectively. We call FC an h–**metric Finsler connection** and v–**metric Finsler connection** if g is h–parallel and v–parallel with respect to FC respectively. In case both (5.17) and (5.18) are satisfied we say that FC is a **metric Finsler connection**. Some classical Finsler connections fall in one of the above categories (see Section 7).

Next, we consider a vector bundle E over M'. A **mixed Finsler tensor field** of type $\begin{pmatrix} p & r \\ q & s \end{pmatrix}$ on $\mathbb{F}^m$ with respect to E is a $(p+q+r+s)$–linear mapping

$$S : (\Gamma(V^*M'))^p \times (\Gamma(VM'))^q \times (\Gamma(E^*))^r \times (\Gamma(E))^s \to \mathcal{F}(M').$$

Then the local components of S have the following general form:

$$S^{k_1\cdots k_p}_{j_1\cdots j_q}{}^{b_1\cdots b_r}_{a_1\cdots a_s}(x, y)$$

and satisfy

$$S^{k_1\cdots k_p\ b_1\cdots b_r}_{j_1\cdots j_q\ a_1\cdots a_s}\frac{\partial \bar{x}^{h_1}}{\partial x^{k_1}}\cdots\frac{\partial \bar{x}^{h_p}}{\partial x^{k_p}}A^{c_1}_{b_1}\cdots A^{c_r}_{b_r} = \bar{S}^{h_1\cdots h_p\ c_1\cdots c_r}_{i_1\cdots i_q\ d_1\cdots d_s}\frac{\partial \bar{x}^{i_1}}{\partial x^{j_1}}\cdots\frac{\partial \bar{x}^{i_q}}{\partial x^{j_q}}A^{d_1}_{a_1}\cdots A^{d_s}_{a_s}, \quad (5.19)$$

with respect to the coordinate transformations $\bar{x}^i = \bar{x}^i(x)$ on M and the transformations $\bar{S}_a = A^b_a S_b$ of local basis in $\Gamma(E)$. Suppose now that $VFC = (HM', \widetilde{\nabla}) = (N^j_i, \widetilde{F}_a{}^b{}_i, \widetilde{C}_a{}^b{}_i)$ and $FC = (HM', \nabla) = (N^k_i, F_i{}^k{}_j, C_i{}^k{}_j)$ are a vectorial Finsler connection on E and a Finsler connection on $\mathbb{F}^m$ respectively. Then by combining the two types of covariant derivatives induced by VFC and FC, we may define some covariant derivatives for mixed Finsler tensor fields. For simplicity, we consider a mixed Finsler tensor field S of type $\begin{pmatrix} 1 & 1 \\ 1 & 1 \end{pmatrix}$ with local components S^{kb}_{ja}. The **relative covariant derivative** of S is defined as follows:

$$\begin{aligned}(\bar{\nabla}_X S)(\ \eta, Y, \omega, Z) &= X\,(S\,(\eta, Y, \omega, Z)) - S\,(\nabla_X \eta, Y, \omega, Z)\\ &\quad - S\,(\eta, \nabla_X Y, \omega, Z) - S(\eta, Y, \widetilde{\nabla}_X \omega, Z) - S(\eta, Y, \omega, \widetilde{\nabla}_X Z),\end{aligned}$$

for any

$$X \in \Gamma(TM'), \eta \in \Gamma(V^*M'), Y \in \Gamma(VM'), \omega \in \Gamma(E^*) \text{ and } Z \in \Gamma(E).$$

Thus locally we obtain the **horizontal relative covariant derivative** and the **vertical relative covariant derivative** of S as follows:

$$S^{kb}_{ja|i} = \frac{\delta S^{kb}_{ja}}{\delta x^i} + S^{hb}_{ja} F_h{}^k{}_i - S^{kb}_{ha} F_j{}^h{}_i + S^{kc}_{ja} \widetilde{F}_c{}^b{}_i - S^{kb}_{jc} \widetilde{F}_a{}^c{}_i, \quad (5.20)$$

and

$$S^{kb}_{ja||i} = \frac{\partial S^{kb}_{ja}}{\partial y^i} + S^{hb}_{ja} C_h{}^k{}_i - S^{kb}_{ha} C_j{}^h{}_i + S^{kc}_{ja} \widetilde{C}_c{}^b{}_i - S^{kb}_{jc} \widetilde{C}_a{}^c{}_i, \quad (5.21)$$

respectively. In the theory of Finsler submanifolds we will meet frequently mixed Finsler tensor fields and their relative covariant derivatives (see Section 2.1).

We now return to the general theory of Finsler connections to introduce their curvature and torsion Finsler tensor fields. First, we define the curvature form of the Finsler connection $FC = (HM', \nabla)$ as being the curvature form Ω of the linear connection ∇. It defines three Finsler tensor fields R, P and S of type (1,3) as follows:

$$R(X,Y)Z = \Omega\,(QX, QY)\,Z; \qquad P(X,Y)Z = \Omega(X, QY)Z;$$

$$S(X,Y)Z = \Omega(X,Y)Z, \tag{5.22}$$

for any $X, Y, Z \in \Gamma(VM')$. Locally, we set:

$$\begin{aligned}
&(a) \quad R\left(\frac{\partial}{\partial y^k}, \frac{\partial}{\partial y^j}\right)\frac{\partial}{\partial y^i} = R_i{}^h{}_{jk}\frac{\partial}{\partial y^h};\\
&(b) \quad P\left(\frac{\partial}{\partial y^k}, \frac{\partial}{\partial y^j}\right)\frac{\partial}{\partial y^i} = P_i{}^h{}_{jk}\frac{\partial}{\partial y^h};\\
&(c) \quad S\left(\frac{\partial}{\partial y^k}, \frac{\partial}{\partial y^j}\right)\frac{\partial}{\partial y^i} = S_i{}^h{}_{jk}\frac{\partial}{\partial y^h}.
\end{aligned} \tag{5.23}$$

Then by using (5.22), (3.11), (5.1), (5.2) and (4.11) we infer that

$$R_i{}^h{}_{jk} = \frac{\delta F_i{}^h{}_j}{\delta x^k} - \frac{\delta F_i{}^h{}_k}{\delta x^j} + F_i{}^r{}_j F_r{}^h{}_k - F_i{}^r{}_k F_r{}^h{}_j + R^r{}_{jk} C_i{}^h{}_r, \tag{5.24}$$

$$P_i{}^h{}_{jk} = \frac{\partial F_i{}^h{}_j}{\partial y^k} - \frac{\delta C_i{}^h{}_k}{\delta x^j} + F_i{}^r{}_j C_r{}^h{}_k - C_i{}^r{}_k F_r{}^h{}_j + \frac{\partial N^r_j}{\partial y^k} C_i{}^h{}_r, \tag{5.25}$$

$$S_i{}^h{}_{jk} = \frac{\partial C_i{}^h{}_j}{\partial y^k} - \frac{\partial C_i{}^h{}_k}{\partial y^j} + C_i{}^r{}_j C_r{}^h{}_k - C_i{}^r{}_k C_r{}^h{}_j. \tag{5.26}$$

We call $R_i{}^h{}_{jk}$, $P_i{}^h{}_{jk}$ and $S_i{}^h{}_{jk}$ the **h–curvature, hv–curvature** and **v–curvature Finsler tensor fields** respectively, of the Finsler connection FC.

In order to define torsion tensor fields for FC we first construct, by using Q and ∇, a linear connection on the manifold M'. Namely, we define the differential operator D by

$$D_X Y = \nabla_X vY + Q\nabla_X QhY, \ \forall X, Y \in \Gamma(TM'). \tag{5.27}$$

It is easy to check that D is a linear connection on M'. We call D the **associate linear connection** to FC on M'. By straightforward calculations, using (3.8), (4.9), and (5.20) we obtain that Q is parallel with respect to D, that is we have

$$(D_X Q)\, Y = 0, \ \forall X, Y \in \Gamma(TM').$$

Remark 5.1. The associate linear connection has been first considered by Oproiu [1]. A different approach to D via Otsuki connections is given in Bejancu [8], p. 38. ■

Next, we denote by R^D and T^D the curvature and torsion tensor fields of D respectively. Then by (5.27) we deduce that R^D and the curvature form Ω of FC are related by

$$R^D(X,Y)Z = \Omega(X,Y)vZ + Q\Omega(X,Y)QhZ, \ \forall X, Y, Z \in \Gamma(TM'). \tag{5.28}$$

Taking into account (5.22), (5.23) and (5.28) we infer that

$$R^D\left(\frac{\delta}{\delta x^k},\frac{\delta}{\delta x^j}\right)\frac{\partial}{\partial y^i} = QR^D\left(\frac{\delta}{\delta x^k},\frac{\delta}{\delta x^j}\right)\frac{\delta}{\delta x^i} = R_i{}^h{}_{jk}\frac{\partial}{\partial y^h}, \tag{5.29}$$

$$R^D\left(\frac{\partial}{\partial y^k},\frac{\delta}{\delta x^j}\right)\frac{\partial}{\partial y^i} = QR^D\left(\frac{\partial}{\partial y^k},\frac{\delta}{\delta x^j}\right)\frac{\delta}{\delta x^i} = P_i{}^h{}_{jk}\frac{\partial}{\partial y^h}, \tag{5.30}$$

$$R^D\left(\frac{\partial}{\partial y^k},\frac{\partial}{\partial y^j}\right)\frac{\partial}{\partial y^i} = QR^D\left(\frac{\partial}{\partial y^k},\frac{\partial}{\partial y^j}\right)\frac{\delta}{\delta x^i} = S_i{}^h{}_{jk}\frac{\partial}{\partial y^h}. \tag{5.31}$$

As the curvature tensor of D determines all curvature Finsler tensor fields of FC we define the torsion Finsler tensor fields of FC by means of T^D. First, by using (5.27) and (3.12) we obtain

$$\begin{aligned} T^D(X,Y) &= (\nabla_X vY - \nabla_Y vX - v[X,Y]) \\ &\quad + Q\left(\nabla_X QhY - \nabla_Y QhX - Qh[X,Y]\right), \end{aligned} \tag{5.32}$$

for any $X, Y \in \Gamma(TM')$. Now we prove the following theorem:

THEOREM 5.2 *The torsion tensor field T^D of the associate linear connection D is completely determined by the following five Finsler tensor fields*

$$T_1(X,Y) = -v\,[QX,QY], \tag{5.33}$$

$$T_2(X,Y) = \nabla_{QX}Y - \nabla_{QY}X - Qh\,[QX,QY], \tag{5.34}$$

$$T_3(X,Y) = -\nabla_{QY}X - v\,[X,QY], \tag{5.35}$$

$$T_4(X,Y) = \nabla_X Y - Qh\,[X,QY], \tag{5.36}$$

$$T_5(X,Y) = \nabla_X Y - \nabla_Y X - [X,Y], \tag{5.37}$$

for any $X, Y \in \Gamma(VM')$, and vice versa.

PROOF. By direct calculations using (5.32) and (4.9) we deduce that

$$\begin{aligned} T^D(QX,QY) &= -v\,[QX,QY] + Q\left(\nabla_{QX}Y - \nabla_{QY}X - Qh\,[QX,QY]\right) \\ &= T_1(X,Y) + QT_2(X,Y), \end{aligned} \tag{5.38}$$

$$\begin{aligned} T^D(X,QY) &= -\left(\nabla_{QY}X + v\,[X,QY]\right) + Q\left(\nabla_X Y - Qh\,[X,QY]\right) \\ &= T_3(X,Y) + QT_4(X,Y), \end{aligned} \tag{5.39}$$

$$T^D(X,Y) = T_5(X,Y), \tag{5.40}$$

for any $X, Y \in \Gamma(VM')$. Thus if T_1–T_4 are given then from (5.38)–(5.40) we obtain T^D. Conversely, if T^D in given then we infer that:

$$T_1(X,Y) = vT^D(QX,QY)\,; \quad T_2(X,Y) = vQT^D(QX,QY);$$

$$T_3(X,Y) = vT^D(X,QY)\,; \quad T_4(X,Y) = vQT^D(X,QY);$$

$$T_5(X,Y) = T^D(X,Y).$$

Thus the proof is complete. ■

We call T_1–T_5 the **torsion Finsler tensor fields** for the Finsler connection FC. We have to note that we have defined T_1 in section 4(see (4.10)) and its local components are given by (4.12). For the other torsion Finsler tensor fields we put:

$$\begin{aligned}&(a)\ \ T_2\left(\frac{\partial}{\partial y^j},\frac{\partial}{\partial y^i}\right) = T_i{}^k{}_j\frac{\partial}{\partial y^k}\ ; \quad (b)\ \ T_3\left(\frac{\partial}{\partial y^j},\frac{\partial}{\partial y^i}\right) = P^k{}_{ij}\frac{\partial}{\partial y^k};\\ &(c)\ \ T_4\left(\frac{\partial}{\partial y^j},\frac{\partial}{\partial y^i}\right) = L_i{}^k{}_j\frac{\partial}{\partial y^k}\ ; \quad (d)\ \ T_5\left(\frac{\partial}{\partial y^j},\frac{\partial}{\partial y^i}\right) = S^k{}_{ij}\frac{\partial}{\partial y^k}.\end{aligned} \tag{5.41}$$

Then by using (5.34)–(5.37), (5.41), (5.1), (5.2) and (4.26) we obtain

$$\begin{cases}(a)\ \ T_i{}^k{}_j = F_i{}^k{}_j - F_j{}^k{}_i\ ; \quad (b)\ \ P^k{}_{ij} = \dfrac{\partial N^k_i}{\partial y^j} - F_j{}^k{}_i;\\ (c)\ \ L_i{}^k{}_j = C_i{}^k{}_j\ ; \qquad\qquad\ (d)\ \ S^k{}_{ij} = C_i{}^k{}_j - C_j{}^k{}_i.\end{cases} \tag{5.42}$$

Thus $\left(R^k{}_{ij}, T_i{}^k{}_j, P^k{}_{ij},\ C_i{}^k{}_j, S^k{}_{ij}\right)$ are the local components of the torsion Finsler tensor fields of the Finsler connection $FC = \left(N^k_i, F_i{}^k{}_j, C_i{}^k{}_j\right)$. Finally, we see that (5.38)–(5.40) yield

$$T^D\left(\frac{\delta}{\delta x^j},\frac{\delta}{\delta x^i}\right) = R^k{}_{ij}\frac{\partial}{\partial y^k} + T_i{}^k{}_j\frac{\delta}{\delta x^k}, \tag{5.43}$$

$$T^D\left(\frac{\partial}{\partial y^j},\frac{\delta}{\delta x^i}\right) = P^k{}_{ij}\frac{\partial}{\partial y^k} + C_i{}^k{}_j\frac{\delta}{\delta x^k}, \tag{5.44}$$

$$T^D\left(\frac{\partial}{\partial y^j},\frac{\partial}{\partial y^i}\right) = S^k{}_{ij}\frac{\partial}{\partial y^k}. \tag{5.45}$$

In order to obtain all the Ricci identities for FC, we consider the identity (cf., (3.10))

$$\nabla_X\nabla_Y vZ - \nabla_Y\nabla_X vZ = \Omega(X,Y)vZ + \nabla_{[X,Y]}vZ,$$

for any $X, Y, Z \in \Gamma(TM')$. Substituting vZ by $Z^i(\partial/\partial y^i)$ and (X,Y) in turn by $\left(\delta/\delta x^k, \delta/\delta x^j\right)$, $\left(\partial/\partial y^k, \delta/\delta x^j\right)$ and $\left(\partial/\partial y^k, \partial/\partial y^j\right)$ and by using (4.11), (4.26), (5.6), (5.7), (5.22), (5.23), and (5.42) we derive

$$Z^i{}_{|j|k} - Z^i{}_{|k|j} = Z^h R_h{}^i{}_{jk} - T_j{}^h{}_k Z^i{}_{|h} - R^h{}_{jk} Z^i{}_{||h}, \tag{5.46}$$

$$Z^i{}_{|j||k} - Z^i{}_{||k|j} = Z^h P_h{}^i{}_{jk} - C_j{}^h{}_k Z^i{}_{|h} - P^h{}_{jk} Z^i{}_{||h}, \tag{5.47}$$

$$Z^i{}_{||j||k} - Z^i{}_{||k||j} = Z^h S_h{}^i{}_{jk} - S^h{}_{jk} Z^i{}_{||h}. \tag{5.48}$$

The indentities (5.46)–(5.48) are known in the literature as the **Ricci identities** of the Finsler connection FC (cf., Matsumoto [6], p. 73).

Next, we consider the Bianchi identities (see Kobayashi–Nomizu [1], p. 135) for the associate linear connection D:

$$\sum_{(X,Y,Z)} \left\{ (D_X T^D)(Y,Z) + T^D(T^D(X,Y),Z) - R^D(X,Y)Z \right\} = 0, \quad (5.49)$$

and

$$\sum_{(X,Y,Z)} \left\{ (D_X R^D)(Y,Z) + R^D(T^D(X,Y),Z) \right\}(U) = 0, \quad (5.50)$$

for any $X, Y, Z,\ U \in \Gamma(TM')$, where $\sum_{(X,Y,Z)}$ denotes the cyclic sum with respect to X, Y, Z. In order to obtain the local expressions of the Bianchi identity (5.49) for the Finsler connection FC we consider the following four cases.

Case I. $\quad X = \dfrac{\delta}{\delta x^k}, \quad Y = \dfrac{\delta}{\delta x^j}, \quad Z = \dfrac{\delta}{\delta x^i}.$

Using (3.7), (5.27), (5.1), (5.2), (5.43), and (5.15) we infer that

$$\left(D_{\delta/\delta x^k} T^D\right)\left(\frac{\delta}{\delta x^j}, \frac{\delta}{\delta x^i}\right) = R^t{}_{ij|k}\frac{\partial}{\partial y^t} + T_i{}^t{}_{j|k}\frac{\delta}{\delta x^t}. \quad (5.51)$$

Also, by using (5.43) and (5.44) we deduce that

$$\begin{aligned} T^D\left(T^D\left(\frac{\delta}{\delta x^k}, \frac{\delta}{\delta x^j}\right), \frac{\delta}{\delta x^i}\right) &= (P^t{}_{ih}R^h{}_{jk} + R^t{}_{ih}T_j{}^h{}_k)\frac{\partial}{\partial y^t} \\ &\quad + (C_i{}^t{}_h R^h{}_{jk} + T_i{}^t{}_h T_j{}^h{}_k)\frac{\delta}{\delta x^t}. \quad (5.52) \end{aligned}$$

Thus, by using (5.51), (5.52), and (5.29) in (5.49) and taking into account that $\{\delta/\delta x^i\}$ and $\{\partial/\partial y^i\}$ are local bases for complementary vector subbundles, we obtain

$$\sum_{(i,j,k)} \{R^t{}_{ij|k} + P^t{}_{ih}R^h{}_{jk} + R^t{}_{ih}T_j{}^h{}_k\} = 0, \quad (5.53)$$

and

$$\sum_{(i,j,k)} \{T_i{}^t{}_{j|k} + C_i{}^t{}_h R^h{}_{jk} + T_i{}^t{}_h T_j{}^h{}_k - R_i{}^t{}_{jk}\} = 0, \quad (5.54)$$

where $\sum_{(i,j,k)}$ denotes the cyclic sum with respect to (i, j, k).

Similarly, we obtain the local expressions of (5.49) for the next three cases.

Case II. $X = \dfrac{\partial}{\partial y^k}, \quad Y = \dfrac{\delta}{\delta x^j}, \quad Z = \dfrac{\delta}{\delta x^i}.$

$$\begin{aligned} R^t{}_{ij\|k} &+ S^t{}_{kh}R^h{}_{ij} - T_i{}^h{}_j P^t{}_{hk} - R_k{}^t{}_{ij} \\ &+ \mathcal{A}_{(ij)}\{P^t{}_{jk|i} + R^t{}_{ih}C_j{}^h{}_k + P^t{}_{ih}P^h{}_{jk}\} = 0, \end{aligned} \tag{5.55}$$

$$T_i{}^t{}_{j\|k} - T_i{}^h{}_j C_h{}^t{}_k + \mathcal{A}_{(ij)}\{C_j{}^t{}_{k|i} + C_i{}^t{}_h P^h{}_{jk} + T_i{}^t{}_h C_j{}^h{}_k + P_j{}^t{}_{ik}\} = 0. \tag{5.56}$$

Case III. $X = \dfrac{\partial}{\partial y^k}, \quad Y = \dfrac{\partial}{\partial y^j}, \quad Z = \dfrac{\delta}{\delta x^i}.$

$$S^t{}_{jk|i} + P^t{}_{ih}S^h{}_{jk} + \mathcal{A}_{(jk)}\{P^t{}_{ij\|k} + P_j{}^t{}_{ik} + S^t{}_{kh}P^h{}_{ij} + C_i{}^h{}_k P^t{}_{hj}\} = 0, \tag{5.57}$$

$$C_i{}^t{}_h S^h{}_{jk} - S_i{}^t{}_{jk} + \mathcal{A}_{(jk)}\{C_i{}^t{}_{j\|k} + C_i{}^h{}_k C_h{}^t{}_j\} = 0. \tag{5.58}$$

Case IV. $X = \dfrac{\partial}{\partial y^k}, \quad Y = \dfrac{\partial}{\partial y^j}, \quad Z = \dfrac{\partial}{\partial y^i}.$

$$\sum_{(i,j,k)} \{S^t{}_{ij\|k} + S^t{}_{ih}S^h{}_{jk} - S_i{}^t{}_{jk}\} = 0. \tag{5.59}$$

Finally we obtain the local expressions for the Bianchi identity (5.50). To this end we again examine four cases.

Case I. $X = \dfrac{\delta}{\delta x^k}, \quad Y = \dfrac{\delta}{\delta x^j}, \quad Z = \dfrac{\delta}{\delta x^i}, \quad U = \dfrac{\partial}{\partial y^\ell}.$

$$\sum_{(i,j,k)} \{R_\ell{}^t{}_{ij|k} + P_\ell{}^t{}_{ih}R^h{}_{jk} + R_\ell{}^t{}_{ih}T_j{}^h{}_k\} = 0. \tag{5.60}$$

Case II. $X = \dfrac{\delta}{\delta x^k}, \quad Y = \dfrac{\delta}{\delta x^j}, \quad Z = \dfrac{\partial}{\partial y^i}, \quad U = \dfrac{\partial}{\partial y^\ell}.$

$$\begin{aligned} R_\ell{}^t{}_{jk\|i} &+ S_\ell{}^t{}_{ih}R^h{}_{jk} - P_\ell{}^t{}_{hi}T_j{}^h{}_k \\ &+ \mathcal{A}_{(jk)}\{P_\ell{}^t{}_{ki|j} + P_\ell{}^t{}_{jh}P^h{}_{ki} + R_\ell{}^t{}_{jh}C_k{}^h{}_i\} = 0. \end{aligned} \tag{5.61}$$

Case III. $X = \dfrac{\delta}{\delta x^k}, \quad Y = \dfrac{\partial}{\partial y^j}, \quad Z = \dfrac{\partial}{\partial y^i}, \quad U = \dfrac{\partial}{\partial y^\ell}.$

$$S_\ell{}^t{}_{ij|k} + P_\ell{}^t{}_{kh}S^h{}_{ij} + \mathcal{A}_{(ij)}\{P_\ell{}^t{}_{ki\|j} + P_\ell{}^t{}_{hi}C_k{}^h{}_j + S_\ell{}^t{}_{jh}P^h{}_{ki}\} = 0. \tag{5.62}$$

Case IV. $X = \frac{\partial}{\partial y^k}, \quad Y = \frac{\partial}{\partial y^j}, \quad Z = \frac{\partial}{\partial y^i}, \quad U = \frac{\partial}{\partial y^\ell}.$

$$\sum_{(i,j,k)} \{S_\ell{}^t{}_{ij\|k} + S_\ell{}^t{}_{ih} S^h{}_{jk}\} = 0. \tag{5.63}$$

The identities (5.53)–(5.63) are known as **Bianchi identities** of the Finsler connection FC (cf., Matsumoto [6], p. 77).

6. The Canonical Non–Linear Connection and the Berwald Connection

As we have seen in Sections 4 and 5, both concepts of vectorial Finsler connection and Finsler connection are defined by means of a non–linear connection on M' and a linear connection on a certain vector bundle over M'. Here we shall construct a canonical non–linear connection on M' and then present the main features of the Berwald connection.

Let $\mathbb{F}^m = (M, M', F^*)$ be a pseudo–Finsler manifold and C be an oriented smooth curve in M. Consider a segment AB of C given by equations

$$x^i = x^i(t), \; t \in [a, b],$$

where A and B have coordinates $(x^i(a))$ and $(x^i(b))$ respectively. Suppose AB is non–degenerate, i.e.,

$$F(x(t), \dot{x}(t)) \neq 0, \quad \forall t \in [a, b], \tag{6.1}$$

where we set $x(t) = (x^i(t))$ and $\dot{x}(t) = \left(\frac{dx^i}{dt}\right)$. Then we define the **length** of AB by

$$\|AB\| = \int_a^b F(x(t), \dot{x}(t))\, dt. \tag{6.2}$$

Taking into account that F^* satisfies condition $(\mathbf{F}_1^*)$ from the definition of a pseudo–Finsler manifold, it is easy to check that $\|AB\|$ does not depend on the choice of the parameter on C. Now we consider the variational problem for the length integral given by (6.2), that is, we look for the curves for which $\|AB\|$ attains its minimum. Such curves, called extremals of the length integral, should satisfy the well known Euler–Lagrange equations:

$$\frac{d}{dt}\left(\frac{\partial F}{\partial y^i}\right) - \frac{\partial F}{\partial x^i} = 0. \tag{6.3}$$

According to (6.2) the **arc length parameter** s on C is given by

$$s = \int_a^t F(x(t), \dot{x}(t))\, dt, \tag{6.4}$$

and therefore

$$\frac{ds}{dt} = F(x(t), \dot{x}(t)). \tag{6.5}$$

Next, we consider s as parameter on C and taking into account the homogeneity of F we infer that

$$F(x(s), x'(s)) = 1; \quad x'(s) = \frac{dx}{ds}. \tag{6.6}$$

Then we have

$$\frac{\partial |F^*|}{\partial y^i} = 2\frac{\partial F}{\partial y^i} \text{ and } \frac{\partial |F^*|}{\partial x^i} = 2\frac{\partial F}{\partial x^i},$$

at any point $(x(s), x'(s))$ where $|F^*| = F^2$. Thus (6.3) becomes

$$\frac{d}{ds}\left(\frac{\partial F^*}{\partial y^i}\right)(x(s), x'(s)) - \frac{\partial F^*}{\partial x^i}(x(s), x'(s)) = 0. \tag{6.7}$$

Further, by direct calculations in (6.7) and using (1.7) we obtain

$$g_{ij}\frac{d^2x^j}{ds^2} + \frac{1}{2}\left(\frac{\partial^2 F^*}{\partial y^i \partial x^j}\frac{dx^j}{ds} - \frac{\partial F^*}{\partial x^i}\right) = 0. \tag{6.8}$$

Finally, denote

$$G^i(x, y) = \frac{1}{4}g^{ih}(x, y)\left(\frac{\partial^2 F^*}{\partial y^h \partial x^j}y^j - \frac{\partial F^*}{\partial x^h}\right)(x, y), \tag{6.9}$$

and deduce from (6.8) that the extremals for the length integral in $\mathbb{F}^m$ are given by equations

$$\frac{d^2x^i}{ds^2} + 2G^i(x(s), x'(s)) = 0. \tag{6.10}$$

Taking into account the homogeneity of both F^* and g^{ih}, we derive from (6.9) that G^i are positively homogeneous of degree two with respect to (y^i). Hence from Euler Theorem with respect to homogeneous functions we infer that

$$\frac{\partial G^i}{\partial y^k}y^k = 2G^i. \tag{6.11}$$

Therefore (6.10) becomes

$$\frac{d^2x^i}{ds^2} + \frac{\partial G^i}{\partial y^k}(x(s), x'(s))\frac{dx^k}{ds} = 0. \tag{6.12}$$

From now on we say that a non–degenerate smooth curve $\mathcal{C}$ in M is a **non–degenerate geodesic** of the pseudo–Finsler manifold $\mathbb{F}^m$ if for any coordinate system $\{(\mathcal{U}, \varphi); x^i\}$ in M there exists a parameter s on $\mathcal{C}$ such that $x^i = x^i(s)$ is a solution of the system of differential equations (6.10). Therefore any extremal for the length integral is a non–degenerate geodesic of $\mathbb{F}^m$.

Now we consider a Riemannian manifold $(M, g = (g_{ij}(x)))$ and recall that the equations of geodesics of M are given by (cf., Do Carmo [1], p. 62)

$$\frac{d^2x^i}{ds^2} + \Gamma_j{}^i{}_k(x(s))\frac{dx^j}{ds}\frac{dx^k}{ds} = 0, \tag{6.13}$$

where $\Gamma_j{}^i{}_k$ are the Christoffel symbols given by (3.30). It is easy to check that the functions

$$N^i_k(x,y) = y^j\Gamma_j{}^i{}_k(x), \tag{6.14}$$

satisfy (4.4) with respect to (1.1), that is, $HTM = (N^i_k(x,y))$ is a non–linear connection on TM. We call it the **canonical non–linear connection (canonical horizontal distribution)** of the Riemannian manifold (M, g). It is interesting to note that the Christoffel symbols can be obtained by differentiating (6.14) with respect to y^j, i.e., we have

$$\Gamma_j{}^i{}_k(x) = \frac{\partial N^i_k}{\partial y^j}(x,y). \tag{6.15}$$

As we shall see later in this section, this formula seems to have inspired Berwald to introduce a parallel transport in Finslerian geometry. Finally, by using (4.12), (6.14), and (3.28) we obtain

$$(a) \quad R^k{}_{ij}(x,y) = y^h R_h{}^k{}_{ij}(x) \quad \text{and} \quad (b) \quad \frac{\partial R^k{}_{ij}}{\partial y^h}(x,y) = R_h{}^k{}_{ij}(x). \tag{6.16}$$

Hence, taking into account Theorem 4.2 and using (6.16) we state that *the canonical horizontal distribution of (M, g) is integrable if and only if (M, g) is locally Euclidean.*

Next, we come back to the equations (6.12) of the geodesics of the pseudo–Finsler manifold $\mathbb{F}^m$. Comparing (6.12) with (6.13) and taking into account (6.14), it is natural to ask whether the functions

$$G^i_k(x,y) = \frac{\partial G^i}{\partial y^k}(x,y), \tag{6.17}$$

define, on each coordinate neighborhood of M', the local coefficients of a non–linear connection on M'. Checking (4.4) for G^i_k and using (6.9), (6.14), (1.2) and (1.3) we conclude that the answer is in the affirmative. As in the case of Riemannian manifolds we call $GM' = (G^i_k(x,y))$ the **canonical non–linear connection (canonical horizontal distribution)** of the pseudo–Finsler manifold $\mathbb{F}^m$. The local frame field $\{\delta^*/\delta^* x^i, \partial/\partial y^i\}$ on M', where we set

$$\frac{\delta^*}{\delta^* x^i} = \frac{\partial}{\partial x^i} - G^j_i \frac{\partial}{\partial y^j}, \tag{6.18}$$

is called the **canonical frame field** on M'.

In order to see how far is Finslerian geometry from the Riemannian geometry we consider the **generalized Christoffel symbols** of $\mathbb{F}^m$ given by (cf., Taylor [1])

$$\begin{cases} \gamma_{ihj}(x,y) &= \dfrac{1}{2}\left(\dfrac{\partial g_{ih}}{\partial x^j} + \dfrac{\partial g_{hj}}{\partial x^i} - \dfrac{\partial g_{ij}}{\partial x^h}\right)(x,y) \\ \gamma_i{}^k{}_j(x,y) &= g^{kh}(x,y)\gamma_{ihj}(x,y). \end{cases} \tag{6.19}$$

Then by direct calculations using (6.19), (1.7), (6.9) and the assertion (ii) of Remark 1.1 we obtain

$$\begin{cases} G^k(x,y) &= \frac{1}{2}\gamma_i{}^k{}_j(x,y)y^iy^j \\ G_h(x,y) &= g_{hk}(x,y)G^k(x,y) = \frac{1}{2}\gamma_{ihj}(x,y)y^iy^j. \end{cases} \tag{6.20}$$

We also need the **Cartan tensor field** whose local components are given by (cf., Cartan [1], p. 11)

$$\begin{cases} g_{ihj}(x,y) &= \dfrac{1}{2}\dfrac{\partial g_{ih}}{\partial y^j}(x,y) \\ g_i{}^k{}_j(x,y) &= g^{kh}(x,y)g_{ihj}(x,y). \end{cases} \tag{6.21}$$

It is easy to see that g_{ihj} and $g_i{}^k{}_j$ are the local components of a covariant symmetric Finsler tensor field of type (0,3) and (1,2) respectively. Any one of these two tensor fields will be refered to as the Cartan tensor field. Moreover, due to (1.17) we infer that

$$(a)\quad y^i g_{ihj} = 0\ ;\quad (b)\quad y^i g_i{}^k{}_j = 0. \tag{6.22}$$

The pseudo–Riemannian manifolds are characterized among the pseudo–Finsler manifolds by means of the Cartan tensor field as in the next theorem.

THEOREM 6.1 *A pseudo–Finsler manifold $\mathbb{F}^m = (M, M', F^*)$ is a pseudo–Riemannian manifold if and only if $M' = TM^0$ and the Cartan tensor field vanishes on TM^0.*

Further, differentiating the first equation in (6.19) with respect to y^k and taking into account (6.21) we derive

$$\frac{\partial \gamma_{ihj}}{\partial y^k} = \frac{\partial g_{ihk}}{\partial x^j} + \frac{\partial g_{khj}}{\partial x^i} - \frac{\partial g_{ikj}}{\partial x^h}. \tag{6.23}$$

Contracting (6.23) by y^i and y^j and using (6.22)a we deduce that

$$\frac{\partial \gamma_{ihj}}{\partial y^k} y^i y^j = 0. \tag{6.24}$$

Finally, we differentiate the second equation in (6.20) with respect to y^t and using (6.22) and (6.24) we obtain

$$G^k_t(x, y) = y^i \gamma_i{}^k{}_t(x, y) - 2G^h(x, y) g_h{}^k{}_t(x, y). \tag{6.25}$$

By Theorem 6.1 we see that (6.25) becomes (6.14) provided $\mathbb{F}^m$ is a pseudo–Riemannian manifold.

In order to construct some special Finsler connections on $\mathbb{F}^m$ we first suppose that, on M', there exists a non–linear connection $HM' = (N^k_j(x, y))$. Then, inspired by (6.15) we consider the functions

$$F_i{}^k{}_j(x, y) = \frac{\partial N^k_j}{\partial y^i}(x, y). \tag{6.26}$$

Taking into account that N^k_j satisfy (4.4), and using (6.26) and (1.3), we check that $F_i{}^k{}_j$ satisfy (5.3). Also, we see that (5.4) is satisfied for $C_i{}^k{}_j = 0$ on each coordinate neighborhood of M'. Hence the triple $FC = (N^k_i, F_i{}^k{}_j, 0)$ is a Finsler connection on $\mathbb{F}^m$. We call it an HM'–**Berwald connection.** If, in particular, HM' is just the canonical non–linear connection GM' of $\mathbb{F}^m$, we say that $\mathbf{BFC} = (G^k_i, G_i{}^k{}_j, 0)$, where we set

$$G_i{}^k{}_j = \frac{\partial G^k_j}{\partial y^i}, \tag{6.27}$$

is the **Berwald connection.** Using (5.42) and (4.12) we deduce that the local components of the torsion Finsler tensor fields of **BFC** satisfy

$$T_i{}^k{}_j = L_i{}^k{}_j = P^k{}_{ij} = S^k{}_{ij} = 0, \tag{6.28}$$

and

$$R^{*k}{}_{ij} = \frac{\delta^* G^k_i}{\delta^* x^j} - \frac{\delta^* G^k_j}{\delta^* x^i}. \tag{6.29}$$

Remark 6.1. Taking into account the homogeneity condition on F^* we may state the following assertions.

(i) G_i^k and $R^{*k}{}_{ij}$ are positively homogeneous of degree one with respect to (y^h).

(ii) $G_i{}^k{}_j$ are positively homogeneous of degree zero with respect to (y^h). ■

Contracting (6.27) by y^i and using the Euler Theorem on the homogeneity of G_j^k we infer that

$$y^i G_i{}^k{}_j = G_j^k. \tag{6.30}$$

Next, differentiating (6.29) with respect to y^h and taking into account (4.26) and (6.27) we obtain

$$\frac{\partial R^{*k}{}_{ij}}{\partial y^h} = \frac{\delta^* G_h{}^k{}_i}{\delta^* x^j} - \frac{\delta^* G_h{}^k{}_j}{\delta^* x^i} + G_h{}^t{}_i G_t{}^k{}_j - G_h{}^t{}_j G_t{}^k{}_i. \tag{6.31}$$

As $C_j{}^i{}_k = 0$, from (5.26) we derive $S_h{}^k{}_{ij} = 0$. Denote by $H_h{}^k{}_{ij}$ and $G_h{}^k{}_{ij}$ the h–curvature and the hv–curvature Finsler tensor field of the Berwald connection. Then by using (5.24), (5.25) and (6.31) we deduce that

$$H_h{}^k{}_{ij} = \frac{\partial R^{*k}{}_{ij}}{\partial y^h}, \tag{6.32}$$

and

$$G_h{}^k{}_{ij} = \frac{\partial G_h{}^k{}_i}{\partial y^j}. \tag{6.33}$$

Moreover, the assertion (i) in Remark 6.1 with respect to $R^{*k}{}_{ij}$ and Euler theorem yield

$$y^h H_h{}^k{}_{ij} = R^{*k}{}_{ij}. \tag{6.34}$$

The horizontal and vertical covariant derivatives of a Finsler tensor field $T^{i\cdots}_{j\cdots}$ with respect to the Berwald connection are denoted by $T^{i\cdots}_{j\cdots|_{\mathbf{b}}k}$ and $T^{i\cdots}_{j\cdots||_{\mathbf{b}}k}$ respectivley. Then by using (5.15) and (5.16) we deduce that

$$g_{ij|_{\mathbf{b}}k} = \frac{\delta^* g_{ij}}{\delta^* x^k} - g_{hj} G_i{}^h{}_k - g_{ih} G_j{}^h{}_k, \tag{6.35}$$

and

$$g_{ij||_{\mathbf{b}}k} = \frac{\partial g_{ij}}{\partial y^k} = 2 g_{ijk}. \tag{6.36}$$

Thus we may say that, in general, the Berwald connection is neither h–metric nor v–metric Finsler connection. Moreover, taking into account (6.36) and Theorem 6.1 we see that $\mathbb{F}^m$ *is a pseudo–Riemannian manifold if and only if the Berwald connection is a v–metric Finsler connection.*

By using (6.28), and the general Bianchi identities for a Finsler connection (see(5.53)–(5.63)) we obtain the following Bianchi identities for the Berwald connection:

$$(a)\ \sum_{(i,j,k)} \{R^{*t}{}_{ij|_{\mathbf{b}}k}\} = 0\,; \qquad (b)\ \sum_{(i,j,k)} \{H_i{}^t{}_{jk}\} = 0\,;$$

$$(c)\ H_k{}^t{}_{ij} = R^{*t}{}_{ij\|_{\mathbf{b}}k}\,; \qquad (d)\ G_i{}^t{}_{jk} = G_j{}^t{}_{ik}\,;$$

$$(e)\ G_i{}^t{}_{jk} = G_k{}^t{}_{ji}\,; \qquad (f)\ \sum_{(i,j,k)} \{H_\ell{}^t{}_{ij|_{\mathbf{b}}k} + G_\ell{}^t{}_{ih} R^{*h}{}_{jk}\} = 0\,;$$

$$(g)\ H_\ell{}^t{}_{jk\|_{\mathbf{b}}i} + \mathcal{A}_{jk}\{G_\ell{}^t{}_{ki|_{\mathbf{b}}j}\} = 0\,; \quad (h)\ G_\ell{}^t{}_{ki\|_{\mathbf{b}}j} = G_\ell{}^t{}_{kj\|_{\mathbf{b}}i}. \tag{6.37}$$

We should note that in case of **BFC** the identities (5.58), (5.59) and (5.63) are trivial. Also, (6.37c) is just (6.32), while (6.37d) and (6.37e) can be easily obtained from (6.33). Moreover, using (6.27) and (6.17) in (6.33) we deduce that

$$G_h{}^k{}_{ij} = \frac{\partial^3 G^k}{\partial y^h \partial y^i \partial y^j}. \tag{6.38}$$

Thus the hv–curvature of the Berwald connection is covariant symmetric. This strange property (especially with respect to the pair (i, j)) is possible because $P_i{}^h{}_{jk}$ from (5.30), in general, is not skew–symmetric with respect to the pair (j, k).

7. The Cartan and Rund Connections

The purpose of this section is to introduce both the Cartan connection and Rund connection on a pseudo–Finsler manifold and to study the main properties of these connections. Starting with a non–linear connection and two Finsler tensor fields, we construct a class of metric Finsler connections which contains the Cartan connection. It is noteworthy that the linear connection of the Cartan connection on the pseudo–Finsler manifold $\mathbb{F}^m = (M, M', F^*)$ is just the projection on VM' of the Levi–Civita connection on M' with respect to the Sasaki–Finsler metric on M'. The Rund connection is presented as a special Finsler connection situated in between the Cartan connection and Berwald connection. We close the section with interelations between the curvature Finsler tensor fields of Berwald, Cartan and Rund connections.

In the first part of this section, by using non–linear connections on M' we construct some metric Finsler connections on $\mathbb{F}^m$. To this end we first note that (5.17), (5.18) and (3.7) imply that a Finsler connection $FC = (HM', \nabla)$ is a metric Finsler connection if and only if

$$\begin{aligned}(\nabla_X g)(vY, vZ) &= X(g(vY, vZ)) - g(\nabla_X vY, vZ) \\ &\quad - g(vY, \nabla_X vZ) = 0, \qquad (7.1)\end{aligned}$$

for any $X, Y, Z \in \Gamma(TM')$. Next, we consider the associate linear connection D to FC on M' (see (5.27)) and denote by T^D its torsion tensor field. Then we may state the following theorem.

THEOREM 7.1 *(Bejancu–Farran [2]). Let HM' be a non–linear connection on M' and S and T be any two skew–symmetric Finsler tensor fields of type (1,2) on $\mathbb{F}^m$. Then there exists a unique linear connection ∇ on VM' satisfying the conditions:*

(i) ∇ is a metric connection, i.e., it satisfies (7.1);

(ii) T^D, S and T satisfy

(a) $T^D(vX, vY) = S(vX, vY)$; *(b)* $hT^D(hX, hY) = T(QhX, QhY)$, (7.2)

for any $X, Y \in \Gamma(TM')$, where Q is the associate almost product structure to HM'.

PROOF. Using g, h, v, Q, S and T we define a linear connection ∇ on VM' by

$$\begin{aligned}&2g(\nabla_{vX} vY, vZ) \\ &= vX(g(vY, vZ)) + vY(g(vZ, vX)) - vZ(g(vX, vY)) \\ &\quad + g(vY, [vZ, vX]) + g(vZ, [vX, vY]) - g(vX, [vY, vZ]) \qquad (7.3) \\ &\quad + g(vY, S(vZ, vX)) + g(vZ, S(vX, vY)) - g(vX, S(vY, vZ)),\end{aligned}$$

and

$$\begin{aligned}&2g(\nabla_{hX} QhY, QhZ) \\ &= hX(g(QhY, QhZ)) + hY(g(QhZ, QhX)) \\ &\quad - hZ(g(QhX, QhY)) + g(QhY, Qh[hZ, hX]) + g(QhZ, Qh[hX, hY]) \\ &\quad - g(QhX, Qh[hY, hZ]) + g(QhY, T(QhZ, QhX)) \\ &\quad + g(QhZ, T(QhX, QhY)) - g(QhX, T(QhY, QhZ)), \qquad (7.4)\end{aligned}$$

for any $X, Y, Z \in \Gamma(TM')$. It is easy to check that ∇ satisfies (i). Moreover, by (5.32) and (4.9b) we obtain

$$T^D(vX, vY) = \nabla_{vX}vY - \nabla_{vY}vX - [vX, vY], \tag{7.5}$$

and

$$hT^D(hX, hY) = Q(\nabla_{hX}QhY - \nabla_{hY}QhX - Qh[hX, hY]), \tag{7.6}$$

for any $X, Y \in \Gamma(TM')$. Then by using (7.3)–(7.6) we deduce (7.2). We now suppose that ∇' is another linear connection on VM' satisfying (i) and (ii). By (7.1) we have

$$\begin{aligned} vX(g(vY, vZ)) &= g(\nabla'_{vX}vY, vZ) + g(vY, \nabla'_{vX}vZ), \\ vY(g(vZ, vX)) &= g(\nabla'_{vY}vZ, vX) + g(vZ, \nabla'_{vY}vX), \\ vZ(g(vX, vY)) &= g(\nabla'_{vZ}vX, vY) + g(vX, \nabla'_{vZ}vY). \end{aligned}$$

Subtracting the third equation from the sum of the first two equations and taking into account (7.5) and (7.2a) we deduce that ∇' satisfies (7.3). Similarly, it follows that ∇' satisfies (7.4). Hence $\nabla' = \nabla$, and the proof is complete. ∎

Thus, starting with a non–linear connection HM' and two skew-symmetric Finsler tensor fields S and T on $\mathbb{F}^m$, by Theorem 7.1 we obtain a unique metric Finsler connection $FC = (HM', \nabla)$ satisfying (7.2), where ∇ is given by (7.3) and (7.4). We call FC the **(HM', S, T)–Cartan connection**. If in particular, HM' is just the cononical non–linear connection GM' of $\mathbb{F}^m$ and $S = T = 0$, we call FC the **Cartan connection** and, from now on, denote it by **FC***. To justify these names we find the local coefficients of the linear connection ∇. First, we set

$$(a)\quad S\left(\frac{\partial}{\partial y^k}, \frac{\partial}{\partial y^j}\right) = S^i{}_{jk}\frac{\partial}{\partial y^i}; \quad (b)\quad T\left(\frac{\partial}{\partial y^k}, \frac{\partial}{\partial y^j}\right) = T_j{}^i{}_k\frac{\partial}{\partial y^i}, \tag{7.7}$$

and

$$(a)\quad S_{hjk} = g_{hi}S^i{}_{jk}\ ; \quad (b)\quad T_{jhk} = g_{hi}T_j{}^i{}_k. \tag{7.8}$$

Then we replace vX, vY and vZ in (7.3) by $\partial/\partial y^k$, $\partial/\partial y^j$ and $\partial/\partial y^h$ respectively, and using (5.2), (2.11), (1.7), (7.7a) and (7.8a) we obtain

$$C_j{}^i{}_k = \frac{1}{2}g^{ih}\left(\frac{\partial g_{jh}}{\partial y^k} + S_{jkh} + S_{hjk} - S_{khj}\right). \tag{7.9}$$

Similarly, we replace hX, hY and hZ in (7.4) by $\delta/\delta x^k$, $\delta/\delta x^j$ and $\delta/\delta x^h$ respectively, and by using (5.1), (2.11), (4.11), (7.7b) and (7.8b) we deduce that

$$F_j{}^i{}_k = \frac{1}{2}g^{ih}\left(\frac{\delta g_{hj}}{\delta x^k} + \frac{\delta g_{hk}}{\delta x^j} - \frac{\delta g_{jk}}{\delta x^h} + T_{kjh} + T_{jhk} - T_{hkj}\right). \tag{7.10}$$

In particular, if we take $HM' = GM'$ and $S = T = 0$ in (7.9) and (7.10) we obtain the local coefficients of the Cartan connection $\mathbf{FC}^* = (GM', \nabla^*) = (G^i_k, F^{*i}_{j\ k}, C^{*i}_{j\ k})$. Thus $C^{*i}_{j\ k} = g_j{}^i{}_k$ (see (6.21)), G^i_k is given by (6.17) and

$$F^{*i}_{j\ k} = \frac{1}{2} g^{ih} \left(\frac{\delta^* g_{hj}}{\delta^* x^k} + \frac{\delta^* g_{hk}}{\delta^* x^j} - \frac{\delta^* g_{jk}}{\delta^* x^h} \right), \tag{7.11}$$

where $\left\{ \delta^* / \delta^* x^k \right\}$ are given by (6.18). It is easy to see that our formulas (6.21) and (7.11) are nothing but the formulas (V) and (XI) in Cartan [1].

In order to obtain the basic properties of the covariant derivative defined by the Cartan connection we need a study of the geometry of the vertical vector bundle VM' of $\mathbb{F}^m$. Since in general g as defined by (2.10) is a pseudo–Riemannian metric on VM', a Finsler vector has a causal character defined as follows. We say that $X \in VM'_u$ is

spacelike if $g_u(X, X) > 0$ or $X = 0$,
timelike if $g_u(X, X) < 0$,
lightlike (null) if $g_u(X, X) = 0$ and $X \neq 0$,

where $g_u = g(u)$. We keep the same terminology for sections of VM'. The **Finsler norm (length)** of X is a non–negative number $\|X\|$ defined by

$$\|X\| = |g_u(X, X)|^{\frac{1}{2}}.$$

If $g_u(X, X) = 1$ or $g_u(X, X) = -1$ we say that X is a **unit spacelike** or **unit timelike Finsler vector**. If X is a unit Finsler vector then $\varepsilon = g_u(X, X)$ is called the **signature** of X.

We now suppose that the Liouville vector field L is not lightlike at any point of M' and denote by $\{L\}$ the line vector sub–bundle spanned by L in VM'. Denote by $\mathcal{L}M'$ the complementary orthogonal vector bundle to $\{L\}$ in VM' with respect to g and call it the **Liouville distribution** of $\mathbb{F}^m$. Then we prove the following theorem:

THEOREM 7.2 *(Bejancu–Farran [1]). The Liouville distribution of a pseudo–Finsler manifold is integrable.*

PROOF. Let $X, Y \in \Gamma(\mathcal{L}M')$. As VM' is an integrable distribution on M' we deduce that $[X, Y] \in \Gamma(VM')$. Hence we need only to show that $[X, Y]$ has no component with respect to L. To this end we note that $X = X^j(\partial/\partial y^j)$ belongs to the Liouville distribution if and only if

$$g_{ij}(x, y) y^i X^j(x, y) = 0. \tag{7.12}$$

Differentiating (7.12) with respect to y^k and using (1.17), we obtain

$$g_{kj}(x,y)X^j(x,y) + g_{ij}(x,y)y^i \frac{\partial X^j}{\partial y^k}(x,y) = 0. \tag{7.13}$$

Then by direct calculations using (2.1), (2.3), (2.11) and (7.13) we derive

$$g([X,Y],L) = g_{ij}y^i \left(\frac{\partial Y^j}{\partial y^k} X^k - \frac{\partial X^j}{\partial y^k} Y^k \right) = 0,$$

which completes the proof of our assertion. ■

Now, in the case of a Finsler manifold the Liouville vector field is automatically spacelike, and hence the Liouville distribution is always defined. Using the above theorem we deduce the following corollary:

COROLLARY 7.1 *On a Finsler manifold $\mathbb{F}^m = (M, M', F)$, the Liouville distribution is integrable and hence, every leaf of VM' is locally a product of a leaf of $\{L\}$ and a leaf of $\mathcal{L}M'$.*

Next, suppose L is spacelike on M'. Then (1.14) implies

$$F^* = g(L,L) > 0, \tag{7.14}$$

on M'. Thus

$$\ell = \frac{1}{F} L = \frac{y^i}{F} \frac{\partial}{\partial y^i}, \tag{7.15}$$

is a unit spacelike Finsler vector field, i.e., we have

$$g(\ell, \ell) = 1. \tag{7.16}$$

By means of g and ℓ we define the global Finsler 1–form η by

$$\eta(X) = g(X, \ell), \quad \forall X \in \Gamma(VM'), \tag{7.17}$$

which by (7.16) satisfies

$$\eta(\ell) = 1. \tag{7.18}$$

We denote by P the projection morphism of VM' on $\mathcal{L}M'$ with respect to the decomposition

$$VM' = \mathcal{L}M' \oplus \{L\}.$$

Then by using (7.17) we infer that

$$X = PX + \eta(X)\ell, \tag{7.19}$$

$$(a)\quad \eta\circ P=0,\qquad (b)\quad P\ell=0, \tag{7.20}$$

and

$$g(X,PY)=g(Y,PX)=g(PX,PY)=g(X,Y)-\eta(X)\eta(Y), \tag{7.21}$$

for any $X,Y\in\Gamma(VM')$. Finally, we define the Finsler tensor field $a\in\Gamma(T_2^0(VM'))$ by

$$a(X,Y)=g(PX,PY),\ \ \forall X,Y\in\Gamma(VM'). \tag{7.22}$$

The local components of the Finsler tensor fields η,ℓ,P and a are given by

$$\eta_i = \eta\left(\frac{\partial}{\partial y^i}\right)=\frac{1}{F}g_{ij}y^j=\frac{\partial F}{\partial y^i}, \tag{7.23}$$

$$\ell^i = \theta^i(\ell)=\frac{1}{F}y^i=g^{ik}\frac{\partial F}{\partial y^k}, \tag{7.24}$$

$$P_i^j = \theta^j\left(P\left(\frac{\partial}{\partial y^i}\right)\right)=\delta_i^j-\eta_i\ell^j, \tag{7.25}$$

$$a_{ij} = g_{ij}-\eta_i\eta_j, \tag{7.26}$$

respectively. Comparing (7.26) with (16.5) in Matsumoto [6] we deduce that the Finsler tensor field a is nothing but the **angular metric** of $\mathbb{F}^m$. Moreover, from (7.22) we deduce the following proposition:

PROPOSITION 7.1 *The angular metric of a pseudo-Finsler manifold $\mathbb{F}^m$ is degenerate on VM', i.e., we have*

$$a(X,\ell)=0,\ \ \forall X\in\Gamma(VM'), \tag{7.27}$$

and coincides with the pseudo-Finsler metric g on the Liouville distribution.

The horizontal and vertical covariant derivatives of a Finsler tensor field $T_{j\cdots}^{i\cdots}$ with respect to the Cartan connection are denoted by $T_{j\cdots|*k}^{i\cdots}$ and $T_{j\cdots||*k}^{i\cdots}$ respectively. Then by condition (i) in Theorem 7.1 we have

$$(a)\quad g_{ij|*k}=0\quad\text{and}\quad (b)\quad g_{ij||*k}=0. \tag{7.28}$$

Contracting (7.11) by y^j and taking into account (6.18), (6.19), (6.21), (6.11), and (6.17) we derive

$$y^jF_j^{*i}{}_k=G_k^i. \tag{7.29}$$

Then by using (7.29) and (6.22b) in (5.8) and (5.9) we obtain

$$y^k{}_{|*j}=0, \tag{7.30}$$

and

$$y^k{}_{\|*j} = \delta^k_j, \tag{7.31}$$

respectively. It is easy to see that

$$\nabla^*_X L = vX, \quad \forall X \in \Gamma(TM'), \tag{7.32}$$

is the coordinate free equality for both (7.30) and (7.31). The formulas stated in the next theorem are obtained by direct calculations using (7.1) and (7.32).

THEOREM 7.3 *Let $\mathbb{F}^m = (M, M', F^*)$ be a pseudo–Finsler manifold such that the Liouville vector field is spacelike on M'. Then we have*

$$X(F) = \eta(vX), \tag{7.33}$$

$$\nabla^*_X \ell = \frac{1}{F} PvX, \tag{7.34}$$

$$(\nabla^*_X \eta)\, vY = \frac{1}{F} a\,(vX, vY)\,, \tag{7.35}$$

$$(\nabla^*_X P)\, vY = -\frac{1}{F}\left(a\,(vX, vY)\,\ell + \eta(vY) PvX\right), \tag{7.36}$$

and

$$(\nabla^*_X a)\,(vY, vZ) = -\frac{1}{F}\left(a\,(vX, vY)\,\eta(vZ) + a\,(vX, vZ)\,\eta(vY)\right), \tag{7.37}$$

for any $X, Y, Z \in \Gamma(TM')$.

In particular, from (7.33) we deduce that

$$(a)\quad \ell(F) = 1 \quad \text{and} \quad (b) \quad X(F) = 0, \;\; \forall X \in \Gamma\left(GM' \oplus \mathcal{L}M'\right). \tag{7.38}$$

Next, we consider the associate almost product structure Q^* to the canonical non–linear connection GM' on M' (see (4.8)). Then by using the pseudo–Finsler metric g and the projection morphisms v^* and h^* of TM' on VM' and GM' respectively, we define

$$\begin{cases} G : \Gamma(TM') \times \Gamma(TM') \longrightarrow \mathcal{F}(M') \\ G(X, Y) = g(v^*X, v^*Y) + g\left(Q^*h^*X, Q^*h^*Y\right), \;\; \forall X, Y \in \Gamma(TM'). \end{cases} \tag{7.39}$$

Clearly G is a symmetric tensor field of type (0,2) on M'. Moreover it is non–degenerate and has a constant index on M'. Indeed, the matrix of local components of G with respect to the frame field $\{\delta^*/\delta^* x^i,\ \partial/\partial y^i\}$ adapted to the decomposition

$$TM' = GM' \oplus VM', \tag{7.40}$$

is given by

$$[G_{AB}] = \begin{bmatrix} g_{ij} & 0 \\ 0 & g_{ij} \end{bmatrix}, \quad \begin{array}{ll} A, B \in & \{1, \cdots, 2m\}, \\ i, j \in & \{1, \cdots, m\}. \end{array} \tag{7.41}$$

Thus G is a pseudo–Riemannian metric on M' of index $2q$ where q is the index of the pseudo–Finsler metric g of $\mathbb{F}^m$. We call G the **Sasaki–Finsler metric** on M'. Denote by ∇' the Levi–Civita connection on M' with respect to G. Then we state the following important result:

THEOREM 7.4 *(Bejancu-Farran [3]). Let* $\mathbf{FC}^* = (GM', \nabla^*)$ *be the Cartan connection of the pseudo–Finsler manifold* $\mathbb{F}^m$*. Then* ∇^* *is the projection of the Levi–Civita connection* ∇' *on the vertical vector bundle, i.e., we have*

$$\nabla^*_X Y = v\nabla'_X Y, \quad \forall X \in \Gamma(TM'), \; Y \in \Gamma(VM'). \tag{7.42}$$

PROOF. According to (3.15), ∇' is given by

$$\begin{aligned} 2G\left(\nabla'_X Y, Z\right) \; = \; & X\left(G(Y,Z)\right) + Y\left(G(Z,X)\right) - Z\left(G(X,Y)\right) \\ & + G\left([X,Y],Z\right) + G\left([Z,X],Y\right) - G\left([Y,Z],X\right), \end{aligned} \tag{7.43}$$

$\forall X, Y, Z \in \Gamma(TM')$.
To prove (7.42) we first set:

$$v\nabla'_{\partial/\partial y^j}\frac{\partial}{\partial y^i} = A_i{}^k{}_j\frac{\partial}{\partial y^k} \quad \text{and} \quad v\nabla'_{\delta^*/\delta^* x^j}\frac{\partial}{\partial y^i} = B_i{}^k{}_j\frac{\partial}{\partial y^k}.$$

Then in (7.43) we replace (X, Y, Z) in turn by $(\partial/\partial y^j,\ \partial/\partial y^i,\ \partial/\partial y^k)$ and $(\delta^*/\delta^* x^j,\ \partial/\partial y^i,\ \partial/\partial y^k)$ and by using (6.21), (4.26) for the canonical non-linear connection and (6.27), we obtain

$$A_i{}^k{}_j = \frac{1}{2}g^{kh}\frac{\partial g_{hi}}{\partial y^j} = g_i{}^k{}_j, \tag{7.44}$$

and

$$B_i{}^k{}_j = \frac{1}{2}g^{kh}\left(\frac{\delta^* g_{hi}}{\delta^* x^j} + G_{jhi} - G_{jih}\right), \tag{7.45}$$

where we set

$$G_{jhi} = g_{hk}G_j{}^k{}_i. \tag{7.46}$$

Next, by direct calculations using (1.7) and (6.6) we infer that

$$\frac{\partial}{\partial y^j}\left(\frac{\partial G_h}{\partial y^i} - \frac{\partial G_i}{\partial y^h}\right) = \frac{\partial g_{hj}}{\partial x^i} - \frac{\partial g_{ij}}{\partial x^h}, \tag{7.47}$$

where G_h is given by (6.20). Further, by using (6.17), (7.47) and (6.18) we derive

$$\begin{aligned} G_{jhi} - G_{jih} &= g_{th}\frac{\partial^2 G^t}{\partial y^i \partial y^j} - g_{ti}\frac{\partial^2 G^t}{\partial y^h \partial y^j} \\ &= \frac{\partial}{\partial y^j}\left(\frac{\partial G_h}{\partial y^i} - \frac{\partial G_i}{\partial y^h}\right) - G_i^t\frac{\partial g_{th}}{\partial y^j} + G_h^t\frac{\partial g_{ti}}{\partial y^j} \\ &= \frac{\delta^* g_{hj}}{\delta^* x^i} - \frac{\delta^* g_{ij}}{\delta^* x^h}. \end{aligned} \tag{7.48}$$

Finally, by using (7.48) in (7.45) and taking into account (7.11) we deduce that $B_i{}^k{}_j = F_i^{*k}{}_j$, which together with (7.44) proves the assertion of the theorem. ∎

THEOREM 7.5 *The associate linear connection D^* to the Cartan connection $\mathbf{FC}^* = (GM', \nabla^*)$ is a metric linear connection.*

PROOF. First, by using (4.8) and (7.39) we deduce that Q^* is both a self–adjoint and a linear isometry with respect to G, i.e., we have

$$(a)\ G(Q^*X, Y) = G(X, Q^*Y) \ \text{ and } \ (b)\ G(Q^*X, Q^*Y) = G(X, Y), \tag{7.49}$$

for any $X, Y \in \Gamma(TM')$. Then by using (5.26) and (7.49) and taking into account that $\mathbf{FC}^*$ is a metric Finsler connection we obtain

$$(D_X^*G)(Y, Z) = (\nabla_X^* g)(v^*Y, v^*Z) + (\nabla_X^* g)(Q^*h^*Y, Q^*h^*Z) = 0,$$

for any $X, Y, Z \in \Gamma(TM')$. ∎

We further present some important identities satisfied by the torsion and curvature Finsler tensor fields of the Cartan connection. First, by means of G and R^{D^*} we define a tensor field of type (0,4) on M', that we still denote by R^{D^*}, as follows

$$R^{D^*}(X, Y, Z, U) = G\left(R^{D^*}(X, Y)Z, U\right), \ \forall X, Y, Z, U \in \Gamma(TM').$$

Then we have the identities:

$$\begin{cases} (a)\ \ R^{D^*}(X, Y, Z, U) + R^{D^*}(Y, X, Z, U) = 0, \\ (b)\ \ R^{D^*}(X, Y, Z, U) + R^{D^*}(X, Y, U, Z) = 0. \end{cases} \tag{7.50}$$

We have to note that (7.50b) is a direct consequence of Theorem 7.5. Locally, by using (5.29)–(5.31) we have:

$$\left\{\begin{array}{lcl} R^*_{ijkh} & = & G\left(R^{D^*}\left(\frac{\delta}{\delta x^h}, \frac{\delta}{\delta x^k}\right)\frac{\partial}{\partial y^i}, \frac{\partial}{\partial y^j}\right) = R^{*t}_{i\ kh} g_{tj}, \\ P^*_{ijkh} & = & G\left(R^{D^*}\left(\frac{\partial}{\partial y^h}, \frac{\delta}{\delta x^k}\right)\frac{\partial}{\partial y^i}, \frac{\partial}{\partial y^j}\right) = P^{*t}_{i\ kh} g_{tj}, \\ S^*_{ijkh} & = & G\left(R^{D^*}\left(\frac{\partial}{\partial y^h}, \frac{\partial}{\partial y^k}\right)\frac{\partial}{\partial y^i}, \frac{\partial}{\partial y^j}\right) = S^{*t}_{i\ kh} g_{tj}. \end{array}\right. \tag{7.51}$$

Then (7.50a) and (7.50b) become

$$(a)\ R^*_{ijkh} + R^*_{ijhk} = 0; \quad (b)\ S^*_{ijkh} + S^*_{ijhk} = 0, \tag{7.52}$$

and

$$(a)\ R^*_{ijkh} + R^*_{jikh} = 0; \quad (b)\ P^*_{ijkh} + P^*_{jikh} = 0;$$
$$(c)\ S^*_{ijkh} + S^*_{jikh} = 0, \tag{7.53}$$

respectively.

As for the Cartan connection **FC*** we have $T^* = S^* = 0$ in (7.2), by using (5.38), (5.40), (5.41a) and (5.41d) we infer that

$$(a)\ T^{*k}_{i\ j} = 0 \quad \text{and} \quad (b)\ S^{*k}_{\ \ ij} = 0. \tag{7.54}$$

We show now that the torsion Finsler tensor fields $R^{*k}_{\ \ ij}$ and $P^{*k}_{\ \ ij}$ are expressed by means of the h–curvature and the hv–curvature Finsler tensor fields of **FC*** respectively. First, by using (6.18) and (7.29) we obtain

$$\frac{\delta^* F_{i\ j}^{\ h}}{\delta^* x^k} y^i = \frac{\delta^* G^h_j}{\delta^* x^k} + G^t_k F^{*h}_{t\ j}. \tag{7.55}$$

Then contracting (5.24) by y^i and taking into account (7.55), (7.29), (6.22) and (6.29) we derive

$$(a)\ y^i R^{*h}_{i\ jk} = R^{*h}_{\ \ jk} \quad \text{and} \quad (b)\ y^i R^*_{itjk} = R^*_{tjk}, \tag{7.56}$$

where we set

$$R^*_{tjk} = g_{th} R^{*h}_{\ \ jk}.$$

Similarly, contracting (5.25) by y^i and using (7.29), (6.22), (6.27), and (5.42b) we obtain

$$(a)\ y^i P^{*h}_{i\ jk} = G_{j\ k}^{\ h} - F^{*h}_{k\ j} = P^{*h}_{\ \ jk} \quad \text{and} \quad (b)\ y^i P^*_{itjk} = P^*_{tjk}, \tag{7.57}$$

where we set

$$P^*_{tjk} = g_{th} P^{*h}_{\ \ jk}.$$

Taking into account (6.27) and (7.11) we deduce that $P^{*h}_{\ \ jk}$ is a symmetric Finsler tensor field. Moreover, (6.30), (7.29) and (7.57a) imply that

$$y^j P^{*h}_{\ \ jk} = 0. \tag{7.58}$$

As $F^{*i}_{j\ k}$ and $g_j{}^i{}_k$ are positively homogeneous of degrees zero and -1 with respect to y^h, by Euler Theorem we have

$$(a)\quad \frac{\partial F^{*i}_{j\ k}}{\partial y^h} y^h = 0 \quad\text{and}\quad (b)\quad \frac{\partial g_j{}^i{}_k}{\partial y^h} y^h = -g_j{}^i{}_k. \tag{7.59}$$

Then, contracting (5.25) by y^k and using (7.59), (6.30) and (6.22) we obtain

$$y^k P^{*h}_{i\ jk} = 0. \tag{7.60}$$

Finally, contracting (5.26) in turn by y^i and y^k and using (6.22) and (6.17) we deduce that

$$(a)\quad y^i S^{*h}_{i\ jk} = 0 \quad\text{and}\quad (b)\quad y^k S^{*h}_{i\ jk} = 0. \tag{7.61}$$

Next, using (7.54) in (5.53)–(5.59) we obtain the first group of Bianchi identities with respect to the Cartan connection:

$$(a)\quad \sum_{(i,j,k)} \{R^{*t}_{\ \ ij|_*k} + P^{*t}_{\ \ ih} R^{*h}_{\ \ jk}\} = 0,$$

$$(b)\quad \sum_{(i,j,k)} \{g_i{}^t{}_h R^{*h}_{\ \ jk} - R^{*t}_{i\ jk}\} = 0,$$

$$(c)\quad R^{*t}_{k\ ij} = R^{*t}_{\ \ ij\|_*k} + \mathcal{A}_{(ij)}\{P^{*t}_{\ \ jk|_*i} + R^{*t}_{\ \ ih} g_j{}^h{}_k + P^{*t}_{\ \ ih} P^{*h}_{\ \ jk}\},$$

$$(d)\quad \mathcal{A}_{(ij)}\{g_i{}^t{}_{k|_*j} + g_j{}^t{}_h P^{*h}_{\ \ ik} + P^{*t}_{i\ jk}\} = 0,$$

$$(e)\quad \mathcal{A}_{(jk)}\{P^{*t}_{\ \ ij\|_*k} + P^{*t}_{j\ ik} + g_i{}^h{}_k P^{*t}_{\ \ hj}\} = 0,$$

$$(f)\quad S^{*t}_{i\ jk} = \mathcal{A}_{(jk)}\{g_i{}^t{}_{j\|_*k} + g_i{}^h{}_k g_h{}^t{}_j\},$$

$$(g)\quad \sum_{(i,j,k)} \{S^{*t}_{i\ jk}\} = 0. \tag{7.62}$$

Similarly, from (5.60)–(5.63) we derive the second group of Bianchi identities with respect to the Cartan connection:

$$(a)\quad \sum_{(i,j,k)} \{R^{*t}_{\ell\ ij|_*k} + P^{*t}_{\ell\ ih} R^{*h}_{\ \ jk}\} = 0,$$

$$(b)\quad R^{*t}_{\ell\ jk||*i} + S^{*t}_{\ell\ ih}R^{*h}{}_{jk} + \mathcal{A}_{(jk)}\{P^{*t}_{\ell\ ki|*j} + P^{*t}_{\ell\ jh}P^{*h}{}_{ki} + R^{*t}_{\ell\ jh}g_k{}^h{}_i\} = 0,$$

$$(c)\quad S^{*t}_{\ell\ ij|*k} + \mathcal{A}_{(ij)}\{P^{*t}_{\ell\ ki||*j} + P^{*t}_{\ell\ hi}g_k{}^h{}_j + S^{*t}_{\ell\ jh}P^{*h}{}_{ki}\} = 0,$$

$$(d)\quad \sum_{(i,j,k)}\{S^{*t}_{\ell\ ij||*k}\} = 0. \tag{7.63}$$

Contracting (7.62b) by g_{ts} and using (7.53a) we deduce that

$$\sum_{(i,j,k)}\{g_{ish}R^{*h}{}_{jk} + R^{*}_{sijk}\} = 0. \tag{7.64}$$

Finally, contracting (7.64) by y^s and taking into account (6.22a) and (7.56b) we infer that

$$\sum_{(i,j,k)}\{R^{*}_{ijk}\} = 0. \tag{7.65}$$

PROPOSITION 7.2 *Let* $\mathbf{FC}^* = (G^i_k, F^{*i}_j{}_k, g_j{}^i{}_k)$ *be the Cartan connection of a pseudo-Finsler manifold* $\mathbb{F}^m$. *Then we have*

$$g_{ih}\frac{\partial F^{*i}_j{}_k}{\partial y^t} = g_{hjt|*k} - g_{hjr}P^{*r}{}_{kt} + \mathcal{A}_{(hj)}\{g_{hkt|*j} + g_{jkr}P^{*r}{}_{ht}\}. \tag{7.66}$$

PROOF. First we rewrite (7.11) as follows

$$F^{*}_{jhk} = g_{ih}F^{*i}_j{}_k = \frac{1}{2}\left(\frac{\delta^* g_{hj}}{\delta^* x^k} + \frac{\delta^* g_{hk}}{\delta^* x^j} - \frac{\delta^* g_{jk}}{\delta^* x^h}\right). \tag{7.67}$$

Then by direct calculations using (7.67), (4.26), (6.27), and (6.21) we deduce that

$$\frac{\partial F^{*}_{jhk}}{\partial y^t} = \frac{\delta^* g_{hjt}}{\delta^* x^k} - g_{hjr}G_t{}^r{}_k + \mathcal{A}_{(hj)}\left\{\frac{\delta^* g_{hkt}}{\delta^* x^j} + G_t{}^r{}_h g_{rjk}\right\}. \tag{7.68}$$

From the general formula for the horizontal covariant derivative (see (5.15)), with respect to $\mathbf{FC}^*$ it follows that

$$\frac{\delta^* g_{hjt}}{\delta^* x^k} = g_{hjt|*k} + g_{sjt}F^{*s}_h{}_k + g_{hst}F^{*s}_j{}_k + g_{hjs}F^{*s}_t{}_k. \tag{7.69}$$

Thus taking into account (7.69) and the second equality of (7.57a) in (7.68) we obtain

$$\frac{\partial F^{*}_{jhk}}{\partial y^t} = g_{hjt|*k} + 2g_{htr}F^{*r}_j{}_k - g_{hjr}P^{*r}{}_{kt} + \mathcal{A}_{(hj)}\{g_{hkt|*j} + g_{jkr}P^{*r}{}_{ht}\}. \tag{7.70}$$

Finally, (7.66) follows by differentiating the first equality in (7.67) with respect to y^t and using (7.70) and (6.21). ■

PROPOSITION 7.3 *The torsion and curvature Finsler tensor fields P^*_{ijk} and P^*_{ijkh} of the Cartan connection are expressed only by means of the Cartan tensor field and its horizontal covariant derivative.*

PROOF. First by using (7.57a) and the horizontal covariant derivative of the Cartan tensor field in (5.25) we deduce that

$$P^{*t}_{i\ hk} = \frac{\partial F^{*t}_{i\ h}}{\partial y^k} - g_{i\ k|*h}^{\ t} + g_{i\ r}^{\ t} P^{*r}_{\ \ hk}. \tag{7.71}$$

Contracting (7.71) by g_{jt} and using (7.66), (6.21), and (7.28a) we infer that

$$P^*_{ijhk} = \mathcal{A}_{(ij)}\{g_{jhk|*i} + g_{ihr}P^{*r}_{\ \ jk}\}. \tag{7.72}$$

Further, contracting (7.72) by y^i and using (7.57b) and (7.58) we obtain

$$P^*_{jhk} = g_{jhk|*i}y^i. \tag{7.73}$$

Thus due to (7.73) we see that (7.72) becomes

$$P^*_{ijhk} = \mathcal{A}_{(ij)}\{g_{jhk|*i} + g_{i\ h}^{\ r} g_{rjk|*s}y^s\}, \tag{7.74}$$

which together with (7.73) completes the proof of our assertion. ■

Finally, we should remark that owing to (6.21) and (1.7), the v–curvature Finsler tensor field of the Cartan connection has a simple expression. More precisely from (5.26) we deduce that

$$S^{*h}_{i\ jk} = g_{i\ j}^{\ r} g_{r\ k}^{\ h} - g_{i\ k}^{\ r} g_{r\ j}^{\ h}. \tag{7.75}$$

As we know (see Theorem 6.1), the Cartan tensor field is an obstruction for a Finsler manifold to be a Riemannian manifold. It is interesting that the vanishing of S on $\mathbb{F}^m$ leads us to the same conclusion which we state in the next theorem.

THEOREM 7.6 *(Brickell [1]). Any Finsler manifold $\mathbb{F}^m = (M, TM^0, F)$ $m \geq 3$, with vanishing v–curvature Finsler tensor field of the Cartan connection, is a Riemannian manifold, provided $F(x, -y) = F(x, y)$ at any point $(x, y) \in TM^0$.*

We have to stress that in the above theorem we have the condition $M' = TM^0$ which can not be removed (see Kikuchi [2]).

Next, we consider a new Finsler connection $\mathbf{RFC} = (G_i^k, F_{i\ j}^{*k}, 0)$ where $F_{i\ j}^{*k}$ are given by (7.11), and call it the **Rund connection** (cf., Matsumoto [6], p. 116). As the h–covariant derivative and the v–covariant derivative defined by **RFC** coincide with the h–covariant derivative and v–covariant derivative defined by $\mathbf{FC}^*$ and **BFC** respectively, we may say that the Rund connection is in between the Cartan connection and the Berwald connection.

In order to be consistent with our previous notations, we denote by $T^{i\cdots}_{j\cdots|\mathbf{r}k}$ and $T^{i\cdots}_{j\cdots||\mathbf{r}k}$ the h–covariant and v–covariant derivatives of the Finsler tensor field $T^{i\cdots}_{j\cdots}$ with respect to **RFC**. Thus, in particular, from (7.28a) and (6.36) we infer that

$$(a) \quad g_{ij|\mathbf{r}k} = 0 \quad \text{and} \quad (b) \quad g_{ij||\mathbf{r}k} = 2g_{ijk}. \tag{7.76}$$

Taking into account (4.12) and (5.42), we see that the only surviving torsion Finsler tensor fields of **RFC** are $R^k{}_{ij}$ and $P^k{}_{ij}$ which coincide with $R^{*k}{}_{ij}$ (see (6.29)) and $P^{*k}{}_{ij}$ (see (7.57a)) of $\mathbf{FC}^*$. The h–curvature, hv–curvature and, v–curvature Finsler tensor fields of **RFC** are denoted by $K_i{}^h{}_{jk}$, $F_i{}^h{}_{jk}$ an $S_i{}^h{}_{jk}$ respectively. Then by using the general formulas (5.24)–(5.26) and taking into account that $C_i{}^k{}_j = 0$, we obtain:

$$\begin{cases} (a) \quad K_i{}^h{}_{jk} = \mathcal{A}_{(jk)}\left\{\dfrac{\delta^* F_{i\ j}^{*h}}{\delta^* x^k} + F_{i\ j}^{*r} F_{r\ k}^{*h}\right\}, \\ (b) \quad F_i{}^h{}_{jk} = \dfrac{\partial F_{i\ j}^{*h}}{\partial y^k}, \quad (c) \quad S_i{}^h{}_{jk} = 0. \end{cases} \tag{7.77}$$

As $T_i{}^k{}_j = C_i{}^k{}_j = S^k{}_{ij} = 0$, the first group of Bianchi identities (5.53)–(5.59) for the Rund connection becomes

$$(a) \quad \sum_{(i,j,k)} \{R^{*t}{}_{ij|\mathbf{r}k} + P^{*t}{}_{ih} R^{*h}{}_{jk}\} = 0,$$

$$(b) \quad \sum_{(i,j,k)} \{K_i{}^t{}_{jk}\} = 0,$$

$$(c) \quad R^{*t}{}_{ij||\mathbf{r}k} - K_k{}^t{}_{ij} + \mathcal{A}_{(ij)}\{P^{*t}{}_{jk|\mathbf{r}i} + P^{*t}{}_{ih} P^{*h}{}_{jk}\} = 0,$$

$$(d) \quad F_i{}^h{}_{jk} - F_j{}^h{}_{ik} = 0,$$

$$(e) \quad \mathcal{A}_{(jk)}\{P^{*t}{}_{ij||\mathbf{r}k} + F_j{}^t{}_{ik}\} = 0, \tag{7.78}$$

since (5.58) and (5.59) are trivial in this case. Similarly the second group of Bianchi indentities (5.60)–(5.63) becomes

$$(a) \qquad \sum_{(i,j,k)} \{K_{\ell\ ij|\mathbf{r}k}^{\ t} + F_{\ell\ ih}^{\ t} R^{*h}_{\ \ jk}\} = 0,$$

$$(b) \qquad K_{\ell\ jk\|\mathbf{r}i}^{\ t} + \mathcal{A}_{(jk)}\{F_{\ell\ ki|\mathbf{r}j}^{\ t} + F_{\ell\ jh}^{\ t} P^{*h}_{\ \ ki}\} = 0,$$

$$(c) \qquad F_{\ell\ ki\|\mathbf{r}j}^{\ t} - F_{\ell\ kj\|\mathbf{r}i}^{\ t} = 0, \tag{7.79}$$

since (5.63) is trivial for **RFC**.

Now we present the interrelations between the curvature Finsler tensor fields of the Cartan, Rund, and Berwald connections. First, we note that by (5.24) the h–curvature Finsler tensor field of type (0, 4) of the Cartan connection is given by

$$R^*_{ijkh} = g_{tj}\mathcal{A}_{(kh)}\left\{\frac{\delta^* F^{*t}_{i\ k}}{\delta^* x^h} + F^{*r}_{i\ k}F^{*t}_{r\ h}\right\} + g_{ijr}R^{*r}_{\ \ kh}. \tag{7.80}$$

Thus, by using (7.77a) in (7.80) we deduce that

$$R^*_{ijkh} = K_{ijkh} + g_{ijr}R^{*r}_{\ \ kh} \tag{7.81}$$

Similarly, using (7.77b) and (7.73) in (7.71) we derive

$$P^*_{ijkh} = F_{ijkh} - g_{ijh|_*k} + g_{ijr}g_{k\ h|_*s}^{\ r}y^s. \tag{7.82}$$

Next, we note that by (7.73) and (7.28a), (7.57a) becomes

$$G_{i\ j}^{\ k} = F^{*k}_{i\ j} + g_{i\ j|_*s}^{\ k}y^s. \tag{7.83}$$

Then by using (7.83) in (5.24) for the Berwald connection and taking into account (7.80) and (7.30) we obtain

$$R^*_{ijkh} = H_{ijkh} + g_{ijr}R^{*r}_{\ \ kh} + y^s\mathcal{A}_{(kh)}\{g_{jkr|_*s}g_{i\ h|_*t}^{\ r}y^t + g_{ijh|_*s|_*k}\}. \tag{7.84}$$

Similarly, using (7.83) in (6.33) and taking into account (7.71) we infer that

$$P^*_{ijkh} = G_{ijkh} - g_{ijh|_*k} + g_{ijr}g_{k\ h|_*s}^{\ r}y^s - g_{tj}\frac{\partial}{\partial y^h}(g_{i\ k|_*s}^{\ t}y^s). \tag{7.85}$$

Finally, comparing (7.81) and (7.82) with (7.84) and (7.85), respectively, we obtain

$$K_{ijkh} = H_{ijkh} + y^s\mathcal{A}_{(kh)}\{g_{jkr|_*s}g_{i\ h|_*t}^{\ r}y^t + g_{ijh|_*s|_*k}\}, \tag{7.86}$$

and

$$F_{ijkh} = G_{ijkh} - g_{tj}\frac{\partial}{\partial y^h}(g_i{}^t{}_{k|*s}y^s), \tag{7.87}$$

which represent the interrelations between the curvature Finsler tensor fields of **RFC** and **BFC**.

As we have seen in the previous section, the Berwald connection is neither h–metric nor v–metric Finsler connection. We are able now to express $g_{ij|_{\mathbf{b}}k}$ in terms of the horizontal covariant derivative of the Cartan tensor field. By using (7.83) in (6.35) and taking into account (7.28a) we deduce that

$$g_{ij|_{\mathbf{b}}k} = -2g_{ijk|*s}y^s. \tag{7.88}$$

8. Special Pseudo–Finsler Manifolds

In the first part of this section we introduce the horizontal flag curvature and by using the Cartan connection we study pseudo–Finsler manifolds of constant curvature. Then we present several characterizations of Landsberg manifolds, Berwald manifolds, and locally Minkowski manifolds. Finally, we introduce generalized Landsberg manifolds and prove that generalized Landsberg manifolds of nonzero constant curvature must be pseudo–Riemannian manifolds.

As is well known, the most interesting examples of Riemannian (pseudo–Riemannian) manifolds are those of constant sectional curvature (see Wolf [1]). Several attempts have been performed to extend the concept to Finsler manifolds (cf., Berwald [5], Matsumoto [6], Akbar–Zadeh [1], [2]). As far as we know, the only attempt that succeeded in producing non–trivial examples of Finsler manifolds of constant curvature was the one based on the Berwald ideas (cf., Matsumoto [6], p. 167). Following Berwald, we shall define pseudo–Finsler manifolds of constant curvature by means of the Cartan connection.

Let $\mathbb{F}^m = (M, M', F^*)$ be a pseudo–Finsler manifold endowed with the Cartan connection $\mathbf{FC}^* = (GM', \nabla^*) = (G_i^k, F_i^{*k}{}_j, g_i{}^k{}_j)$. Consider the almost product structure Q^* associated to GM' (see (4.8)) and the associate linear connection D^* to $\mathbf{FC}^*$ on M' (see (5.27)). We assume that the Liouville vector field L is spacelike on M' and denote by the same symbol L the Liouville vector at the point $(x, y) \in M'$. Then we take a Finsler vector X at (x, y) such that $\mathrm{Span}\{L, X\}$ is a 2–dimensional non–degenerate subspace of $VM'_{(x,y)}$ with respect to the Finsler metric g. Then by using the curvature tensor field R^{D^*} of type $(0, 4)$ of D^* and the Sasaki–Finsler metric G on M' (see (7.39)) we define

$$K\left(x, y; Q^*L, Q^*X\right) = \frac{R^{D^*}\left(Q^*L, Q^*X, Q^*X, Q^*L\right)}{G(Q^*L, Q^*L)G(Q^*X, Q^*X) - G(Q^*L, Q^*X)^2}.$$

The plane

$$\Pi = \mathrm{Span}\{Q^*L, Q^*X\},$$

is non–degenerate with respect to G and it is called a **horizontal flag** at (x, y) with Q^*L as flagpole and Q^*X as transverse edge. By straightforward calculations using (7.50) we obtain

$$K(x, y;\; Q^*L, Q^*X') = K(x, y;\; Q^*L, Q^*X)$$

for any $Q^*X' \in \Pi$, such that $\{Q^*L, Q^*X'\}$ is a basis for Π. Hence we may define $K(x, y; \Pi) = K(x, y; Q^*L, Q^*X)$ and call it the **horizontal flag curvature** of $\mathbb{F}^m$ with respect to the horizontal flag Π at point (x, y). Taking into account (7.49b), (7.15), (7.17), (7.21), and (7.22) we infer that

$$K\left(x, y; \Pi\right) = \frac{R^{D^*}\left(Q^*L, Q^*X, Q^*X, Q^*L\right)}{F^*a(X, X)}, \tag{8.1}$$

where a is the angular metric of $\mathbb{F}^m$.

Locally, by using (4.8), (2.3), and (2.1) we obtain $Q^*L = y^i(\delta^*/\delta^*x^i)$ and $Q^*X = X^i(x, y)(\delta^*/\delta^*x^i)$. Then taking into account (5.29), (7.39), (7.52), (7.53), (7.56a), and (7.26) we deduce that

$$R^*_{ijk}X^iy^jX^k = K(x, y; \Pi)F^*a_{ik}X^iX^k. \tag{8.2}$$

If the horizontal flag curvature of $\mathbb{F}^m$ does not depend on the horizontal flag Π, that is, K is a scalar field $K(x, y)$ on M', then $\mathbb{F}^m$ is said to be a pseudo–Finsler manifold of **scalar curvature** $K(x, y)$. Thus from (8.2) we deduce that $\mathbb{F}^m$ is of scalar curvature K if and only if

$$R^*_{ijk}y^j = KF^*a_{ik}. \tag{8.3}$$

Now we prove the following proposition.

PROPOSITION 8.1 *The torsion Finsler tensor field* R^*_{ijk} *of* **FC*** *satisfies the identity*

$$3R^*_{ijk} = \mathcal{A}_{(jk)}\left\{\frac{\partial(R^*_{ihk}y^h)}{\partial y^j} + 2g_i{}^s{}_kR^*_{shj}y^h\right\}. \tag{8.4}$$

PROOF. By using (6.21) we see that (6.32) becomes

$$\frac{\partial(R^*_{ihk})}{\partial y^j} = H_{jihk} + 2g_i{}^s{}_j R^*_{shk}. \tag{8.5}$$

Then contracting (8.5) by y^h we find

$$\frac{\partial(R^*_{ihk}y^h)}{\partial y^j} = R^*_{ijk} + H_{jihk}y^h + 2g_i{}^s{}_j R^*_{shk}y^h. \tag{8.6}$$

On the other hand, the identity (6.37b) becomes

$$\sum_{(jhk)} \{H_{jihk}\} = 0. \tag{8.7}$$

Contracting (8.7) by y^h and taking into account (6.34) we infer that

$$R^*_{ijk} = (H_{jihk} - H_{kihj})\,y^h. \tag{8.8}$$

Thus (8.4) is a consequence of (8.6) and (8.8). ■

The next theorem is a generalization of the Schur Theorem for pseudo–Finsler manifolds.

THEOREM 8.1 *Let $\mathbb{F}^m = (M, M', F^*)$ be a pseudo–Finsler manifold with a spacelike Liouville vector field and $m > 2$. If the horizontal flag curvature $K(x, y; \Pi)$ depends only on $x \in M$, then it is a constant on M, and the Cartan tensor field and the h–curvature Finsler tensor field of* **FC*** *satisfy*

$$\mathcal{A}_{(kh)}\{FK g_{ijh}\eta_k + g_{ijh|*s|*k}y^s\} = 0, \tag{8.9}$$

and

$$R^*_{ijkh} = \mathcal{A}_{(kh)}\{K g_{ik}g_{jh} + g_{jkr|*s}y^s g_i{}^r{}_{h|*t}y^t\}. \tag{8.10}$$

PROOF. First we prove (8.9) and (8.10). Since K does not depend on the pair (y, X), from (8.3) we deduce that

$$R^*_{ijk}y^j = K(x)F^*a_{ik}. \tag{8.11}$$

By using (8.11) in the right part of (8.4) and taking into account (7.26), (6.21) and (1.12) we obtain

$$R^*_{ijk} = KF(g_{ik}\eta_j - g_{ij}\eta_k). \tag{8.12}$$

Contracting the Bianchi identity (7.62c) by g_{th} and using (7.28), (5.16), (7.73) and (7.30) we infer that

$$\begin{aligned} R^*_{ijkh} &= \frac{\partial R^*_{jkh}}{\partial y^i} - R^*_{rkh}g_i{}^r{}_j \\ &\quad + \mathcal{A}_{(hk)}\{g_{ijh|*s|*k}y^s + g_{jkr|*s}y^s g_i{}^r{}_{h|*t}y^t\}. \end{aligned} \tag{8.13}$$

Differentiating (8.12) with respect to y^h and taking into account (5.23), (1.12) and (6.21) we derive

$$\frac{\partial R^*_{ijk}}{\partial y^h} = K\mathcal{A}_{(jk)}\left\{g_{ik}g_{hj} + 2Fg_{ihk}\eta_j\right\}. \tag{8.14}$$

Further, by using (8.12) and (8.14) in (8.13) we find that

$$\begin{aligned} R^*_{ijkh} &= \mathcal{A}_{(kh)}\{Kg_{ik}g_{jh} + g_{jkr|*s}y^s g_i{}^r{}_{h|*t}y^t \\ &\quad + FKg_{ijh}\eta_k + g_{ijh|*s|*k}y^s\}. \end{aligned} \tag{8.15}$$

Thus (8.9) and (8.10) follow from (8.15) via (7.53a).

Next, to prove that K is constant on M we first note that from (7.35) it follows that

$$(a)\ \ \eta_{i|*j} = 0 \quad \text{and} \quad (b)\ \ \eta_{i||*j} = \frac{1}{F}a_{ij}. \tag{8.16}$$

We then take the horizontal covariant derivative of (8.12) with respect to **FC*** and by using (8.16a) and (7.28a) we obtain

$$R^*_{ijk|*h} = K_{|*h}F(g_{ik}\eta_j - g_{ij}\eta_k). \tag{8.17}$$

Taking into account (8.12) and (8.17) we see that (7.62a) becomes

$$\sum_{(j,k,h)}\left\{K_{|*h}(g_{ik}\eta_j - g_{ij}\eta_k)\right\} = 0. \tag{8.18}$$

Contracting (8.18) by $g^{ik}y^h$ we infer that

$$(m-2)\left(K_{|*h}y^h\eta_j - FK_{|*j}\right) = 0,$$

which together with condition $m > 2$ implies that

$$FK_{|*j} = K_{|*h}y^h\eta_j. \tag{8.19}$$

As K is independent of (y^i) we have $K_{|*j} = \partial K/\partial x^j$, and therefore $K_{|*j}$ is also independent of (y^i). Thus by (8.19) we have

$$K_{|*j||*i} = -K_{|*r}g_j{}^r{}_i = -\frac{1}{F}K_{|*h}y^h\eta_r g_j{}^r{}_i = 0. \tag{8.20}$$

We now take the vertical covariant derivative of (8.19) and by using (8.20), (7.31) and (8.16b) we derive

$$\eta_i K_{|*j} = K_{|*i}\eta_j + \frac{1}{F}K_{|*h}y^h a_{ij}. \tag{8.21}$$

Taking into account (8.19) we see that (8.21) becomes

$$K_{|*h}y^h a_{ij} = 0. \tag{8.22}$$

Since L is assumed to be spacelike we deduce that the Liouville distribution is non–degenerate. Hence there exists $X \in \Gamma(\mathcal{L}M')$ such that $g(X, X) \neq 0$. By using (7.22), (7.21) and (7.17) we obtain

$$a(X, X) = a_{ij}X^i X^j = g(X, X) \neq 0.$$

Hence from (8.22) we obtain $K_{|*h}y^h = 0$, which together with (8.19) implies $K_{|*j} = \partial K/\partial x^j = 0$. Hence K is constant on M. ∎

When the horizontal flag curvature of $\mathbb{F}^m$ is constant we say that $\mathbb{F}^m$ is a **pseudo–Finsler manifold of constant curvature**. The next theorem states characterizations of pseudo–Finsler manifolds of constant curvature by means of h–curvature tensor fields of **FC***, **RFC** and **BFC**.

THEOREM 8.2 *Let $\mathbb{F}^m$ be a pseudo–Finsler manifold with a spacelike Liouville vector field and $m > 2$. Then the following assertions are equivalent:*

(i) $\mathbb{F}^m$ is a pseudo–Finsler manifold of constant curvature.

(ii) The Cartan tensor field and the h–curvature Finsler tensor field of **FC*** *satisfy (8.9) and (8.10).*

(iii) The h–curvature Finsler tensor field of **RFC** *satisfies*

$$K_{ijkh} = \mathcal{A}_{(kh)}\{K g_{ik}g_{jh} + y^s(g_{jkr|*s}g_i{}^r{}_{h|*t}y^t + g_{ijh|*s|*k})\}. \tag{8.23}$$

(iv) The h–curvature Finsler tensor field of **BFC** *satisfies*

$$H_{ijkh} = K(g_{ik}g_{jh} - g_{ih}g_{jk}). \tag{8.24}$$

(v) The torsion Finsler tensor field R^*_{ijk} satisfies

$$R^*_{ijk} = KF\,(g_{ik}\eta_j - g_{ij}\eta_k), \tag{8.25}$$

where K from assertions (iii)–(v) is a function on M.

PROOF. $(i) \Longrightarrow (ii)$. It is proved by Theorem 8.1.

$(ii) \Longrightarrow (iii)$. Contracting (8.10) by $y^i g^{jr}$ and using (6.22) and (7.23) we obtain

$$R^{*r}{}_{kh} = KF(\delta^r_h \eta_k - \delta^r_k \eta_h). \tag{8.26}$$

Taking into account (8.26), (8.9) and (8.10) in (7.81) we deduce (8.23).

$(iii) \Longrightarrow (iv)$. Substituting K_{ijkh} from (8.23) into (7.86) we obtain (8.24).

$(iv) \Longrightarrow (v)$ Contracting (8.24) by y^i and using (6.34) and (7.23) we derive (8.25).

$(v) \Longrightarrow (i)$. Contracting (8.25) by y^j and using (7.23) and (7.26) we obtain (8.11). Comparing (8.11) with (8.3) we deduce that $K(x, y; \Pi) = K(x)$, that is the horizontal flag curvature of $\mathbb{F}^m$ depends only on $x \in M$. Then by Theorem 8.1 we obtain the assertion (i). ∎

Yasuda and Shimada [1] determined all the Randers manifolds of constant curvature. However, the geometry of Finsler (pseudo–Finsler) manifolds of constant curvature is far from being settled. By using the Sasaki–Finsler metric on M', we shall present here a geometric characterization of pseudo–Finsler manifolds of constant curvature $K = 1$.

First, by direct calculations, using (7.43), we obtain all the local coefficients of the Levi–Civita connection ∇' with respect to G as in the following proposition.

PROPOSITION 8.2 *The Levi–Civita connection ∇' on M' with respect to the Sasaki–Finsler metric G is locally expressed as follows:*

$$\begin{aligned}
&(a)\ \nabla'_{\delta^*/\delta^* x^j} \frac{\delta^*}{\delta^* x^i} = -(g_i{}^k{}_j + \frac{1}{2} R^{*k}{}_{ij}) \frac{\partial}{\partial y^k} + F^{*k}_{i\ j} \frac{\delta^*}{\delta^* x^k}, \\
&(b)\ \nabla'_{\partial/\partial y^j} \frac{\partial}{\partial y^i} = g_i{}^k{}_j \frac{\partial}{\partial y^k} + 2 g_i{}^k{}_{j|*h} y^h \frac{\delta^*}{\delta^* x^k}, \\
&(c)\ \nabla'_{\delta^*/\delta^* x^j} \frac{\partial}{\partial y^i} = F^{*k}_{i\ j} \frac{\partial}{\partial y^k} + (g_i{}^k{}_j + \frac{1}{2} R^*_{ihj} g^{hk}) \frac{\delta^*}{\delta^* x^k} \\
&\qquad = \nabla_{\partial/\partial y^i} \frac{\delta^*}{\delta^* x^j} + G_i{}^k{}_j \frac{\partial}{\partial y^k},
\end{aligned} \tag{8.27}$$

where $G_i{}^k{}_j$ are the local coefficients of the Berwald connection (see (6.27)).

Since L is supposed to be spacelike on M' we may consider the unit vector field $\xi = Q^*\ell$ (see (7.15)). Clearly ξ is a globally defined vector field on M' which lies in GM'. For this reason we call it the **horizontal unit Liouville vector field** of $\mathbb{F}^m$. Further, we study some of the properties of ξ. First we prove the following proposition:

PROPOSITION 8.3 *The Lie derivative of the Sasaki–Finsler metric with respect to the horizontal unit Liouville vector field satisfies the equations:*

$$(\mathcal{L}_\xi G)(v^*X, v^*Y) = (\mathcal{L}_\xi G)(h^*X, h^*Y) = 0, \tag{8.28}$$

and

$$(\mathcal{L}_\xi G)(h^*X, v^*Y) = \frac{1}{F}(a_{ij} - R^*_{ihj}y^h)X^i\dot{Y}^j, \tag{8.29}$$

for any $X, Y \in \Gamma(TM')$, where $h^*X = X^i(\delta^*/\delta^*x^i)$ and $v^*Y = \dot{Y}^j(\partial/\partial y^j)$ (cf., (4.6)).

PROOF. First, by using (8.27c), (7.41), (7.29), and (6.30) we obtain

$$G\left(\nabla'_{\partial/\partial y^j}\xi, \frac{\partial}{\partial y^i}\right) = \frac{y^k}{F}(F^{*h}_{k\ j} - G_k{}^h{}_j)g_{hi} = 0. \tag{8.30}$$

Next, taking into account (8.27a), (7.41), (7.23), and (8.16a) we deduce that

$$G\left(\nabla'_{\delta^*/\delta^* x^j}\xi, \frac{\delta^*}{\delta^* x^i}\right) = \eta_{i|_{*j}} = 0. \tag{8.31}$$

Thus, by using (8.30) and (8.31) and the expression of the Lie derivative of G with respect to ξ (see (3.16)), we obtain (8.28). Further, from (7.56b) and (7.53a) we deduce that $y^t R^*_{tjk} = 0$, which together with (7.65) implies

$$R^*_{ijk}y^j = R^*_{kji}y^j. \tag{8.32}$$

Finally, using (3.16), (8.27a), (8.27c), (7.41), and (8.32) we infer that

$$\begin{aligned}(\mathcal{L}_\xi G)\left(\frac{\delta^*}{\delta^* x^i}, \frac{\partial}{\partial y^j}\right) &= \eta_{i||_{*j}} - \frac{1}{F}R^*_{jhi}y^h \\ &= \frac{1}{F}(a_{ij} - R^*_{ihj}y^h),\end{aligned}$$

which proves (8.29). ∎

As L is assumed to be spacelike on M', from (7.14) it follows that the positive indicatrix bundle IM^+ is the only one surviving over M. Moreover, the total space of IM^+ is a hypersurface of M' given by the equation $F(x, y) = 1$. It is easy to show that $Q^*\xi = \ell$ is the unit normal vector field to IM^+ with respect to G. Indeed, if the local equations of IM^+ in M' are

$$x^i = x^i(t^\alpha); \quad y^i = y^i(t^\alpha); \quad \alpha \in \{1, \cdots, 2m-1\},$$

then we have

$$\frac{\partial F}{\partial x^i}\frac{\partial x^i}{\partial t^\alpha} + \frac{\partial F}{\partial y^i}\frac{\partial y^i}{\partial t^\alpha} = 0.$$

As the horizontal covariant derivative of F with respect to $\mathbf{FC}^*$ vanishes (see (7.38)), by using (6.18) and (7.23) we obtain

$$\left(G^i_k\frac{\partial x^k}{\partial t^\alpha} + \frac{\partial y^i}{\partial t^\alpha}\right)\eta_i = 0. \tag{8.33}$$

Taking into account that the natural frame field on IM^+ is represented by

$$\frac{\partial}{\partial t^\alpha} = \frac{\partial x^i}{\partial t^\alpha}\frac{\delta^*}{\delta^* x^i} + \left(G^i_k\frac{\partial x^k}{\partial t^\alpha} + \frac{\partial y^i}{\partial t^\alpha}\right)\frac{\partial}{\partial y^i},$$

and using (8.33) we derive

$$G\left(\frac{\partial}{\partial t^\alpha}, \ell\right) = \left(G^i_k\frac{\partial x^k}{\partial t^\alpha} + \frac{\partial y^i}{\partial t^\alpha}\right)\eta_i = 0.$$

Thus ℓ is orthogonal to the tangent bundle of IM^+. Moreover, the horizontal unit vector field is tangent to IM^+ since it lies in GM' and therefore it is orthogonal to ℓ (see (7.41)). Then by Proposition 8.3 we may state the following corollary.

COROLLARY 8.1 *Let $\mathbb{F}^m = (M, M', F^*)$ be a pseudo–Finsler manifold whose Liouville vector field is spacelike. Then the horizontal unit Liouville vector field is a Killing vector field on the hypersurface IM^+ of M', if and only if,*

$$a_{ij}(x,y) = R^*_{ihj}(x,y)y^h, \quad \forall (x,y) \in IM^+. \tag{8.34}$$

Finally, we obtain the following geometric characterization of pseudo–Finsler manifolds of constant curvature $K = 1$.

THEOREM 8.3 *(Bejancu–Farran [3]). Let $\mathbb{F}^m = (M, M', F^*)$ be a pseudo-Finsler manifold whose Liouville vector field is spacelike. Then $\mathbb{F}^m$ has constant curvature $K = 1$ if and only if the horizontal unit Liouville vector field is a Killing vector field on IM^+.*

PROOF. Assume $K = 1$, and from (8.3) we deduce (8.34) since $F^*(x,y) = 1$ on IM^+. Conversely, suppose ξ is a Killing vector field on IM^+. Then by using (8.34) in (8.3) we deduce that $K(x,y;\Pi) = 1$ for any horizontal flag and any $(x,y) \in IM^+$. Now, we take a point $(x_0,y_0) \in M' \setminus IM^+$. Then there exists $a \in (0,\infty)\setminus\{1\}$ such that $F(x_0,y_0) = a$. As F is positively homogeneous of degree one with respect to (y^i), we have $F(x_0, y_0/a) = 1$. Hence $(x_0, y_0/a) \in IM^+$ and by (8.34) we have

$$a_{ij}(x_0, \frac{1}{a}y_0) = R_{ihj}(x_0, \frac{1}{a}y_0)\frac{1}{a}y_0^h.$$

Taking into account that a_{ij} and R_{ihj} are positively homogeneous of degrees zero and one respectively, we infer that

$$R_{ihj}(x_0, y_0)y_0^h = F^2(x_0, y_0)a_{ij}(x_0, y_0).$$

Thus from (8.3) we deduce $K(x_0, y_0, \Pi) = 1$, which completes the proof of the theorem. ■

Next, following Antonelli et al. [1], p. 98, we say that the pseudo-Finsler manifold $\mathbb{F}^m$ is a **Landsberg manifold (space)** if the Berwald connection coincides with the Rund connection. Some useful tensorial characterizations for this special class of pseudo-Finsler manifolds are given in the theorem which follows.

THEOREM 8.4 *Let $\mathbb{F}^m = (M, M', F^*)$ be a pseudo-Finsler manifold. Then the following assertions are equivalent:*

(i) $\mathbb{F}^m$ is a Landsberg manifold.

(ii) The Cartan tensor field satisfies

$$g_{ijk|_*h}y^h = 0. \tag{8.35}$$

(iii) The Berwald connection is an h–metric Finsler connection.

PROOF. (*i*) $\Longrightarrow$ (*ii*). As **BFC** = **RFC**, we have $G_i{}^k{}_j = F_i^*{}^k_j$. Thus (8.35) follows from (7.83).

(*ii*) $\Longrightarrow$ (*iii*). To prove this we first note that by using (7.83) and (7.28a) in (6.35) we infer that, in general,

$$g_{ij|_{\mathbf{b}}k} = -2g_{ijk|_*h}y^h. \tag{8.36}$$

Thus (8.35) implies $g_{ij|_{\mathbf{b}}k} = 0$, that is, **BFC** is h–metric.

(*iii*) $\Longrightarrow$ (*i*). As $g_{ij|_{\mathbf{b}}k} = 0$, from (8.36) we obtain (8.35) which together with (7.83) implies **BFC** = **RFC**. ■

In 1926 Berwald [1] introduced a special class of Finsler manifolds which later took his name. We say that the pseudo-Finsler manifold $\mathbb{F}^m$ is a **Berwald manifold (space)** if the local coefficients $G_i{}^k{}_j$ of the Berwald connection depend on (x^i) only.

THEOREM 8.5 *Let $\mathbb{F}^m = (M, M', F^*)$ be a pseudo–Finsler manifold. Then the following assertions are equivalent:*

(i) $\mathbb{F}^m$ is a Berwald manifold.

(ii) The hv–curvature tensor field of the Berwald connection vanishes.

(iii) The Cartan tensor field is h–parallel with respect to $\mathbf{FC}^*$, *that is, we have*

$$g_{ijk|*h} = 0. \tag{8.37}$$

(iv) The hv–curvature Finsler tensor field of the Rund connection vanishes.

(v) The local coefficients of the Rund connection depend on (x^i) only.

PROOF. (i) $\Longrightarrow$ (ii). It is a consequence of (6.33).

(ii) $\Longrightarrow$ (iii). By (7.87) we have

$$F_{ijkh} + \frac{\partial}{\partial y^h}(g_{ijk|*s}y^s) - 2g_{jhr}g_i{}^r{}_{k|*s}y^s = 0. \tag{8.38}$$

On the other hand, by using (7.53b) and (7.82) we obtain

$$F_{ijkh} + F_{jikh} - 2g_{ijh|*k} + 2g_{ijr}g_k{}^r{}_{h|*s}y^s = 0. \tag{8.39}$$

By (8.38) we see that (8.39) becomes

$$\begin{aligned} \frac{\partial}{\partial y^h}(g_{ijk|*s}y^s) \quad &- \quad g_{jhr}g_i{}^r{}_{k|*s}y^s - g_{ihr}g_j{}^r{}_{k|*s}y^s \\ &+ \quad g_{ijr}g_k{}^r{}_{h|*s}y^s - g_{ijh|*k} = 0. \end{aligned} \tag{8.40}$$

Contracting (8.40) by y^i and using (6.22) and (7.30) we deduce that $g_{hjk|*s}y^s = 0$, which together with (8.40) implies (8.37).

(iii) $\Longrightarrow$ (iv). By (8.37) and (7.73) we infer that $P^{*i}{}_{jk} = 0$. Then from (7.66) and (7.77b) it follows that $F_i{}^h{}_{jk} = 0$.

(iv) $\Longrightarrow$ (v). It is a consequence of (7.77b).

(v) $\Longrightarrow$ (i). In this case (7.29) becomes

$$G_k^i(x, y) = y^j F_j^{*i}{}_k(x).$$

Differentiating this equation with respect to (y^h) we derive $G_h{}^i{}_k(x, y) = F_h^{*i}{}_k(x)$. Thus the proof is complete. ■

Finally, we say that a pseudo–Finsler manifold $\mathbb{F}^m = (M, M', F^*)$ is a **locally pseudo–Minkowski manifold** if on each coordinate neighborhood of M', F^* is a function of (y^i) only. As in the case of Finsler manifolds (see Matsumoto [6], p. 159) we have the following characterization of locally pseudo–Minkowski manifolds.

THEOREM 8.6 *Let $\mathbb{F}^m$ be a pseudo–Finsler manifold. Then the following assertions are equivalent:*

(i) $\mathbb{F}^m$ is a locally pseudo–Minkowski manifold.

(ii) $\mathbb{F}^m$ is a Berwald manifold and the h–curvature Finsler tensor field of the Cartan connection vanishes.

(iii) The h–curvature and hv–curvature Finsler tensor fields of the Berwald connection vanish.

(iv) The h–curvature and hv–curvature Finsler tensor fields of the Rund connection vanish.

Remark 8.1. By the assertions (iii) and (iv) of Theorem 8.6 we may say that locally pseudo–Minkowski manifolds are flat pseudo–Finsler manifolds with respect to both the Berwald and Rund connections. Moreover, by (ii) and (7.71) we see that $R^*_{ijkh} = P^*_{ijkh} = 0$. So, S^*_{ijkh} is the only surviving curvature Finsler tensor field on a locally pseudo–Minkowski manifold. ■

Example 8.1. Let $\mathbb{F}^2 = (D, M', F^*)$ be the pseudo–Finsler manifold in Example 1.5. Suppose $\mathbb{F}^2$ is not pseudo–Riemannian, that is, $m \neq 1$ and $n \neq 1$ in (1.28). Then by direct calculations using (6.9) and (6.17) we deduce that

$$\left[G_i^k(x,y)\right] = \begin{bmatrix} \dfrac{y^1}{m}\dfrac{\partial f}{\partial x^1} & 0 \\ 0 & \dfrac{y^2}{m}\dfrac{\partial f}{\partial x^2} \end{bmatrix}.$$

Hence by (6.27), ${G_i}^k{}_j$ are functions of (x^i) only. Thus for any non constant smooth function f in (1.28) we obtain an example of a Berwald manifold which is not locally pseudo–Minkowski. ■

The relationship between Berwald manifolds and Landsberg manifolds is interesting. Since (8.37) yields (8.35) we conclude that every Berwald manifold is a Landsberg manifold. On the other hand, Matsumoto's results [8] indicate that there are many situations when a Landsberg manifold becomes a Berwald manifold. According to Matsumoto [8] and Bao–Chern–Shen [1], the problem of finding explicite examples of Landsberg manifolds that are not Berwald is still open. This shows that, although Landsberg manifolds were introduced a long time ago, (see Landsberg [1]-[3]), and inspite of the great deal of research done on Landsberg manifolds, this area of research is still active, and very far from being exhausted.

For the above reasons we introduce here a new class of pseudo–Finsler manifolds which contains, in particular, Landsberg manifolds. More precisely, we say that the pseudo–Finsler manifold $\mathbb{F}^m$ is a **generalized Landsberg manifold** if the h–curvature Finsler tensor fields of **BFC** and **RFC** coincide, that is, we have

$$H_{ijkh} = K_{ijkh}. \tag{8.41}$$

By using (7.86) we may state the following theorem:

THEOREM 8.7 *A pseudo–Finsler manifold is a generalized Landsberg manifold if and only if the Cartan tensor field satisfies:*

$$(a)\ \mathcal{A}_{(kh)}\{g_{jkr|*s}y^s g_i{}^r{}_{h|*t}y^t\} = 0 \text{ and } (b)\ \mathcal{A}_{(kh)}\{g_{ijh|*s|*k}y^s\} = 0. \tag{8.42}$$

Certainly, the class of generalized Landsberg manifolds contains the three classes of Landsberg manifolds, Berwald manifolds and locally pseudo–Minkowski manifolds. On the other hand, we have seen that the vanishing of the h–curvature of the Cartan connection is one of the conditions (see (ii) of Theorem 8.6) for $\mathbb{F}^m$ to be locally Minkowski. Therefore we must find the right place of an $\mathbb{F}^m$ with $R^*_{ijkh} = 0$ among the above four classes of pseudo–Finsler manifolds. We call such an $\mathbb{F}^m$ an h–**flat** pseudo–Finsler manifold with respect to **FC***. The following theorem is related to the above discussion.

THEOREM 8.8 *(Matsumoto [6], p. 164). If $\mathbb{F}^m$ is h–flat with respect to FC^* then $P^*_{ijk|*h}$ is completely symmetric and*

$$\mathcal{A}_{(kh)}\{P^*_{jkr}P^{*r}{}_{ih}\} = 0.$$

Thus, by using (7.73) and Theorems 8.7 and 8.8 we may state the following important corollory.

COROLLARY 8.2 *Any h–flat pseudo–Finsler manifold with respect to the Cartan connection is a generalized Landsberg manifold.*

We are now able to prove the following theorem for generalized Landsberg manifolds of constant curvature.

THEOREM 8.9 *(Bejancu–Farran [4]) Let $\mathbb{F}^m = (M, M', F^*)$ be a generalized Landsberg manifold of constant curvature, whose Liouville vector field is spacelike. Then $\mathbb{F}^m$ is an h–flat pseudo–Finsler manifold with respect to the Cartan connection or $g_{ijk} = 0$ on M'.*

PROOF. Taking into account (8.42b) and (8.9) we obtian

$$FK\left(g_{ijh}\eta_k - g_{ijk}\eta_h\right) = 0. \tag{8.43}$$

Contracting (8.43) by y^k and using (7.23), (1.14), and (6.22a) we deduce that

$$Kg_{ijh} = 0. \tag{8.44}$$

If $K = 0$ then from (8.10) and (8.42a) we deduce that $R^*_{ijkh} = 0$, so $\mathbb{F}^m$ is h–flat with respect to FC^*. When $K \neq 0$, then from (8.44) it follows that $g_{ijh} = 0$, which completes the proof of the theorem. ■

Taking into account the Theorems 6.1 and 8.9, we may state the following corollary.

COROLLARY 8.3 *Any generalized Landsberg manifold $\mathbb{F}^m = (M, TM^0, F^*)$ of non–zero constant curvature is a pseudo–Riemannian manifold.*

Remark 8.2. We learned from Bao–Chern–Shen [2] that Akbar–Zadeh [2] proved that any Landsberg manifold of positive constant curvature is a Riemannian manifold. Also, we should mention that while writing the book, the authors proved that Corollary 8.3 is also valid for generalized Landsberg manifolds of scalar curvature (cf., Bejancu–Farran [5]). In this way we have extended a well known result of Numata [1] on Landsberg manifolds to generalized Landsberg manifolds. ■

Now, following Matsumoto [1], we say that $\mathbb{F}^m = (M, M', F)$ is a C–**reducible Finsler manifold** if the Cartan tensor field is expressed in the form

$$g_{ijk} = \frac{1}{m+1}\left(a_{ij}g_k + a_{jk}g_i + a_{ki}g_j\right), \tag{8.45}$$

where a_{ij} are the local components of the angular metric and

$$g_i = g_i{}^j{}_j. \tag{8.46}$$

An important result of Matsumoto and Hōjō [1] establishes that the only C–reducible Finsler manifolds are Randers manifolds and Kropina manifolds. Thus there exist on M a Riemannian metric and a differential 1–form with local components $b_{ij}(x)$ and $c_i(x)$ respectively, such that the fundamental function has one of the following forms

$$(a)\ \ F(x,y)=b(x,y)+c(x,y) \quad \text{or} \quad (b)\ \ F(x,y)=\frac{b^2(x,y)}{c(x,y)}, \tag{8.47}$$

according as $\mathbb{F}^m$ is a Randers manifold or a Kropina manifold, where we set

$$(a)\ \ b(x,y)=(b_{ij}(x)y^iy^j)^{\frac{1}{2}} \quad \text{and} \quad (b)\ \ c(x,y)=c_i(x)y^i. \tag{8.48}$$

Moreover, in case $\mathbb{F}^m$ is a Randers manifold or a Kropina manifold we have (Shibata–Shimada–Azuma–Yasuda [1])

$$(a)\ \ g_i=\frac{m+1}{2F}\left(\mu\eta_i-\tau' c_i\right) \quad \text{or} \quad (b)\ \ g_i=\frac{m+1}{2F}\left(\eta_i-\tau c_i\right), \tag{8.49}$$

where we set

$$(a)\ \ \mu=\frac{c}{b};\quad (b)\ \ \tau'=1+\mu;\quad (c)\ \ \tau=\frac{b^2}{c^2}. \tag{8.50}$$

The Finsler metrics on a Randers manifold and a Kropina manifold are given by (see (30.16) in Matsumoto [6] and (1.1) in Shibata–Shimada–Azuma–Yasuda [1])

$$g_{ij}=\frac{F}{b}b_{ij}+c_ic_j+\frac{1}{b}\left(c_iY_j+c_jY_i\right)-\frac{c}{b^3}Y_iY_j, \tag{8.51}$$

and

$$g_{ij}=\tau(2b_{ij}-\eta_ic_j-\eta_jc_i)+\eta_i\eta_j, \tag{8.52}$$

respectively, where we have put

$$Y_i=b_{ij}(x)y^j. \tag{8.53}$$

Moreover, Shibata–Shimada–Azuma–Yasuda [1] gave explicit expressions for the local coefficients for **BFC** and **FC*** of a Randers manifold. They are too complicated, so we omit them here. We only give the local functions

$$2G^i={\Gamma_j}^i{}_k y^jy^k+\left(E_{jk}y^jy^k-2bF_jy^j\right)\frac{y^i}{F}+2bF^i_jy^j, \tag{8.54}$$

where ${\Gamma_j}^i{}_k$ are the Christoffel symbols with respect to the Riemannian metric $(b_{ij}(x))$, and we set:

$$(a)\ \ E_{jk}\ =\ \frac{1}{2}(c_{j;k}+c_{k;j});\quad (b)\ \ c_{j;k}=\frac{\partial c_j}{\partial x^k}-c_h{\Gamma_j}^h{}_k; \tag{8.55}$$

$$(c)\ \ F_{jk}\ =\ \frac{1}{2}(c_{j;k}-c_{k;j});\quad (d)\ \ F^i_k=b^{ij}F_{jk};\quad (e)\ \ F_k=c_iF^i_k.$$

CHAPTER 2

PSEUDO–FINSLER SUBMANIFOLDS

In the present chapter we develop a general theory of pseudo–Finsler submanifolds using the vertical vector bundle. First, we consider an arbitrary Finsler connection on the ambient pseudo–Finsler manifold $\widetilde{\mathbb{F}}^{m+n}$ and study the induced geometric objects on the pseudo–Finsler submanifold $\mathbb{F}^m$. In particular, we deduce the equations of Gauss, Codazzi, and Ricci for $\mathbb{F}^m$ in $\widetilde{\mathbb{F}}^{m+n}$. Finally, we present a comparison between the induced and intrinsic Finsler connections in each of the cases of Cartan, Berwald, and Rund connections.

1. The pseudo–Finsler Normal Bundle and the Induced Geometric Objects

This section sets the stage for studying the geometry of pseudo–Finsler submanifolds. By means of the pseudo–Finsler normal bundle, we develop some of the machinery needed later in the book. This includes geometric objects such as: induced connections, normal curvature forms, second fundamental forms and shape operators.

Let M and $\widetilde{M}$ be two manifolds of dimension m and $m+n$ respectively, and $f : M \to \widetilde{M}$ be a smooth mapping. Suppose that locally f is given by the equations

$$x^i = x^i(u^1, \cdots, u^m) \;\; ; \;\; 1 \leq i \leq m+n. \tag{1.1}$$

The **tangent mapping (differential)** $f_* : TM \to T\widetilde{M}$ of f is locally defined as follows. A point of TM with coordinates (u^α, v^α) is carried by f_* into a point of $T\widetilde{M}$ with coordinates $(x^i(u),\ y^i(u,v))$, where x^i are the functions in (1.1) and

$$(a) \quad y^i(u,v) = B^i_\alpha v^\alpha; \qquad (b) \quad B^i_\alpha = \frac{\partial x^i}{\partial u^\alpha}. \tag{1.2}$$

We say that f is an **immersion** of M into $\widetilde{M}$ if $(f_*)_u : T_uM \to T_{f(u)}\widetilde{M}$ is injective for every point u of M. In this case we say that M is an **immersed submanifold** (or simply, **submanifold**) of $\widetilde{M}$. The mapping $f : M \to \widetilde{M}$ is called an **imbedding** if it is an injective immersion. We say, in this case, that M is an **imbedded submanifold** of $\widetilde{M}$. By the definition of f_* we see that rank $[B^i_\alpha] = m$ whenever f is an immersion. Thus any immersion f is a local imbedding, that is, any point $u \in M$ has a coordinate neighborhood $\mathcal{U}$ such that f is injective on $\mathcal{U}$. As is well known, any open subset of $\widetilde{M}$ is an $(m+n)$–dimensional manifold which is called an **open submanifold** of $\widetilde{M}$.

Throughout this chapter we shall use the following ranges for indices: $i, j, k, \cdots \in \{1, \cdots, m+n\}$; $\alpha, \beta, \gamma, \cdots \in \{1, \cdots, m\}$; $a, b, c, \cdots \in \{m+1, \cdots, m+n\}$. To simplify the equations involved in the study we use the notations:

$$B^{ij\cdots}_{\alpha\beta\cdots} = B^i_\alpha B^{j\cdots}_{\beta\cdots}; \quad B^i_{\alpha\beta} = \frac{\partial^2 x^i}{\partial u^\alpha \partial u^\beta}; \quad B^i_{\alpha 0} = B^i_{\alpha\beta} v^\beta; \quad B^i_{00} = B^i_{\alpha\beta} v^\alpha v^\beta .$$

Next, we consider an $(m+n)$–dimensional pseudo–Finsler manifold $\widetilde{\mathbb{F}}^{m+n} = (\widetilde{M}, \widetilde{M}', \widetilde{F}^*)$ of index $0 \leq q < m+n$, and an m–dimensional submanifold M of $\widetilde{M}$. Suppose $M'_u = (f_*)^{-1}_u \left(\widetilde{M}'_{f(u)}\right)$ is non empty for any $u \in M$. Then M'_u is a positive conic set in T_uM, since $\widetilde{M}'_{f(u)}$ is so in $T_{f(u)}\widetilde{M}$. Moreover, $M' = (f_*)^{-1}(\widetilde{M}')$ is an open submanifold of TM such that $\pi(M') = M$ and $\theta(M) \cap M' = \phi$, where π and θ are the canonical projection of TM on M and the zero section of TM respectively. Also, $\widetilde{F}^*$ induces on M' the function F^* locally given by

$$F^*(u, v) = \widetilde{F}^*\left(x(u), y(u, v)\right). \tag{1.3}$$

It is easy to see that F^* satisfies condition $\mathbf{F}^*_1$ of Section 1.1. Moreover, by direct calculations using (1.3) and (1.2) we infer that

$$g_{\alpha\beta} = \frac{1}{2} \frac{\partial^2 F^*}{\partial v^\alpha \partial v^\beta} = \tilde{g}_{ij}\left(x(u), y(u, v)\right) B^{ij}_{\alpha\beta}. \tag{1.4}$$

If for any coordinate neighborhood $\mathcal{U}'$ in M' the functions $g_{\alpha\beta}$ define at every point of $\mathcal{U}'$ a quadratic form with r negative eigenvalues and $m - r$ positive eigenvalues (for a fixed $0 \leq r < m$), we say that $\mathbb{F}^m = (M, M', F^*)$ is a **pseudo–Finsler submanifold** of $\widetilde{\mathbb{F}}^{m+n}$. If, in particular, $\widetilde{\mathbb{F}}^{m+n}$ is a Finsler manifold then (1.4) implies that on each $\mathcal{U}' \subset M'$, $g_{\alpha\beta}$ define a positive definite quadratic form. In this case we say that $\mathbb{F}^m$ is a **Finsler submanifold** of $\widetilde{\mathbb{F}}^{m+n}$.

We have to note that all the geometric objects defined in Chapter 1 for the ambient pseudo–Finsler manifolds $\widetilde{\mathbb{F}}^{m+n}$ will have here a tilde, as for example, $\tilde{g}, \tilde{g}_{ij}, \widetilde{N}^i_j, \widetilde{F}_j{}^i{}_k, \tilde{g}_j{}^i{}_k$, etc.

Now we consider a pseudo–Finsler submanifold $\mathbb{F}^m = (M, M', F^*)$ of the pseudo–Finsler manifold $\widetilde{\mathbb{F}}^{m+n} = (\widetilde{M}, \widetilde{M}', \widetilde{F}^*)$ and take the coordinate systems $\{\widetilde{\mathcal{U}}; x^i, y^i\}$ and $\{\mathcal{U}; u^\alpha, v^\alpha\}$ on $\widetilde{M}'$ and M' respectively, where $\widetilde{\mathcal{U}}$ is a coordinate neighborhood on $\widetilde{M}'$ and $\mathcal{U} = \widetilde{\mathcal{U}} \cap f_*(M')$. Then the natural fields of frames $\{\partial/\partial u^\alpha,\ \partial/\partial v^\alpha\}$ and $\{\partial/\partial x^i,\ \partial/\partial y^i\}$ on M' and $\widetilde{M}'$ are related on $\mathcal{U}$ by

$$\frac{\partial}{\partial u^\alpha} = B^i_\alpha \frac{\partial}{\partial x^i} + B^i_{\alpha 0} \frac{\partial}{\partial y^i}, \tag{1.5}$$

and

$$\frac{\partial}{\partial v^\alpha} = B^i_\alpha \frac{\partial}{\partial y^i}. \tag{1.6}$$

By (1.6) one deduces that the vertical vector bundle VM' is a vector sub–bundle of $V\widetilde{M}'_{|M'}$. Hence the pseudo–Riemannian metric $\tilde{g}$ on $V\widetilde{M}'$ (see (1.2.10)) induces a pseudo–Riemannian metric g on VM'. Actually, g is just the pseudo–Finsler metric of $\mathbb{F}^m$, that is, $g_{\alpha\beta}$ from (1.4) are given by

$$g_{\alpha\beta} = g\left(\frac{\partial}{\partial v^\alpha}, \frac{\partial}{\partial v^\beta}\right). \tag{1.7}$$

Further, we denote by $VM'^\perp$ the orthogonal complementary vector bundle of VM' in $V\widetilde{M}'_{|M'}$ with respect to $\tilde{g}$ and call it the **pseudo–Finsler normal bundle** of the pseudo–Finsler submanifold $\mathbb{F}^m$ in $\widetilde{\mathbb{F}}^{m+n}$. Thus we have the orthogonal decomposition

$$V\widetilde{M}'_{|M'} = VM' \oplus VM'^\perp. \tag{1.8}$$

To define the local components of some induced geometric objects on $\mathbb{F}^m$ we consider a local field of orthonormal frames $\{B_a = B^i_a \partial/\partial y^i\}$ in $VM'^\perp$ with respect to $\tilde{g}$. In order to simplify the presentation we assume that $B_a, a \in \{m+1, \cdots, m+n\}$ are spacelike. Otherwise, the signature of B_a should appear in some formulas involving these Finsler normal vector fields. Hence $\tilde{g}\,(\partial/\partial v^\alpha, B_a) = 0$ and $\tilde{g}\,(B_a, B_b) = \delta_{ab}$, which are locally equivalent to

$$\tilde{g}_{ij} B^i_\alpha B^j_a = 0, \quad \forall \alpha \in \{1, \cdots, m\},\ a \in \{m+1, \cdots, m+n\}, \tag{1.9}$$

and

$$\tilde{g}_{ij} B^i_a B^j_b = \delta_{ab}, \quad \forall a, b \in \{m+1, \cdots, m+n\}, \tag{1.10}$$

respectively.

Here, and in the sequel, we shortly write $\tilde{g}_{ij}, B^i_\alpha, B^i_a$, etc., instead of $\tilde{g}_{ij}\left(x(u), y(u,v)\right)$, $B^i_\alpha(u)$, $B^i_a(u,v)$, etc.

Denote by $[B^i_\alpha B^i_a]$ the transition matrix from the natural field of frames $\{\partial/\partial y^1, \cdots, \partial/\partial y^{m+n}\}$ on $V\widetilde{M}'_{|M'}$ to the field of frames

$\{\partial/\partial v^1, \cdots, \partial/\partial v^m, B_{m+1}, \cdots, B_{m+n}\}$ adapted to the decomposition (1.8). If $\left[\widetilde{B}^\alpha_i \widetilde{B}^a_i\right]$ is the inverse of the matrix $[B^i_\alpha B^i_a]$, then we have:

$$(a)\ \widetilde{B}^\alpha_i B^i_\beta = \delta^\alpha_\beta;\ \ (b)\ \widetilde{B}^\alpha_i B^i_a = 0;\ \ (c)\ \widetilde{B}^a_i B^i_\alpha = 0;\ \ (d)\ \widetilde{B}^a_i B^i_b = \delta^a_b, \tag{1.11}$$

and

$$B^i_\alpha \widetilde{B}^\alpha_j + B^i_a \widetilde{B}^a_j = \delta^i_j. \tag{1.12}$$

Contracting (1.4) by $g^{\beta\gamma}\widetilde{B}^\alpha_k$ and taking into account (1.12) and (1.9) we deduce that

$$\widetilde{B}^\gamma_k = \tilde{g}_{kj} B^j_\beta g^{\beta\gamma}. \tag{1.13}$$

Locally, the decomposition (1.8) is expressed by

$$\frac{\partial}{\partial y^j} = \widetilde{B}^\alpha_j \frac{\partial}{\partial v^\alpha} + \widetilde{B}^a_j B_a, \tag{1.14}$$

which is obtained by contracting (1.6) by $\widetilde{B}^\alpha_j$ and using (1.12). Finally, contracting (1.10) by $\widetilde{B}^b_k$ and using (1.12) and (1.9) we infer that

$$\widetilde{B}^a_i = \tilde{g}_{ij} B^j_b \delta^{ba}. \tag{1.15}$$

Next, we introduce the Finsler tensor fields:

$$(a)\quad C^i_j = \widetilde{B}^a_j B^i_a;\quad (b)\quad C_{ij} = \tilde{g}_{ik} C^k_j = \sum_{a=m+1}^{m+n} \widetilde{B}^a_i \widetilde{B}^a_j;$$

$$(c)\quad C^{ij} = \tilde{g}^{ik} C^j_k = \sum_{a=m+1}^{m+n} B^i_a B^j_a. \tag{1.16}$$

Then we see that (1.12) becomes

$$B^i_\alpha \widetilde{B}^\alpha_j + C^i_j = \delta^i_j. \tag{1.17}$$

Moreover, we derive

$$(a)\quad C^i_j B^j_\alpha = 0;\quad (b)\quad C_{ij} B^j_\alpha = 0;\quad (c)\quad C^{ij}\widetilde{B}^\alpha_j = 0. \tag{1.18}$$

Contracting (1.4) by $\widetilde{B}^\alpha_k \widetilde{B}^\beta_h$ and using (1.17) and (1.18b) we obtain

$$\tilde{g}_{kh} = g_{\alpha\beta}\widetilde{B}^\alpha_k \widetilde{B}^\beta_h + C_{kh}.$$

Similarly, we deduce that

$$g^{\alpha\beta} = \tilde{g}^{ij}\widetilde{B}^\alpha_i \widetilde{B}^\beta_j \quad \text{and} \quad \tilde{g}^{kh} = g^{\alpha\beta}B^k_\alpha B^h_\beta + C^{kh}.$$

We now consider a non–linear connection $H\widetilde{M}' = (\widetilde{N}^j_i)$ on $\widetilde{M}'$ and the Sasaki–Finsler metric $\widetilde{G}$ defined on $\widetilde{M}'$ (cf., (1.7.39)). Then on each coordinate neighborhood of M' we define the functions

$$N^\beta_\alpha = g^{\beta\gamma}\widetilde{G}\left(\frac{\partial}{\partial u^\alpha}, \frac{\partial}{\partial v^\gamma}\right). \tag{1.19}$$

By using (1.1.2), (1.1.3) for M', and taking into account that $g^{\beta\gamma}$ and $g_{\alpha\mu}$ are the components of Finsler tensor fields on $\mathbb{F}^m$, we conclude that N^β_α satisfy (1.4.4) on M'. Hence, owing to Theorem 1.4.1, we obtain a non–linear connection $HM' = (N^\beta_\alpha)$ on M'. As a distribution, HM' is locally spanned by

$$\frac{\delta}{\delta u^\alpha} = \frac{\partial}{\partial u^\alpha} - N^\beta_\alpha \frac{\partial}{\partial v^\beta}. \tag{1.20}$$

We call it the **induced non–linear connection** on $\mathbb{F}^m$. By the definition of $\widetilde{G}$, we know that $H\widetilde{M}'$ is orthogonal to $V\widetilde{M}'$ with respect to $\widetilde{G}$. It is interesting to note that HM' inherits this property.

PROPOSITION 1.1 *Let $\mathbb{F}^m = (M, M', F^*)$ be a pseudo–Finsler submanifold of $\widetilde{\mathbb{F}}^{m+n} = (\widetilde{M}, \widetilde{M}', \widetilde{F}^*)$ and $H\widetilde{M}' = (\widetilde{N}^j_i)$ be a non–linear connection on $\widetilde{M}'$. Then the induced non–linear connection $HM' = (N^\beta_\alpha)$ on M' is orthogonal to VM' with respect to $\widetilde{G}$ and has the local coefficients given by*

$$N^\beta_\alpha = \widetilde{B}^\beta_i (B^i_{\alpha 0} + B^j_\alpha \widetilde{N}^i_j). \tag{1.21}$$

PROOF. Substituting the expression for N^β_α from (1.19) into (1.20) and using (1.7) we deduce that $\widetilde{G}(\delta/\delta u^\alpha,\ \partial/\partial v^\beta) = 0$, that is, HM' is orthogonal to VM'. Then by using (1.5), (1.6), (1.4.3), (1.2.11) and (1.13) and taking into account that $H\widetilde{M}'$ is orthogonal to $V\widetilde{M}'$ with respect to $\widetilde{G}$ we obtain (1.21). ■

By direct calculations using (1.1) and (1.2) we infer that

$$dx^i = B^i_\alpha du^\alpha, \tag{1.22}$$

and

$$dy^i = B^i_{\alpha 0} du^\alpha + B^i_\alpha dv^\alpha. \tag{1.23}$$

Then, following (1.5.10) we define on M' the local 1–forms

$$\delta v^\alpha = dv^\alpha + N^\alpha_\beta du^\beta. \tag{1.24}$$

Thus (1.21)–(1.24) enable us to state the following result.

PROPOSITION 1.2 *A complementary distribution $HM' = (N^\beta_\alpha)$ to VM' in TM' is the induced non–linear connection on $\mathbb{F}^m$ by the non–linear connection $H\widetilde{M}' = (\widetilde{N}^j_i)$ of $\widetilde{\mathbb{F}}^{m+n}$, if and only if,*

$$\delta v^\alpha = \widetilde{B}^\alpha_j \delta y^j, \tag{1.25}$$

where we set

$$\delta y^j = dy^j + \widetilde{N}^j_i dx^i.$$

Remark 1.1. We have to note that relations of the type of (1.25) were used in general for Finsler vector fields by Matsumoto [4], Miron [2], Rund ([3], p.159), in order to get the local coefficients of the induced Finsler connection. As we shall see later in this section, by using the orthogonal decomposition (1.8) and our approach of pseudo–Finsler submanifolds via the vertical vector bundle, we reach the same goal. ■

As $H\widetilde{M}'_{|M'} \oplus VM'^{\perp}$ is the orthogonal complementary vector bundle to VM' in $T\widetilde{M}'_{|M'}$, and HM' is orthogonal to VM', we deduce that HM' is a vector subbundle of $H\widetilde{M}'_{|M'} \oplus VM'^{\perp}$. Then using (1.20), (1.5), (1.6), (1.21) and (1.12) we obtain

$$\frac{\delta}{\delta u^\alpha} = B^i_\alpha \frac{\delta}{\delta x^i} + H^a_\alpha B_a, \tag{1.26}$$

where we set

$$H^a_\alpha = \widetilde{B}^a_i (B^i_{\alpha 0} + B^j_\alpha \widetilde{N}^i_j). \tag{1.27}$$

We now check that H^a_α are the components of a mixed Finsler tensor field. First, we note that $\delta/\delta u^\alpha$ satisfy (see (1.4.5))

$$\frac{\delta}{\delta u^\alpha} = \frac{\partial \bar{u}^\beta}{\partial u^\alpha} \frac{\delta}{\delta \bar{u}^\beta}. \tag{1.28}$$

Also we have

$$\bar{B}^i_a \frac{\partial x^j}{\partial \bar{x}^i} = A^b_a B^j_b. \tag{1.29}$$

Thus by using (1.26), (1.28), and (1.29) we obtain that H^a_α are the local components of a mixed Finsler tensor field of type $\begin{pmatrix} 0 & 1 \\ 1 & 0 \end{pmatrix}$ on $\mathbb{F}^m$ with respect to $VM'^\perp$ (see(1.5.19)). We call $\omega^a = H^a_\alpha du^\alpha$ the **normal curvature form** of $\mathbb{F}^m$ with respect to the Finsler normal direction B_a. The above name will be justified in Section 1 of Chapter 3.

Next, we consider a Finsler connection $\widetilde{FC} = (H\widetilde{M}', \widetilde{\nabla})$ on the pseudo–Finsler manifold $\widetilde{\mathbb{F}}^{m+n} = (\widetilde{M}, \widetilde{M}', \widetilde{F}^*)$ and look for the induced geometric objects on the pseudo–Finsler submanifold $\mathbb{F}^m = (M, M', F^*)$. We have already defined the induced non–linear connection $HM' = (N^\beta_\alpha)$ with local coefficients given by (1.21). Thus we proceed with the study of the geometric objects induced by $\widetilde{\nabla}$ on $\mathbb{F}^m$. To this end we denote by h and v the projection morphisms of TM' on HM' and VM' respectively. Then according to the orthogonal decomposition (1.8) we set

$$\widetilde{\nabla}_X vY = \nabla_X vY + B(X, vY), \ \forall X, Y \in \Gamma(TM'), \tag{1.30}$$

where $\nabla_X vY \in \Gamma(VM')$ and $B(X, vY) \in \Gamma(VM'^\perp)$. It is easy to check that ∇ is a linear connection on the vertical vector bundle of $\mathbb{F}^m$. Thus $IFC = (HM', \nabla)$ is a Finsler connection on $\mathbb{F}^m$ which we call the **induced Finsler connection**. Also B is an $\mathcal{F}(M')$–bilinear mapping on $\Gamma(TM') \times \Gamma(VM')$ and $\Gamma(VM'^\perp)$–valued, which we call the **second fundamental form** of $\mathbb{F}^m$. Taking into account that, in the classical theory of Riemannian submanifolds, such formula comes from the work of Gauss, we are entitled to call (1.30) the **Gauss formula** for the immersion of $\mathbb{F}^m$ in $(\widetilde{\mathbb{F}}^{m+n}, \widetilde{FC})$.

Using B we define the $\mathcal{F}(M')$–bilinear mappings:

$$\begin{cases} H : \Gamma(HM') \times \Gamma(HM') \longrightarrow \Gamma(VM'^\perp) \\ H(hX, hY) = B\,(hX, QhY)\,, \end{cases} \tag{1.31}$$

and

$$\begin{cases} V : \Gamma(VM') \times \Gamma(VM') \longrightarrow \Gamma(VM'^\perp) \\ V(vX, vY) = B\,(vX, vY)\,, \end{cases} \tag{1.32}$$

for any $X, Y \in \Gamma(TM')$, where Q is the associate almost product structure to the induced non–linear connection HM' (see (1.4.8)). We call H and V the h–**second fundamental form** and the v–**second fundamental form** of $\mathbb{F}^m$ respectively. Finally, from Gauss formula we derive:

$$\widetilde{\nabla}_{hX} QhY = \nabla_{hX} QhY + H(hX, hY), \tag{1.33}$$

and

$$\widetilde{\nabla}_{vX} vY = \nabla_{vX} vY + V\,(vX, vY)\,, \tag{1.34}$$

for any $X, Y \in \Gamma(TM')$.

Now, for any $X \in \Gamma(TM')$ and $W \in \Gamma(VM'^{\perp})$ we set

$$\widetilde{\nabla}_X W = -A_W X + \nabla^{\perp}_X W, \tag{1.35}$$

where $A_W X \in \Gamma(VM')$ and $\nabla^{\perp}_X W \in \Gamma(VM'^{\perp})$. It follows that $\nabla^{\perp}$ is a linear connection on the pseudo–Finsler normal bundle $VM'^{\perp}$ of $\mathbb{F}^m$. Therefore according to the theory we developed is section 1.4, the pair $NFC = (HM', \nabla^{\perp})$ is a vectorial Finsler connection on $VM'^{\perp}$. We call NFC the **normal Finsler connection** with respect to $\widetilde{FC}$ (cf., Bejancu [2]). We also note that

$$\begin{cases} A : \Gamma\left(VM'^{\perp}\right) \times \Gamma(TM') \longrightarrow \Gamma(VM') \\ A(W, X) = A_W X, \end{cases} \tag{1.36}$$

is an $\mathcal{F}(M')$–bilinear mapping. For any $W \in \Gamma(VM'^{\perp})$ we call A_W the **shape operator** (the **Weingarten operator**) with respect to W. Finally, (1.35) is named the **Weingarten formula** for the immersion of $\mathbb{F}^m$ in $\left(\widetilde{\mathbb{F}}^{m+n}, \widetilde{FC}\right)$.

As in the case of the second fundamental form, by means of A we define for any $W \in \Gamma(VM'^{\perp})$ the $\mathcal{F}(M')$–linear mappings:

$$\begin{cases} A^h_W : \Gamma(HM') \longrightarrow \Gamma(HM'), \\ A^h_W hX = QA_W hX, \end{cases} \tag{1.37}$$

and

$$\begin{cases} A^v_W : \Gamma(VM') \longrightarrow \Gamma(VM'), \\ A^v_W vX = A_W vX, \end{cases} \tag{1.38}$$

and call them the h–**shape operator** and v–**shape operator** respectively. Thus from the Weingarten formula we deduce that

$$\widetilde{\nabla}_{hX} W = -QA^h_W hX + \nabla^{\perp}_{hX} W, \tag{1.39}$$

and

$$\widetilde{\nabla}_{vX} W = -A^v_W vX + \nabla^{\perp}_{vX} W, \tag{1.40}$$

for any $X \in \Gamma(TM')$ and $W \in \Gamma(VM'^{\perp})$.

We close the section with local expressions for both the geometric objects and the equations induced on $\mathbb{F}^m$. To achieve this, we suppose that the Finsler connection $\widetilde{FC}$ on $\widetilde{\mathbb{F}}^{m+n}$ is locally given by the triple $(\widetilde{N}^k_i, \widetilde{F}_i{}^k{}_j, \widetilde{C}_i{}^k{}_j)$ (see Theorem 1.5.1). Thus according to (1.5.1) and (1.5.2) we have

$$\widetilde{\nabla}_{\delta/\delta x^j} \frac{\partial}{\partial y^i} = \widetilde{F}_i{}^k{}_j \frac{\partial}{\partial y^k}, \tag{1.41}$$

and

$$\widetilde{\nabla}_{\partial/\partial y^j}\frac{\partial}{\partial y^i} = \widetilde{C}_i{}^k{}_j\frac{\partial}{\partial y^k}. \tag{1.42}$$

Now, for any $X \in \Gamma(TM')$ we define the differential operator

$$\begin{cases} \bar{\nabla}_X : \Gamma(V\widetilde{M}'_{|M'}) \longrightarrow \Gamma(V\widetilde{M}'_{|M'}), \\ \bar{\nabla}_X\widetilde{Y} = \widetilde{\nabla}_X\widetilde{Y}, \ \ \forall\, \widetilde{Y} \in \Gamma(V\widetilde{M}'_{|M'}). \end{cases} \tag{1.43}$$

Clearly $\bar{\nabla}$ is a linear connection on the vector bundle $V\widetilde{M}'_{|M'}$ and thus $\bar{FC} = (HM', \bar{\nabla})$ is a vectorial Finsler connection on $V\widetilde{M}'_{|M'}$. Locally we put

$$\bar{\nabla}_{\delta/\delta u^\alpha}\frac{\partial}{\partial y^i} = \bar{F}_i{}^k{}_\alpha\frac{\partial}{\partial y^k} \tag{1.44}$$

and

$$\bar{\nabla}_{\partial/\partial v^\alpha}\frac{\partial}{\partial y^i} = \bar{C}_i{}^k{}_\alpha\frac{\partial}{\partial y^k}. \tag{1.45}$$

Then we replace $\delta/\delta u^\alpha$ and $\partial/\partial v^\alpha$ from (1.26) and (1.6) in (1.44) and (1.45), respectively, and using (1.41) and (1.42) we derive

$$\bar{F}_i{}^k{}_\alpha = \widetilde{F}_i{}^k{}_j B^j_\alpha + \widetilde{C}_i{}^k{}_j B^j_a H^a_\alpha, \tag{1.46}$$

and

$$\bar{C}_i{}^k{}_\alpha = \widetilde{C}_i{}^k{}_j B^j_\alpha. \tag{1.47}$$

Further, we introduce the local coefficients $F_\alpha{}^\gamma{}_\beta$ and $C_\alpha{}^\gamma{}_\beta$ of the linear connection ∇ from IFC by:

$$(a)\ \nabla_{\delta/\delta u^\beta}\frac{\partial}{\partial v^\alpha} = F_\alpha{}^\gamma{}_\beta\frac{\partial}{\partial v^\gamma} \ \text{ and } \ (b)\ \nabla_{\partial/\partial v^\beta}\frac{\partial}{\partial v^\alpha} = C_\alpha{}^\gamma{}_\beta\frac{\partial}{\partial v^\gamma}, \tag{1.48}$$

and the local components $H^a{}_{\alpha\beta}$ and $V^a{}_{\alpha\beta}$ of H and V by:

$$(a)\ H\left(\frac{\delta}{\delta u^\beta}, \frac{\delta}{\delta u^\alpha}\right) = H^a{}_{\alpha\beta}B_a \ \text{ and } \ (b)\ V\left(\frac{\partial}{\partial v^\beta}, \frac{\partial}{\partial v^\alpha}\right) = V^a{}_{\alpha\beta}B_a. \tag{1.49}$$

Then replace (hX, hY) from (1.33) by $(\delta/\delta u^\beta, \delta/\delta u^\alpha)$ and (vX, vY) from (1.34) by $(\partial/\partial v^\beta,\ \partial/\partial v^\alpha)$ and using (1.48) and (1.49) obtain

$$\widetilde{\nabla}_{\delta/\delta u^\beta}\frac{\partial}{\partial v^\alpha} = F_\alpha{}^\gamma{}_\beta\frac{\partial}{\partial v^\gamma} + H^a{}_{\alpha\beta}\, B_a, \tag{1.50}$$

and

$$\widetilde{\nabla}_{\partial/\partial v^\beta}\frac{\partial}{\partial v^\alpha} = C_\alpha{}^\gamma{}_\beta\frac{\partial}{\partial v^\gamma} + V^a{}_{\alpha\beta}\, B_a. \tag{1.51}$$

On the other hand, using (1.6), (1.20), (1.44), (1.45), and (1.14) we deduce that

$$\widetilde{\nabla}_{\delta/\delta u^{\beta}}\frac{\partial}{\partial v^{\alpha}} = \left(B^{k}_{\alpha\beta} + B^{i}_{\alpha}\bar{F}_{i\ \beta}^{\ k}\right)\left(\widetilde{B}^{\alpha}_{k}\frac{\partial}{\partial v^{\alpha}} + \widetilde{B}^{a}_{k}B_{a}\right), \tag{1.52}$$

and

$$\widetilde{\nabla}_{\partial/\partial v^{\beta}}\frac{\partial}{\partial v^{\alpha}} = B^{i}_{\alpha}\bar{C}_{i\ \beta}^{\ k}\left(\widetilde{B}^{\alpha}_{k}\frac{\partial}{\partial v^{\alpha}} + \widetilde{B}^{a}_{k}B_{a}\right). \tag{1.53}$$

Finally, comparing (1.50) and (1.51) with (1.52) and (1.53), respectively, and using (1.46) and (1.47) we obtain the following theorem:

THEOREM 1.1 *Let $\widetilde{\mathbb{F}}^{m+n}$ be a pseudo–Finsler manifold endowed with the Finsler connection $\widetilde{FC} = (\widetilde{N}^{k}_{i}, \widetilde{F}_{i\ j}^{\ k}, \widetilde{C}_{i\ j}^{\ k})$ and $\mathbb{F}^{m}$ be a pseudo–Finsler submanifold of $\widetilde{\mathbb{F}}^{m+n}$. Then we have the assertions:*

(i) The local coefficients of the induced Finsler connection $IFC = (N^{\gamma}_{\alpha}, F_{\alpha\ \beta}^{\ \gamma}, C_{\alpha\ \beta}^{\ \gamma})$ are given by (1.21) and the following formulas:

$$\begin{aligned} F_{\alpha\ \beta}^{\ \gamma} &= (B^{k}_{\alpha\beta} + B^{i}_{\alpha}\bar{F}_{i\ \beta}^{\ k})\widetilde{B}^{\gamma}_{k} \\ &= (B^{k}_{\alpha\beta} + B^{ij}_{\alpha\beta}\widetilde{F}_{i\ j}^{\ k} + B^{i}_{\alpha}\widetilde{C}_{i\ j}^{\ k}B^{j}_{a}H^{a}_{\beta})\widetilde{B}^{\gamma}_{k}, \end{aligned} \tag{1.54}$$

and

$$C_{\alpha\ \beta}^{\ \gamma} = B^{i}_{\alpha}\ \bar{C}_{i\ \beta}^{\ k}\widetilde{B}^{\gamma}_{k} = B^{ij}_{\alpha\beta}\widetilde{C}_{i\ j}^{\ k}\widetilde{B}^{\gamma}_{k}. \tag{1.55}$$

(ii) The local components of the second fundamental forms H and V are given by

$$\begin{aligned} H^{a}_{\ \alpha\beta} &= (B^{k}_{\alpha\beta} + B^{i}_{\alpha}\bar{F}_{i\ \beta}^{\ k})\widetilde{B}^{a}_{k} \\ &= (B^{k}_{\alpha\beta} + B^{ij}_{\alpha\beta}\widetilde{F}_{i\ j}^{\ k} + B^{i}_{\alpha}\widetilde{C}_{i\ j}^{\ k}B^{j}_{b}H^{b}_{\beta})\widetilde{B}^{a}_{k}, \end{aligned} \tag{1.56}$$

and

$$V^{a}_{\ \alpha\beta} = B^{i}_{\alpha}\bar{C}_{i\ \beta}^{\ k}\widetilde{B}^{a}_{k} = B^{ij}_{\alpha\beta}\widetilde{C}_{i\ j}^{\ k}\widetilde{B}^{a}_{k}, \tag{1.57}$$

respectively.

Next we note that B^{i}_{α} are the local components of a mixed Finsler tensor field of type $\begin{pmatrix} 0 & 1 \\ 1 & 0 \end{pmatrix}$ on $\mathbb{F}^{m}$ with respect to the vector bundle $V\widetilde{M}'_{|M'}$. Moreover, the pair $(IFC, \bar{FC})$ defines the horizontal and vertical relative covariant derivatives of B^{i}_{α} given by (see (1.5.20) and (1.5.21))

$$\begin{aligned} B^{i}_{\alpha|\beta} &= \frac{\delta B^{i}_{\alpha}}{\delta u^{\beta}} + B^{j}_{\alpha}\bar{F}_{j\ \beta}^{\ i} - B^{i}_{\gamma}F_{\alpha\ \beta}^{\ \gamma} \\ &= B^{i}_{\alpha\beta} + B^{j}_{\alpha}\bar{F}_{j\ \beta}^{\ i} - B^{i}_{\gamma}F_{\alpha\ \beta}^{\ \gamma}, \end{aligned} \tag{1.58}$$

and

$$\begin{aligned} B^i_{\alpha\|\beta} &= \frac{\partial B^i_\alpha}{\partial v^\beta} + B^j_\alpha \bar{C}_j{}^i{}_\beta - B^i_\gamma C_\alpha{}^\gamma{}_\beta \\ &= B^j_\alpha \bar{C}_j{}^i{}_\beta - B^i_\gamma C_\alpha{}^\gamma{}_\beta, \end{aligned} \tag{1.59}$$

respectively. By straightforward calculations using (1.54)–(1.59) and (1.12) we deduce that

$$(a) \quad B^i_{\alpha|\beta} = H^a{}_{\alpha\beta} B^i_a \quad \text{and} \quad (b) \quad B^i_{\alpha\|\beta} = V^a{}_{\alpha\beta} B^i_a. \tag{1.60}$$

We now consider the normal Finsler connection $NFC = (HM', \nabla^\perp)$ and set

$$(a) \quad \nabla^\perp_{\delta/\delta u^\alpha} B_a = F_a{}^b{}_\alpha B_b \quad \text{and} \quad (b) \quad \nabla^\perp_{\partial/\partial v^\alpha} B_a = C_a{}^b{}_\alpha B_b. \tag{1.61}$$

Thus by Theorem 1.4.3 we see that NFC is locally determined by the triple $(N^\beta_\alpha, F_a{}^b{}_\alpha, C_a{}^b{}_\alpha)$. The shape operators with respect to the orthonormal field of frames $\{B_a\}$ are denoted by A_a.

Then we put

$$(a) \quad A_a\left(\frac{\delta}{\delta u^\alpha}\right) = H'^\beta_a{}_\alpha \frac{\delta}{\delta u^\beta} \quad \text{and} \quad (b) \quad A_a\left(\frac{\partial}{\partial v^\alpha}\right) = V'^\beta_a{}_\alpha \frac{\partial}{\partial v^\beta}. \tag{1.62}$$

Next we replace X from (1.39) and (1.40) by $\delta/\delta u^\alpha$ and $\partial/\partial v^\alpha$ respectively and W by B_a, and using (1.61) and (1.62) we infer that

$$\widetilde{\nabla}_{\delta/\delta u^\alpha} B_a = -H'^\beta_a{}_\alpha \frac{\partial}{\partial v^\beta} + F_a{}^b{}_\alpha B_b, \tag{1.63}$$

and

$$\widetilde{\nabla}_{\partial/\partial v^\alpha} B_a = -V'^\beta_a{}_\alpha \frac{\partial}{\partial v^\beta} + C_a{}^b{}_\alpha B_b. \tag{1.64}$$

On the other hand, using (1.44), (1.45), and (1.14) we derive

$$\widetilde{\nabla}_{\delta/\delta u^\alpha} B_a = \left(\frac{\delta B^k_a}{\delta u^\alpha} + B^j_a \bar{F}_j{}^k{}_\alpha\right)\left(\widetilde{B}^\beta_k \frac{\partial}{\partial v^\beta} + \widetilde{B}^b_k B_b\right), \tag{1.65}$$

and

$$\widetilde{\nabla}_{\partial/\partial v^\alpha} B_a = \left(\frac{\partial B^k_a}{\partial v^\alpha} + B^j_a \bar{C}_j{}^k{}_\alpha\right)\left(\widetilde{B}^\beta_k \frac{\partial}{\partial v^\beta} + \widetilde{B}^b_k B_b\right). \tag{1.66}$$

Comparing (1.63) and (1.64) with (1.65) and (1.66), respectively, we obtain

$$(a) \quad \left(\frac{\delta B^k_a}{\delta u^\alpha} + B^j_a \bar{F}_j{}^k{}_\alpha\right) \widetilde{B}^\beta_k = -H'^\beta_a{}_\alpha;$$

$$(b)\quad \left(\frac{\partial B^k_a}{\partial v^\alpha} + B^j_a \bar{C}_j{}^k{}_\alpha\right) \widetilde{B}^\beta_k = -V'^\beta_a{}_\alpha, \tag{1.67}$$

and

$$(a)\quad \left(\frac{\delta B^k_a}{\delta u^\alpha} + B^j_a \bar{F}_j{}^k{}_\alpha\right) \widetilde{B}^b_k = F_a{}^b{}_\alpha;$$

$$(b)\quad \left(\frac{\partial B^k_a}{\partial v^\alpha} + B^j_a \bar{C}_j{}^k{}_\alpha\right) \widetilde{B}^b_k = C_a{}^b{}_\alpha. \tag{1.68}$$

Contracting (1.67a) by B^i_β and (1.68a) by B^i_b and then adding them and using (1.12) we obtain

$$B^i_{a|\alpha} = -H'^\beta_a{}_\alpha B^i_\beta, \tag{1.69}$$

where we set

$$B^i_{a|\alpha} = \frac{\delta B^i_a}{\delta u^\alpha} + B^j_a \bar{F}_j{}^i{}_\alpha - B^i_b F_a{}^b{}_\alpha. \tag{1.70}$$

Similarly, from (1.67b) and (1.68b) we deduce that

$$B^i_{a\|\alpha} = -V'^\beta_a{}_\alpha B^i_\beta, \tag{1.71}$$

where we set

$$B^i_{a\|\alpha} = \frac{\partial B^i_a}{\partial v^\alpha} + B^j_a \bar{C}_j{}^i{}_\alpha - B^i_b C_a{}^b{}_\alpha. \tag{1.72}$$

Remark 1.2. It is easy to see that $B^i_{a|\alpha}$ and $B^i_{a\|\alpha}$ are the horizontal and vertical relative covariant derivatives of the mixed Finsler tensor field B^i_a on $\mathbb{F}^m$ with respect to the vector bundle $V\widetilde{M}'_{|M'} \oplus VM'^\perp$ on which we consider a linear connection defined in a standard way by the pair $(\bar{\nabla}, \nabla^\perp)$. ■

According to (1.5.20) and (1.5.21) the horizontal and vertical relative covariant derivatives of $\tilde{g}_{ij}$ with respect to the pair $(IFC, \bar{FC})$ are given by:

$$\tilde{g}_{ij|\alpha} = \frac{\delta \tilde{g}_{ij}}{\delta u^\alpha} - \tilde{g}_{kj}\bar{F}_i{}^k{}_\alpha - \tilde{g}_{ik}\bar{F}_j{}^k{}_\alpha, \tag{1.73}$$

and

$$\tilde{g}_{ij\|\alpha} = \frac{\partial \tilde{g}_{ij}}{\partial v^\alpha} - \tilde{g}_{kj}\bar{C}_i{}^k{}_\alpha - \tilde{g}_{ik}\bar{C}_j{}^k{}_\alpha, \tag{1.74}$$

respectively. Then, using (1.26), (1.6), (1.46) and (1.47) we infer that

$$\tilde{g}_{ij|\alpha} = \tilde{g}_{ij|k} B^k_\alpha + \tilde{g}_{ij\|k} H^a_\alpha B^k_a, \tag{1.75}$$

and

$$\tilde{g}_{ij\|\alpha} = \tilde{g}_{ij\|k} B^k_\alpha, \tag{1.76}$$

where $\tilde{g}_{ij|k}$ and $\tilde{g}_{ij||k}$ are the horizontal and vertical covariant derivatives of $\tilde{g}_{ij}$ with respect to the Finsler connection $\widetilde{FC}$. Finally, using (1.4), (1.60), (1.9), (1.75) and (1.76) we obtain

$$g_{\alpha\beta|\gamma} = \tilde{g}_{ij|k} B^{ijk}_{\alpha\beta\gamma} + \tilde{g}_{ij||k} B^{ij}_{\alpha\beta} B^{k}_{a} H^{a}_{\gamma}, \tag{1.77}$$

and

$$g_{\alpha\beta||\gamma} = \tilde{g}_{ij||k} B^{ijk}_{\alpha\beta\gamma}. \tag{1.78}$$

As a direct consequence of (1.77) and (1.78) we state the following result.

PROPOSITION 1.3 *Let $\mathbb{F}^m$ be a pseudo-Finsler submanifold of $\widetilde{\mathbb{F}}^{m+n}$ and IFC be the induced Finsler connection on $\mathbb{F}^m$ by the Finsler connection $\widetilde{FC}$ of $\widetilde{\mathbb{F}}^{m+n}$. Then we have the assertions:*

(i) If $\widetilde{FC}$ is a v–metric Finsler connection, then IFC is a v–metric Finsler connection too.

(ii) If $\widetilde{FC}$ is a metric Finsler connection, then IFC is a metric Finsler connection too.

It is important to note that if $\widetilde{FC}$ is an h–metric Finsler connection then IFC is not necessarily so (see the induced Finsler connection by the Rund connection in Section 5).

Now, from (1.33) and (1.39) we deduce that

$$\tilde{g}(\widetilde{\nabla}_{hX} QhY, W) + \tilde{g}(QhY, \widetilde{\nabla}_{hX} W) = \tilde{g}(H(hX, hY), W) - \tilde{g}(QhY, QA^{h}_{W} hX),$$

for any $X, Y \in \Gamma(TM')$ and $W \in \Gamma(VM'^{\perp})$. As $\tilde{g}(QhY, W) = 0$, by (1.3.8) we derive that

$$\tilde{g}(QhY, QA^{h}_{W} hX) = (\widetilde{\nabla}_{hX} \tilde{g})(QhY, W) + \tilde{g}(H(hX, hY), W). \tag{1.79}$$

Substituting hX, hY and W by $\delta/\delta u^{\beta}, \delta/\delta u^{\alpha}$ and B_a respectively, and by using (1.5.20), (1.73), (1.49a) and (1.62a) we obtain

$$H'_{a\alpha\beta} = H_{a\alpha\beta} + \tilde{g}_{ij|\beta} B^{i}_{\alpha} B^{j}_{a}, \tag{1.80}$$

where we set

$$(a)\quad H'_{a\alpha\beta} = H'^{\ \gamma}_{a\ \ \beta} g_{\gamma\alpha} \quad \text{and} \quad (b)\quad H_{a\alpha\beta} = \delta_{ab} H^{b}_{\ \alpha\beta}. \tag{1.81}$$

Similarly, by using (1.34) and (1.40) we infer that

$$(a)\quad V'_{a\alpha\beta} = V_{a\alpha\beta} + \tilde{g}_{ij||\beta} B^{i}_{\alpha} B^{j}_{a}, \tag{1.82}$$

where we put

$$(a)\ \ V'_{a\alpha\beta} = V'_{a}{}^{\gamma}{}_{\beta} g_{\gamma\alpha} \ \ \text{and} \ \ (b)\ \ V_{a\alpha\beta} = \delta_{ab} V^{b}{}_{\alpha\beta}. \tag{1.83}$$

2. The Equations of Gauss, Codazzi, and Ricci for $\mathbb{F}^m$ in $\widetilde{\mathbb{F}}^{m+n}$

As is well known (cf., Chen [1], p. 44, and Yano–Kon [1], p. 67) the Gauss, Codazzi, and Ricci equations for a Riemannian submanifold are easily obtained by a direct calculation of the curvature tensor field of the Levi–Civita connection on the ambient manifold using the Gauss and Weingarten formulas and the induced geometric objects. Our approach of Finsler geometry via the vertical vector bundle will enable us to obtain a generalization of such equations to pseudo–Finsler submanifolds for the general case when the ambient pseudo–Finsler manifold is endowed with an arbitrary Finsler connection.

Let $\mathbb{F}^m = (M, M', F^*)$ be a pseudo–Finsler submanifold of the pseudo–Finsler manifold $\widetilde{\mathbb{F}}^{m+n} = (\widetilde{M}, \widetilde{M}', \widetilde{F}^*)$. Suppose $\widetilde{FC} = (H\widetilde{M}', \widetilde{\nabla}) = (\widetilde{N}_i^k, \widetilde{F}_i{}^k{}_j, \widetilde{C}_i{}^k{}_j)$ is a Finsler connection on $\widetilde{\mathbb{F}}^{m+n}$. In the first section we have defined the induced Finsler connection $IFC = (HM', \nabla) = (N_\alpha^\gamma, F_\alpha{}^\gamma{}_\beta, C_\alpha{}^\gamma{}_\beta)$ on $\mathbb{F}^m$ and the vectorial Finsler connections $\bar{FC} = (HM', \bar{\nabla}) = (N_\alpha^\gamma, \bar{F}_i{}^k{}_\alpha, \bar{C}_i{}^k{}_\alpha)$ and $NFC = (HM', \nabla^\perp) = (N_\alpha^\gamma, F_a{}^b{}_\alpha, C_a{}^b{}_\alpha)$ on the vector bundles $V\widetilde{M}'_{|M'}$ and $VM'^\perp$, respectively.

We now denote by $\widetilde{R}$ and R the curvature forms of the linear connections $\widetilde{\nabla}$ and ∇ respectively. Then by direct calculations using (1.3.11), (1.30), (1.35) and (1.36) we obtain

$$\begin{aligned}\widetilde{R}(X,Y)vZ &= R(X,Y)vZ + A(B(X,vZ),Y) - A(B(Y,vZ),X)\\ &\quad + \nabla^\perp_X(B(Y,vZ)) - \nabla^\perp_Y(B(X,vZ))\\ &\quad + B(X,\nabla_Y vZ) - B(Y,\nabla_X vZ) - B([X,Y],vZ),\end{aligned} \tag{2.1}$$

for any $X, Y, Z \in \Gamma(TM')$. Consider the associate linear connection D to the induced Finsler connection IFC (see (1.5.27)) and denote by T its torsion tensor field. Thus (1.3.12) yields

$$[X,Y] = D_X Y - D_Y X - T(X,Y),\ \forall X, Y \in \Gamma(TM'). \tag{2.2}$$

Next, by means of the triple $(D, \nabla, \nabla^\perp)$ of linear connections, we define a covariant derivative of the second fundamental form B of $\mathbb{F}^m$ by (see (1.3.9))

$$(\nabla'_X B)(Y, vZ) = \nabla^\perp_X(B(Y,vZ)) - B(D_X Y, vZ) - B(Y, \nabla_X vZ). \tag{2.3}$$

Then using (2.2) and (2.3) in (2.1) and taking into account that $D_X vY = \nabla_X vY$, for any $X, Y \in \Gamma(TM')$, we deduce that

$$\begin{aligned} \widetilde{R}(X,Y)vZ &= R(X,Y)vZ + A(B(X,vZ),Y) - A(B(Y,vZ),X) \quad (2.4) \\ &+(\nabla'_X B)(Y,vZ) - (\nabla'_Y B)(X,vZ) + B(T(X,Y),vZ). \end{aligned}$$

Now we take the components of (2.4) in VM' and $VM'^{\perp}$ and obtain

$$\begin{aligned} \tilde{g}(\widetilde{R}(X,Y)vZ, vU) &= g(R(X,Y)vZ, vU) \\ &+g(A_{B(X,vZ)}Y - A_{B(Y,vZ)}X, vU), \quad (2.5) \end{aligned}$$

and

$$\begin{aligned} \tilde{g}(\widetilde{R}(X,Y)vZ, W) &= \tilde{g}((\nabla'_X B)(Y,vZ) - (\nabla'_Y B)(X,vZ), W) \\ &+\tilde{g}(B(T(X,Y),vZ), W), \quad (2.6) \end{aligned}$$

respectively, for any $X, Y, Z, U \in \Gamma(TM')$ and $W \in \Gamma(VM'^{\perp})$. Following the theory of pseudo–Riemannian submanifolds we call (2.5) and (2.6) the **Gauss equation** and the **B–Codazzi equation**, respectively, for the pseudo–Finsler submanifold $\mathbb{F}^m$ with respect to the Finsler connection $\widetilde{FC}$ on $\widetilde{\mathbb{F}}^{m+n}$.

Next, we define a covariant derivative of A given by (1.36) as follows:

$$(\nabla'_X A)(W,Y) = \nabla_X(A(W,Y)) - A(\nabla^{\perp}_X W, Y) - A(W, D_X Y), \quad (2.7)$$

for any $X, Y \in \Gamma(TM')$ and $W \in \Gamma(VM'^{\perp})$. According to (1.3.10) the curvature form $R^{\perp}$ of the normal Finsler connection is given by

$$R^{\perp}(X,Y)W = \nabla^{\perp}_X(\nabla^{\perp}_Y W) - \nabla^{\perp}_Y(\nabla^{\perp}_X W) - \nabla^{\perp}_{[X,Y]}W, \quad (2.8)$$

for any $X, Y \in \Gamma(TM')$ and $W \in \Gamma(VM'^{\perp})$. Then using (1.3.11), (1.30) (1.35), (2.2), (2.7) and (2.8) we derive

$$\begin{aligned} \widetilde{R}(X,Y)W &= R^{\perp}(X,Y)W + B(Y, A_W X) - B(X, A_W Y) \quad (2.9) \\ &+(\nabla'_Y A)(W,X) - (\nabla'_X A)(W,Y) - A_W(T(X,Y)). \end{aligned}$$

Finally, taking the components of (2.9) in VM' and $VM'^{\perp}$ we obtain the equations

$$\begin{aligned} \tilde{g}(\widetilde{R}(X,Y)W, vZ) &= g((\nabla'_Y A)(W,X) - (\nabla'_X A)(W,Y), vZ) \\ &-g(A_W(T(X,Y)), vZ), \quad (2.10) \end{aligned}$$

and

$$\begin{aligned}\tilde{g}(\widetilde{R}(X,Y)W_1,W_2) &= \tilde{g}(R^{\perp}(X,Y)W_1,W_2)\\ &+\tilde{g}\left(B(Y,A_{W_1}X)-B(X,A_{W_1}Y),W_2\right), \quad (2.11)\end{aligned}$$

respectively, for any $X,Y,Z \in \Gamma(TM')$ and $W,W_1,W_2 \in \Gamma(VM'^{\perp})$. We call (2.10) and (2.11) the A–**Codazzi equation** and the **Ricci equation** respectively, for the pseudo–Finsler submanifold $\mathbb{F}^m$ with respect to the Finsler connection $\widetilde{FC}$ on $\widetilde{\mathbb{F}}^{m+n}$.

Summing up the above results we may state the following theorem:

THEOREM 2.1 *Let $\mathbb{F}^m$ be a pseudo–Finsler submanifold of a pseudo–Finsler manifold $\widetilde{\mathbb{F}}^{m+n}$ endowed with a Finsler connection $\widetilde{FC}$. Then the Gauss equation, the A–Codazzi equation, the B–Codazzi equation and the Ricci equation are given by (2.5), (2.10), (2.6), and (2.11), respectively.*

We now seek local expressions for the above structure equations of the immersion of $\mathbb{F}^m$ in $\widetilde{\mathbb{F}}^{m+n}$. Moreover, we shall find the inter–relations between the torsion tensor fields of the Finsler connections $\widetilde{FC}$ and IFC.

First, we introduce the local components of the curvature forms $\widetilde{R}$ and R as follows:

$$\begin{aligned}
&(a) \quad \widetilde{R}\left(\frac{\delta}{\delta x^k},\frac{\delta}{\delta x^j}\right)\frac{\partial}{\partial y^i}=\widetilde{R}_i{}^h{}_{jk}\frac{\partial}{\partial y^h},\\
&(b) \quad \widetilde{R}\left(\frac{\partial}{\partial y^k},\frac{\delta}{\delta x^j}\right)\frac{\partial}{\partial y^i}=\widetilde{P}_i{}^h{}_{jk}\frac{\partial}{\partial y^h},\\
&(c) \quad \widetilde{R}\left(\frac{\partial}{\partial y^k},\frac{\partial}{\partial y^j}\right)\frac{\partial}{\partial y^i}=\widetilde{S}_i{}^h{}_{jk}\frac{\partial}{\partial y^h},
\end{aligned} \qquad (2.12)$$

and

$$\begin{aligned}
&(a) \quad R\left(\frac{\delta}{\delta u^\gamma},\frac{\delta}{\delta u^\beta}\right)\frac{\partial}{\partial v^\alpha}=R_\alpha{}^\mu{}_{\beta\gamma}\frac{\partial}{\partial v^\mu},\\
&(b) \quad R\left(\frac{\partial}{\partial v^\gamma},\frac{\delta}{\delta u^\beta}\right)\frac{\partial}{\partial v^\alpha}=P_\alpha{}^\mu{}_{\beta\gamma}\frac{\partial}{\partial v^\mu},\\
&(c) \quad R\left(\frac{\partial}{\partial v^\gamma},\frac{\partial}{\partial v^\beta}\right)\frac{\partial}{\partial v^\alpha}=S_\alpha{}^\mu{}_{\beta\gamma}\frac{\partial}{\partial v^\mu}.
\end{aligned} \qquad (2.13)$$

Next, by (1.43) we see that the curvature forms $\bar{R}$ and $\widetilde{R}$ of $\bar{\nabla}$ and $\widetilde{\nabla}$ are related by

$$\bar{R}(X,Y)Z=\widetilde{R}(X,Y)Z,\ \forall X,Y\in\Gamma(TM'),\ Z\in\Gamma(V\widetilde{M}'_{|M'}). \qquad (2.14)$$

Thus locally we set

$$\begin{aligned}
&(a) \quad \widetilde{R}\left(\frac{\delta}{\delta u^\gamma}, \frac{\delta}{\delta u^\beta}\right)\frac{\partial}{\partial y^i} = \bar{R}_i{}^h{}_{\beta\gamma}\frac{\partial}{\partial y^h},\\
&(b) \quad \widetilde{R}\left(\frac{\partial}{\partial v^\gamma}, \frac{\delta}{\delta u^\beta}\right)\frac{\partial}{\partial y^i} = \bar{P}_i{}^h{}_{\beta\gamma}\frac{\partial}{\partial y^h},\\
&(c) \quad \widetilde{R}\left(\frac{\partial}{\partial v^\gamma}, \frac{\partial}{\partial v^\beta}\right)\frac{\partial}{\partial y^i} = \bar{S}_i{}^h{}_{\beta\gamma}\frac{\partial}{\partial y^h}.
\end{aligned} \tag{2.15}$$

Then using (1.6) and (1.26) in (2.15) and taking into account (2.12) we deduce that

$$\bar{R}_i{}^h{}_{\beta\gamma} = \widetilde{R}_i{}^h{}_{jk}B^{jk}_{\beta\gamma} + \widetilde{P}_i{}^h{}_{jk}B^k_a(B^j_\beta H^a_\gamma - B^j_\gamma H^a_\beta) + \widetilde{S}_i{}^h{}_{jk}B^j_a B^k_b H^a_\beta H^b_\gamma, \tag{2.16}$$

$$\bar{P}_i{}^h{}_{\beta\gamma} = \widetilde{P}_i{}^h{}_{jk}B^{jk}_{\beta\gamma} + \widetilde{S}_i{}^h{}_{jk}B^j_a H^a_\beta B^k_\gamma, \tag{2.17}$$

$$\bar{S}_i{}^h{}_{\beta\gamma} = \widetilde{S}_i{}^h{}_{jk}B^{jk}_{\beta\gamma}. \tag{2.18}$$

Further, we shall show in detail the whole process of obtaining local formulas from (2.4) in case $X = \delta/\delta u^\gamma$, $Y = \delta/\delta u^\beta$, $Z = \partial/\partial v^\alpha$. First, using (1.6) and (2.15a) we deduce that the left hand side of (2.4) is given by

$$\widetilde{R}\left(\frac{\delta}{\delta u^\gamma}, \frac{\delta}{\delta u^\beta}\right)\frac{\partial}{\partial v^\alpha} = B^i_\alpha \bar{R}_i{}^h{}_{\beta\gamma}\frac{\partial}{\partial y^h}. \tag{2.19}$$

Next, taking into account (1.31), (1.36), (1.49a) and (1.62a) we infer that

$$A\left(B\left(\frac{\delta}{\delta u^\gamma}, \frac{\partial}{\partial v^\alpha}\right), \frac{\delta}{\delta u^\beta}\right) = H^a{}_{\alpha\gamma}H'^\mu_a{}_\beta\frac{\partial}{\partial v^\mu}. \tag{2.20}$$

Using (1.5.27), (1.49a), (1.61a) and (1.48a) we see that (2.3) becomes

$$\left(\nabla'_{\delta/\delta u^\gamma}B\right)\left(\frac{\delta}{\delta u^\beta}, \frac{\partial}{\partial v^\alpha}\right) = H^a{}_{\alpha\beta|\gamma}B_a, \tag{2.21}$$

where $H^a{}_{\alpha\beta|\gamma}$ is the horizontal relative covariant derivative of the mixed Finsler tensor field $H^a{}_{\alpha\beta}$ with respect to the pair (IFC, NFC) (see (1.5.20)).

In Section 1.5 we have seen that all the local components of torsion Finsler tensor fields of a Finsler connection are given by the torsion tensor field of its associate linear connection (see (1.5.43)–(1.5.45)). Hence the torsion Finsler tensor fields of IFC are given by

$$T\left(\frac{\delta}{\delta u^\gamma}, \frac{\delta}{\delta u^\beta}\right) = R^\mu{}_{\beta\gamma}\frac{\partial}{\partial v^\mu} + T_\beta{}^\mu{}_\gamma\frac{\delta}{\delta u^\mu}, \tag{2.22}$$

$$T\left(\frac{\partial}{\partial v^\gamma}, \frac{\delta}{\delta u^\beta}\right) = P^\mu{}_{\beta\gamma}\frac{\partial}{\partial v^\mu} + C_\beta{}^\mu{}_\gamma\frac{\delta}{\delta u^\mu}, \tag{2.23}$$

$$T\left(\frac{\partial}{\partial v^\gamma}, \frac{\partial}{\partial v^\beta}\right) = S^\mu{}_{\beta\gamma}\frac{\partial}{\partial v^\mu}. \tag{2.24}$$

Thus by (2.22) and (1.49) we obtain

$$B\left(T\left(\frac{\delta}{\delta u^{\gamma}}, \frac{\delta}{\delta u^{\beta}}\right), \frac{\partial}{\partial v^{\alpha}}\right) = (H^{a}{}_{\alpha\mu}T_{\beta}{}^{\mu}{}_{\gamma} + V^{a}{}_{\alpha\mu}R^{\mu}{}_{\beta\gamma})B_{a}. \tag{2.25}$$

Therefore, taking into account (2.13a), (2.19)–(2.21) and (2.25) we see that (2.4) becomes

$$\begin{aligned} B^{i}_{\alpha}\bar{R}_{i}{}^{h}{}_{\beta\gamma}\frac{\partial}{\partial y^{h}} &= (R_{\alpha}{}^{\mu}{}_{\beta\gamma} + \mathcal{A}_{(\beta\gamma)}\{H^{a}{}_{\alpha\gamma}H'^{\mu}_{a}{}_{\beta}\})\frac{\partial}{\partial v^{\mu}} \\ &\quad +(H^{a}{}_{\alpha\mu}T_{\beta}{}^{\mu}{}_{\gamma} + V^{a}{}_{\alpha\mu}R^{\mu}{}_{\beta\gamma} + \mathcal{A}_{(\beta\gamma)}\{H^{a}{}_{\alpha\beta|\gamma}\})B_{a}. \end{aligned} \tag{2.26}$$

Finally, using (1.14) in the left hand side of (2.26) and then taking the VM' and $VM'^{\perp}$ components we deduce that

$$B^{i}_{\alpha}\bar{R}_{i}{}^{h}{}_{\beta\gamma}\widetilde{B}^{\mu}_{h} = R_{\alpha}{}^{\mu}{}_{\beta\gamma} + \mathcal{A}_{(\beta\gamma)}\{H^{a}{}_{\alpha\gamma}H'^{\mu}_{a}{}_{\beta}\}, \tag{2.27}$$

and

$$B^{i}_{\alpha}\bar{R}_{i}{}^{h}{}_{\beta\gamma}\widetilde{B}^{a}_{h} = H^{a}{}_{\alpha\mu}T_{\beta}{}^{\mu}{}_{\gamma} + V^{a}{}_{\alpha\mu}R^{\mu}{}_{\beta\gamma} + \mathcal{A}_{(\beta\gamma)}\{H^{a}{}_{\alpha\beta|\gamma}\}. \tag{2.28}$$

By similar calculations for $\{X = \partial/\partial v^{\gamma},\ Y = \delta/\delta u^{\beta},\ Z = \partial/\partial v^{\alpha}\}$ and $\{X = \partial/\partial v^{\gamma},\ Y = \partial/\partial v^{\beta},\ Z = \partial/\partial v^{\alpha}\}$ in (2.4) we obtain

$$B^{i}_{\alpha}\bar{P}_{i}{}^{h}{}_{\beta\gamma}\widetilde{B}^{\mu}_{h} = P_{\alpha}{}^{\mu}{}_{\beta\gamma} + V^{a}{}_{\alpha\gamma}H'^{\mu}_{a}{}_{\beta} - H^{a}{}_{\alpha\beta}V'^{\mu}_{a}{}_{\gamma}, \tag{2.29}$$

$$B^{i}_{\alpha}\bar{P}_{i}{}^{h}{}_{\beta\gamma}\widetilde{B}^{a}_{h} = H^{a}{}_{\alpha\beta||\gamma} - V^{a}{}_{\alpha\gamma|\beta} + H^{a}{}_{\alpha\mu}C_{\beta}{}^{\mu}{}_{\gamma} + V^{a}{}_{\alpha\mu}P^{\mu}{}_{\beta\gamma}, \tag{2.30}$$

and

$$B^{i}_{\alpha}\bar{S}_{i}{}^{h}{}_{\beta\gamma}\widetilde{B}^{\mu}_{h} = S_{\alpha}{}^{\mu}{}_{\beta\gamma} + \mathcal{A}_{(\beta\gamma)}\{V^{a}{}_{\alpha\gamma}V'^{\mu}_{a}{}_{\beta}\}, \tag{2.31}$$

$$B^{i}_{\alpha}\bar{S}_{i}{}^{h}{}_{\beta\gamma}\widetilde{B}^{a}_{h} = V^{a}{}_{\alpha\mu}S^{\mu}{}_{\beta\gamma} + \mathcal{A}_{(\beta\gamma)}\{V^{a}{}_{\alpha\beta||\gamma}\}, \tag{2.32}$$

respectively, where $V^{a}{}_{\alpha\gamma|\beta}$ and $\{H^{a}{}_{\alpha\beta||\gamma}, V^{a}{}_{\alpha\beta||\gamma}\}$ are horizontal and vertical relative covariant derivatives of the mixed Finsler tensor fields $V^{a}{}_{\alpha\gamma}$ and $\{H^{a}{}_{\alpha\beta}, V^{a}{}_{\alpha\beta}\}$ on $\mathbb{F}^{m}$ with respect to the pair (IFC, NFC) (see (1.5.20) and (1.5.21)).

In order to get local expressions for (2.9) we set

$$\begin{aligned} &(a)\quad R^{\perp}\left(\frac{\delta}{\delta u^{\beta}}, \frac{\delta}{\delta u^{\alpha}}\right)B_{a} = R^{\perp b}_{a}{}_{\alpha\beta}B_{b}, \\ &(b)\quad R^{\perp}\left(\frac{\partial}{\partial v^{\beta}}, \frac{\delta}{\delta u^{\alpha}}\right)B_{a} = P^{\perp b}_{a}{}_{\alpha\beta}B_{b}, \\ &(c)\quad R^{\perp}\left(\frac{\partial}{\partial v^{\beta}}, \frac{\partial}{\partial v^{\alpha}}\right)B_{a} = S^{\perp b}_{a}{}_{\alpha\beta}B_{b}. \end{aligned} \tag{2.33}$$

We shall now evaluate (2.9) for $X = \delta/\delta u^\beta$, $Y = \delta/\delta u^\alpha$ and $W = B_a$. Using (2.15a) we derive

$$\widetilde{R}\left(\frac{\delta}{\delta u^\beta}, \frac{\delta}{\delta u^\alpha}\right) B_a = B^i_a \bar{R}_i{}^h{}_{\alpha\beta} \frac{\partial}{\partial y^h}. \tag{2.34}$$

Then (1.49a) and (1.62a) yield

$$B\left(\frac{\delta}{\delta u^\alpha}, A_a(\frac{\delta}{\delta u^\beta})\right) = H'^\mu_{a\ \beta} H^b{}_{\mu\alpha} B_b. \tag{2.35}$$

Further, using (2.7), (1.62a), (1.61a), (1.48a) and the general formula (1.5.20) for the horizontal relative covariant derivative of $H'^\mu_{a\ \beta}$ with respect to the pair (IFC, NFC) we infer that

$$\left(\nabla'_{\delta/\delta u^\alpha} A\right)\left(B_a, \frac{\delta}{\delta u^\beta}\right) = H'^\mu_{a\ \beta|\alpha} \frac{\partial}{\partial v^\mu}. \tag{2.36}$$

Also, (2.22) and (1.62) imply that

$$A_a\left(T\left(\frac{\delta}{\delta u^\beta}, \frac{\delta}{\delta u^\alpha}\right)\right) = (H'^\mu_{a\ \gamma} T_\alpha{}^\gamma{}_\beta + V'^\mu_{a\ \gamma} R^\gamma{}_{\alpha\beta}) \frac{\partial}{\partial v^\mu}. \tag{2.37}$$

Finally, using (2.34)–(2.37) and (2.33a) we easily see that (2.9) becomes

$$\begin{aligned} B^i_a \bar{R}_i{}^h{}_{\alpha\beta} \frac{\partial}{\partial y^h} \quad &= \quad (R^{\perp b}_{a\ \alpha\beta} + \mathcal{A}_{(\alpha\beta)}\{H'^\mu_{a\ \beta} H^b{}_{\mu\alpha}\}) B_b \\ &+ (\mathcal{A}_{(\alpha\beta)}\{H'^\mu_{a\ \beta|\alpha}\} - H'^\mu_{a\ \gamma} T_\alpha{}^\gamma{}_\beta - V'^\mu_{a\ \gamma} R^\gamma{}_{\alpha\beta}) \frac{\partial}{\partial v^\mu}. \end{aligned} \tag{2.38}$$

Taking the VM' and $VM'^\perp$ components in (2.38) we obtain

$$B^i_a \bar{R}_i{}^h{}_{\alpha\beta} \widetilde{B}^\mu_h = \mathcal{A}_{(\alpha\beta)}\{H'^\mu_{a\ \beta|\alpha}\} - H'^\mu_{a\ \gamma} T_\alpha{}^\gamma{}_\beta - V'^\mu_{a\ \gamma} R^\gamma{}_{\alpha\beta}, \tag{2.39}$$

and

$$B^i_a \bar{R}_i{}^h{}_{\alpha\beta} \widetilde{B}^b_h = R^{\perp b}_{a\ \alpha\beta} + \mathcal{A}_{(\alpha\beta)}\{H'^\mu_{a\ \beta} H^b{}_{\mu\alpha}\}, \tag{2.40}$$

respectively.

By similar calculations in (2.9) for $\{X = \partial/\partial v^\beta,\ Y = \delta/\delta u^\alpha,\ W = B_a\}$ and $\{X = \partial/\partial v^\beta,\ Y = \partial/\partial v^\alpha,\ W = B_a\}$ we derive that

$$B^i_a \bar{P}_i{}^h{}_{\alpha\beta} \widetilde{B}^\mu_h = V'^\mu_{a\ \beta|\alpha} - H'^\mu_{a\ \alpha\|\beta} - H'^\mu_{a\ \gamma} C_\alpha{}^\gamma{}_\beta - V'^\mu_{a\ \gamma} P^\gamma{}_{\alpha\beta}, \tag{2.41}$$

$$B^i_a \bar{P}_i{}^h{}_{\alpha\beta} \widetilde{B}^b_h = P^{\perp b}_{a\ \alpha\beta} + V'^\mu_{a\ \beta} H^b{}_{\mu\alpha} - H'^\mu_{a\ \alpha} V^b{}_{\mu\beta}, \tag{2.42}$$

and

$$B^i_a \bar{S}_i{}^h{}_{\alpha\beta} \widetilde{B}^\mu_h = \mathcal{A}_{(\alpha\beta)}\{V'^\mu_a{}_{\beta\|\alpha}\} - V'^\mu_a{}_\gamma S^\gamma{}_{\alpha\beta}, \tag{2.43}$$

$$B^i_a \bar{S}_i{}^h{}_{\alpha\beta} \widetilde{B}^b_h = S^{\perp b}_a{}_{\alpha\beta} + \mathcal{A}_{(\alpha\beta)}\{V'^\gamma_a{}_\beta V^b{}_{\gamma\alpha}\}, \tag{2.44}$$

respectively, where we use (1.5.20) and (1.5.21) for horizontal and vertical relative covariant derivatives of $V'^\mu_a{}_\beta$ and $H'^\mu_a{}_\beta$ with respect to the pair (IFC, NFC).

The above calculations enable us to state the following theorem:

THEOREM 2.2 *Let $\widetilde{\mathbb{F}}^{m+n}$ be a pseudo-Finsler manifold endowed with an arbitrary Finsler connection $\widetilde{FC} = (\widetilde{N}^k_i, \widetilde{F}_i{}^k{}_j, \widetilde{C}_i{}^k{}_j)$ and $\mathbb{F}^m$ be a pseudo-Finsler submanifold of $\widetilde{\mathbb{F}}^{m+n}$. Then we have the following assertions:*

(i) The local expressions of the Gauss equation are given by (2.27), (2.29) and (2.31).

(ii) The local expressions of the A–Codazzi equation and B–Codazzi equation are given by the groups of equations (2.39), (2.41), (2.43) and (2.28), (2.30), (2.32) respectively.

(iii) The local expressions of the Ricci equation are given by (2.40), (2.42) and (2.44).

As in general, Finsler connections have torsions, we close this section with some structure equations about the torsion Finsler tensor fields.

First, we consider the torsion tensor field $\widetilde{T}$ of the associate linear connection $\widetilde{D}$ of $\widetilde{FC} = (H\widetilde{M}', \widetilde{\nabla})$ and set

$$\widetilde{T}\left(\frac{\delta}{\delta u^\beta}, \frac{\delta}{\delta u^\alpha}\right) = \bar{R}^i{}_{\alpha\beta}\frac{\partial}{\partial y^i} + \bar{T}_\alpha{}^i{}_\beta \frac{\delta}{\delta x^i}, \tag{2.45}$$

$$\widetilde{T}\left(\frac{\partial}{\partial v^\beta}, \frac{\delta}{\delta u^\alpha}\right) = \bar{P}^i{}_{\alpha\beta}\frac{\partial}{\partial y^i} + \bar{C}_\alpha{}^i{}_\beta \frac{\delta}{\delta x^i}, \tag{2.46}$$

$$\widetilde{T}\left(\frac{\partial}{\partial v^\beta}, \frac{\partial}{\partial v^\alpha}\right) = \bar{S}^i{}_{\alpha\beta}\frac{\partial}{\partial y^i}. \tag{2.47}$$

Then using (1.26), (1.6) and (1.5.43)–(1.5.45) in the left hand side of (2.45)–(2.47) and taking components from $V\widetilde{M}'_{|M'}$ and $H\widetilde{M}'_{|M'}$ we deduce that

$$\bar{R}^i{}_{\alpha\beta} = \widetilde{R}^i{}_{jk} B^{jk}_{\alpha\beta} + \mathcal{A}_{(\alpha\beta)}\{B^j_\alpha H^a_\beta\}\widetilde{P}^i{}_{jk} B^k_a + \widetilde{S}^i{}_{jk} B^j_a B^k_b H^a_\alpha H^b_\beta, \tag{2.48}$$

$$\bar{P}^i{}_{\alpha\beta} = \widetilde{P}^i{}_{jk} B^{jk}_{\alpha\beta} + \widetilde{S}^i{}_{jk} B^j_a B^k_\beta H^a_\alpha, \tag{2.49}$$

$$\bar{S}^i{}_{\alpha\beta} = \widetilde{S}^i{}_{jk} B^{jk}_{\alpha\beta}, \tag{2.50}$$

and

$$\bar{T}_{\alpha\ \beta}^{\ i} = \widetilde{T}_{j\ k}^{\ i}B^{jk}_{\alpha\beta} + \mathcal{A}_{(\alpha\beta)}\{B^j_\alpha H^a_\beta\}\widetilde{C}_{j\ k}^{\ i}B^k_a, \tag{2.51}$$

$$\bar{C}_{\alpha\ \beta}^{\ i} = \widetilde{C}_{j\ k}^{\ i}B^{jk}_{\alpha\beta}, \tag{2.52}$$

respectively.

On the other hand, the torsion tensor field $\widetilde{T}$ of $\widetilde{D}$ is expressed (see (1.5.32)) by

$$\begin{aligned}\widetilde{T}(X,Y) &= \widetilde{\nabla}_X\tilde{v}Y - \widetilde{\nabla}_Y\tilde{v}X - \tilde{v}[X,Y] \\ &\quad +\widetilde{Q}(\widetilde{\nabla}_X\widetilde{Q}\tilde{h}Y - \widetilde{\nabla}_Y\widetilde{Q}\tilde{h}X - \widetilde{Q}\tilde{h}[X,Y]), \\ &\quad \forall X,Y \in \Gamma(T\widetilde{M}'),\end{aligned} \tag{2.53}$$

where $\tilde{v}$ and $\tilde{h}$ are the projection morphisms of $T\widetilde{M}'$ on $V\widetilde{M}'$ and $H\widetilde{M}'$ respectively, and $\widetilde{Q}$ stands for the associate almost product structure to $H\widetilde{M}'$. Next, from (1.6) and (1.26) we obtain:

$$\begin{aligned}&(a)\ \ \tilde{h}(\frac{\partial}{\partial v^\alpha}) = 0; &&(b)\ \ \tilde{v}(\frac{\partial}{\partial v^\alpha}) = \frac{\partial}{\partial v^\alpha};\\ &(c)\ \ \tilde{v}(\frac{\delta}{\delta u^\alpha}) = H^a_\alpha B_a; &&(d)\ \ \widetilde{Q}\tilde{h}(\frac{\delta}{\delta u^\alpha}) = Qh(\frac{\delta}{\delta u^\alpha}).\end{aligned} \tag{2.54}$$

Also, by using (1.4.11) and (2.54) for $\mathbb{F}^m$ we derive

$$(a)\ \ \tilde{h}\left[\frac{\delta}{\delta u^\beta}, \frac{\delta}{\delta u^\alpha}\right] = 0; \quad (b)\ \ \tilde{v}\left[\frac{\delta}{\delta u^\beta}, \frac{\delta}{\delta u^\alpha}\right] = R^\gamma_{\ \beta\alpha}\frac{\partial}{\partial v^\gamma}. \tag{2.55}$$

Replace (X,Y) from (2.53) by $(\delta/\delta u^\beta,\ \delta/\delta u^\alpha)$ and using (2.54c), (2.54d) and (2.55) we infer that

$$\begin{aligned}\widetilde{T}\left(\frac{\delta}{\delta u^\beta}, \frac{\delta}{\delta u^\gamma}\right) &= \widetilde{\nabla}_{\delta/\delta u^\beta}(H^a_\alpha B_a) - \widetilde{\nabla}_{\delta/\delta u^\alpha}(H^a_\beta B_a) \\ &\quad +R^\gamma_{\ \alpha\beta}\frac{\partial}{\partial v^\gamma} + \widetilde{Q}\left(\widetilde{\nabla}_{\delta/\delta u^\beta}\frac{\partial}{\partial v^\alpha} - \widetilde{\nabla}_{\delta/\delta u^\alpha}\frac{\partial}{\partial v^\beta}\right).\end{aligned} \tag{2.56}$$

Finally, using (1.50), (1.65) and (1.4.8) for $\widetilde{Q}$, it is easily seen that (2.56) becomes

$$\begin{aligned}\widetilde{T}\left(\frac{\delta}{\delta u^\beta}, \frac{\delta}{\delta u^\alpha}\right) &= (T_{\alpha\ \beta}^{\ \gamma}B^i_\gamma + \mathcal{A}_{(\alpha\beta)}\{H^a_{\ \alpha\beta}B^i_a\})\frac{\delta}{\delta x^i} \\ &\quad +(T_{\alpha\ \beta}^{\ \gamma}H^a_\gamma B^i_a + R^\gamma_{\ \alpha\beta}B^i_\gamma + \mathcal{A}_{(\alpha\beta)}\{H^a_{\alpha|\beta}B^i_a \\ &\quad +H^a_\beta H'^{\gamma}_{a\ \alpha}B^i_\gamma\})\frac{\partial}{\partial y^i},\end{aligned} \tag{2.57}$$

where

$$T_{\alpha\ \beta}^{\ \gamma} = F_{\alpha\ \beta}^{\ \gamma} - F_{\beta\ \alpha}^{\ \gamma}, \tag{2.58}$$

is one of the torsions of IFC expressed in (2.22) and $H^a_{\alpha|\beta}$ stands for the horizontal relative covariant derivative of H^a_α with respect to (IFC, NFC). Comparing (2.45) with (2.57) we deduce that

$$\bar{R}^i{}_{\alpha\beta} = T_{\alpha\ \beta}^{\ \gamma} H^a_\gamma B^i_a + R^\gamma{}_{\alpha\beta} B^i_\gamma + \mathcal{A}_{(\alpha\beta)}\{H^a_{\alpha|\beta} B^i_a + H^a_\beta H'^{\gamma}_{a\ \alpha} B^i_\gamma\}, \tag{2.59}$$

and

$$\bar{T}_{\alpha\ \beta}^{\ i} = T_{\alpha\ \beta}^{\ \gamma} B^i_\gamma + \mathcal{A}_{(\alpha\beta)}\{H^a{}_{\alpha\beta} B^i_a\}. \tag{2.60}$$

Contracting (2.59) and (2.60) by both $\widetilde{B}^\mu_i$ and $\widetilde{B}^b_i$ and using (1.11), (2.51) and (2.48) we obtain

$$T_{\alpha\ \beta}^{\ \mu} = (\widetilde{T}_{j\ k}^{\ i} B^{jk}_{\alpha\beta} + \mathcal{A}_{(\alpha\beta)}\{B^j_\alpha H^a_\beta\}\widetilde{C}_{j\ k}^{\ i} B^k_a)\widetilde{B}^\mu_i, \tag{2.61}$$

$$\begin{aligned} R^\mu{}_{\alpha\beta} \ = \ & \widetilde{R}^i{}_{jk} B^{jk}_{\alpha\beta} \widetilde{B}^\mu_i + \mathcal{A}_{(\alpha\beta)}\{B^j_\alpha H^a_\beta \widetilde{P}^i{}_{jk} B^k_a \widetilde{B}^\mu_i - H^a_\beta H'^{\mu}_{a\ \alpha}\} \\ & + \widetilde{S}^i{}_{jk} B^j_a B^k_b H^a_\alpha H^b_\beta \widetilde{B}^\mu_i, \end{aligned} \tag{2.62}$$

and

$$(a)\ \ H^b{}_{\alpha\beta} - H^b{}_{\beta\alpha} = \bar{T}_{\alpha\ \beta}^{\ i} \widetilde{B}^b_i; \ \ (b)\ \ T_{\alpha\ \beta}^{\ \gamma} H^b_\gamma + H^b_{\alpha|\beta} - H^b_{\beta|\alpha} = \bar{R}^i{}_{\alpha\beta} \widetilde{B}^b_i. \tag{2.63}$$

According to (1.4.26), (1.5.42b) and (1.5.42d) we have

$$\left[\frac{\delta}{\delta u^\alpha}, \frac{\partial}{\partial v^\beta}\right] = \frac{\partial N^\gamma_\alpha}{\partial v^\beta}\frac{\partial}{\partial v^\gamma}, \tag{2.64}$$

$$P^\gamma{}_{\alpha\beta} = \frac{\partial N^\gamma_\alpha}{\partial v^\beta} - F_{\beta\ \alpha}^{\ \gamma}, \tag{2.65}$$

and

$$S^\gamma{}_{\alpha\beta} = C_{\alpha\ \beta}^{\ \gamma} - C_{\beta\ \alpha}^{\ \gamma}, \tag{2.66}$$

respectively. Then by using (2.53), (2.54), (1.50), (1.51), (1.64) and (2.64) –(2.66) we infer that

$$\bar{P}^i{}_{\alpha\beta} = (P^\gamma{}_{\alpha\beta} - H^a_\alpha V'^{\gamma}_{a\ \beta}) B^i_\gamma + (H^a_{\alpha\|\beta} + H^a_\gamma C_{\alpha\ \beta}^{\ \gamma} - H^a{}_{\beta\alpha}) B^i_a, \tag{2.67}$$

$$\bar{C}_{\alpha\ \beta}^{\ i} = C_{\alpha\ \beta}^{\ \gamma} B^i_\gamma + V^a{}_{\alpha\beta} B^i_a, \tag{2.68}$$

and

$$\bar{S}^i{}_{\alpha\beta} = S^\gamma{}_{\alpha\beta} B^i_\gamma + (V^a_{\alpha\beta} - V^a_{\beta\alpha}) B^i_a, \tag{2.69}$$

where $H^a_{\alpha\|\beta}$ is the vertical relative covariant derivative of H^a_α with respect to (IFC, NFC). Finally, contracting in turn (2.67)–(2.69) by both $\widetilde{B}^\mu_i$ and $\widetilde{B}^b_i$ and taking into account (2.49), (2.50) and (2.52) we derive (1.55), (1.57) and

$$P^{\mu}{}_{\alpha\beta} = H^a_\alpha V'^{\mu}_{a\ \beta} + (\widetilde{P}^i{}_{jk} B^{jk}_{\alpha\beta} + \widetilde{S}^i{}_{jk} B^j_a B^k_\beta H^a_\alpha)\widetilde{B}^\mu_i, \qquad (2.70)$$

$$S^{\mu}{}_{\alpha\beta} = \widetilde{S}^i{}_{jk} B^{jk}_{\alpha\beta} \widetilde{B}^\mu_i, \qquad (2.71)$$

and

$$\begin{aligned} V^b{}_{\alpha\beta} - V^b{}_{\beta\alpha} &= \widetilde{S}^i{}_{jk} B^{jk}_{\alpha\beta} \widetilde{B}^a_i & (a) \\ H^b_{\alpha\|\beta} + H^b_\gamma C_\alpha{}^\gamma{}_\beta - H^b{}_{\beta\alpha} &= \bar{P}^i{}_{\alpha\beta} \widetilde{B}^b_i. & (b) \end{aligned} \qquad (2.72)$$

As a result of the above calculations we may state the following theorem:

THEOREM 2.3 *Let $\widetilde{\mathbb{F}}^{m+n}$ be a pseudo-Finsler manifold endowed with an arbitrary Finsler connection $\widetilde{FC} = (\widetilde{N}^k_i, \widetilde{F}_i{}^k{}_j, \widetilde{C}_i{}^k{}_j)$ and $\mathbb{F}^m$ be a pseudo-Finsler submanifold of $\widetilde{\mathbb{F}}^{m+n}$. Then the torsion Finsler tensor fields $T_\alpha{}^\mu{}_\beta$, $R^\mu{}_{\alpha\beta}$, $C_\alpha{}^\mu{}_\beta$, $P^\mu{}_{\alpha\beta}$ and $S^\mu{}_{\alpha\beta}$ of the induced Finsler connection $IFC = (N^\gamma_\alpha, F_\alpha{}^\gamma{}_\beta, C_\alpha{}^\gamma{}_\beta)$ are given by structure equations (2.61), (2.62), (1.55), (2.70), and (2.71), respectively.*

The theory we have developed in these two sections of the chapter can be considered as a foundation for the study of the differential geometry of pseudo–Finsler submanifolds using the vertical vector bundle. From now on, for any special Finsler connection on the ambient manifold we may apply the above theory and obtain all induced geometric objects and structure equations. This is done in the next three sections for the Cartan, Berwald, and Rund connections.

Many authors have been concerned with the extension of the Gauss–Codazzi–Ricci equations to Finsler submanifolds. Important results about this matter have been obtained by Comic [1], [2], Davies [1], Dhawan and Prakash [1], Dragomir [1], Eliopoulos [1], Haimovici [2], Hassan [1], Matsumoto [4], Miron [2], Miron and Bejancu [1], [2], Prakash and Behari [1], Rastogi [1], Rund [1]–[5], Sinha [1], Verma [1], etc.. The structure equations presented in our Theorem 2.2 have been obtained in 1985 by Bejancu [1] for Finsler submanifolds of arbitrary codimension. Later on, Sakaguchi [1], obtained the above structure equations by applying the methods developed by Miron [2] and Matsumoto [4] for Finsler hypersurfaces. The study of the geometry of Finsler submanifolds via the vertical vector bundle has been initiated by Bejancu (cf., [2], [3]).

Finally, we should stress that in these two sections there is no need for the fundamental function of $\widetilde{\mathbb{F}}^{m+n}$. The main geometric objects of the ambient manifold are the pseudo–Riemannian metric on the vertical bundle and the Finsler connection. Thus our theory works fully for $\widetilde{M}'$ whose vertical bundle is endowed with a pseudo–Riemannian metric $\tilde{g} = \tilde{g}_{ij}(x, y)$ (cf., Bejancu [4], [5]).

3. Induced Cartan Connection

Let $\mathbb{F}^m = (M, M', F^*)$ be a pseudo–Finsler submanifold of the pseudo–Finsler manifold $\widetilde{\mathbb{F}}^{m+n} = (\widetilde{M}, \widetilde{M}', \widetilde{F}^*)$. In order to study the induced Finsler connection on $\mathbb{F}^m$ by the Cartan connection on $\widetilde{F}^{m+n}$, we introduce some mixed Finsler tensor fields on $\mathbb{F}^m$ and give their basic properties.

First, we consider the Cartan tensor fields of type (0, 3) (see (1.6.21)) $g_{\alpha\beta\gamma}$ and $\tilde{g}_{ijk}$ of $\mathbb{F}^m$ and $\widetilde{\mathbb{F}}^{m+n}$ respectively. Then differentiating (1.4) with respect to v^α and using (1.2) we obtain

$$g_{\alpha\beta\gamma} = \tilde{g}_{ijk} B^{ijk}_{\alpha\beta\gamma}. \tag{3.1}$$

Contracting (3.1) by $g^{\mu\beta}$ and taking into account (1.13) we infer that

$$g_{\alpha}{}^{\mu}{}_{\gamma} = \tilde{g}_{i}{}^{h}{}_{k} B^{ik}_{\alpha\gamma} \widetilde{B}^{\mu}_{h}. \tag{3.2}$$

Next, we locally define the functions

$$(a)\quad g_{a\alpha\beta} = \tilde{g}_{ijk} B^i_a B^{jk}_{\alpha\beta};\quad (b)\quad g_{ab\alpha} = \tilde{g}_{ijk} B^i_a B^j_b B^k_\alpha;$$

$$(c)\quad g_{abc} = \tilde{g}_{ijk} B^i_a B^j_b B^k_c. \tag{3.3}$$

It is easy to check that $(g_{a\alpha\beta})$, $(g_{ab\alpha})$ and (g_{abc}) are the local components of three mixed Finsler tensor fields of type $\begin{pmatrix} 0 & 0 \\ 2 & 1 \end{pmatrix}$, $\begin{pmatrix} 0 & 0 \\ 1 & 2 \end{pmatrix}$, and $\begin{pmatrix} 0 & 0 \\ 0 & 3 \end{pmatrix}$ respectively, on $\mathbb{F}^m$ with respect to $VM'^{\perp}$. Contracting (3.3) by $\widetilde{B}^a_h$ and using (1.12) we derive

$$\begin{aligned} (a)\quad & g_{a\alpha\beta}\widetilde{B}^a_h = \tilde{g}_{hjk} B^{jk}_{\alpha\beta} - g_{\alpha\beta\gamma}\widetilde{B}^\gamma_h, \\ (b)\quad & g_{ab\alpha}\widetilde{B}^a_h = \tilde{g}_{hjk} B^j_\alpha B^k_b - g_{b\alpha\gamma}\widetilde{B}^\gamma_h, \\ (c)\quad & g_{abc}\widetilde{B}^a_h = \tilde{g}_{hjk} B^j_b B^k_c - g_{bc\gamma}\widetilde{B}^\gamma_h. \end{aligned} \tag{3.4}$$

Remark 3.1. Starting with this section we shall frequently use, throughout the book, the mixed Finsler tensor fields $\delta_{ab}, \delta^{ab}, \tilde{g}_{ij}, \tilde{g}^{ij}, g_{\alpha\beta}$ and $g^{\alpha\beta}$

to lower or raise the indices of the local components $T^{ai\alpha\cdots}_{bj\beta\cdots}$ of some mixed Finsler tensor fields on $\mathbb{F}^m$ with respect to the vector bundle $V\widetilde{M}'_{|M'} \oplus VM'^{\perp}$. ■

Certainly, B^i_α are independent of (v^β). On the contrary, $\widetilde{B}^\alpha_i$, B^i_a, $\widetilde{B}^a_i$ and C^j_i (see 1.16), as we will show in the next proposition, depend on (v^β).

PROPOSITION 3.1 *Let $\mathbb{F}^m$ be a pseudo–Finsler submanifold of $\widetilde{\mathbb{F}}^{m+n}$. Then we have:*

$$\frac{\partial \widetilde{B}^\alpha_i}{\partial v^\beta} = 2g_a{}^\alpha{}_\beta \widetilde{B}^a_i, \tag{3.5}$$

$$\frac{\partial C^j_i}{\partial v^\beta} = 2g_a{}^\alpha{}_\beta B^j_\alpha \widetilde{B}^a_i, \tag{3.6}$$

$$\frac{\partial B^i_a}{\partial v^\beta} = -2g_a{}^\alpha{}_\beta B^i_\alpha - B^j_a B^i_b \frac{\partial \widetilde{B}^b_j}{\partial v^\beta}, \tag{3.7}$$

$$\frac{\partial \widetilde{B}^a_i}{\partial v^\beta} = -\widetilde{B}^a_j \widetilde{B}^b_i \frac{\partial B^j_b}{\partial v^\beta}. \tag{3.8}$$

PROOF. First, from (1.13) one obtains

$$g_{\varepsilon\gamma} \widetilde{B}^\varepsilon_j = \tilde{g}_{jk} B^k_\gamma. \tag{3.9}$$

Differentiating (3.9) with respect to v^β and taking into account (1.6.21) for both $\mathbb{F}^m$ and $\widetilde{F}^{m+n}$, we infer that

$$2g_{\varepsilon\gamma\beta} \widetilde{B}^\varepsilon_j + g_{\varepsilon\gamma} \frac{\partial \widetilde{B}^\varepsilon_j}{\partial v^\beta} = 2\tilde{g}_{jkh} B^{kh}_{\gamma\beta}. \tag{3.10}$$

Contracting (3.10) by $B^j_a g^{\gamma\alpha}$ and using (1.11b) and (3.4a) we derive

$$B^j_a \frac{\partial \widetilde{B}^\alpha_j}{\partial v^\beta} = 2g_a{}^\alpha{}_\beta. \tag{3.11}$$

On the other hand, differentiating (1.11a) with respect to v^β we obtain

$$B^j_\gamma \frac{\partial \widetilde{B}^\alpha_j}{\partial v^\beta} = 0. \tag{3.12}$$

Contracting (3.11) and (3.12) by $\widetilde{B}^a_i$ and $\widetilde{B}^\gamma_i$ respectively and adding them we obtain (3.5) via (1.12). Next, differentiating (1.17) with respect to v^β and using (3.5) we deduce (3.6). Further, we differentiate (1.11b), (1.11c) and (1.11d) and using (3.5) we obtain

$$2g_a{}^\alpha{}_\beta + \widetilde{B}^\alpha_j \frac{\partial B^j_a}{\partial v^\beta} = 0, \tag{3.13}$$

$$B^j_\alpha \frac{\partial \widetilde{B}^a_j}{\partial v^\beta} = 0, \tag{3.14}$$

and

$$B^j_a \frac{\partial \widetilde{B}^b_j}{\partial v^\beta} + \widetilde{B}^b_j \frac{\partial B^j_a}{\partial v^\beta} = 0, \tag{3.15}$$

respectively. Contracting (3.13) and (3.15) by B^i_α and B^i_b respectively and adding the results we obtain (3.7). Similarly, from (3.14) and (3.15) one deduces (3.8). ■

Next, we consider the Cartan connection

$$\widetilde{\mathbf{FC}}^* = (G\widetilde{M}', \widetilde{\nabla}^*) = (\widetilde{G}^k_i, \widetilde{F}^{*k}_{i\ j}, \tilde{g}_i{}^k{}_j)$$

on $\widetilde{\mathbb{F}}^{m+n}$ given (see (1.6.17), (1.7.11), (1.6.21)) by

$$\widetilde{G}^k_i = \frac{\partial \widetilde{G}^k}{\partial y^i}, \tag{3.16}$$

$$\widetilde{F}^{*k}_{i\ j} = \frac{1}{2}\tilde{g}^{kh}\left(\frac{\delta^* \tilde{g}_{hi}}{\delta^* x^j} + \frac{\delta^* \tilde{g}_{hj}}{\delta^* x^i} - \frac{\delta^* \tilde{g}_{ij}}{\delta^* x^h}\right), \tag{3.17}$$

$$\tilde{g}_i{}^k{}_j = \frac{1}{2}\tilde{g}^{kh}\frac{\partial \tilde{g}_{hi}}{\partial y^j}, \tag{3.18}$$

where we used the functions $\widetilde{G}^k$ and the differential operators $\delta^*/\delta^* x^i$ given (see (1.6.9) and (1.6.18)) by

$$\widetilde{G}^k = \frac{1}{4}\tilde{g}^{kh}\left(\frac{\partial^2 \widetilde{F}^*}{\partial y^h \partial x^j} y^j - \frac{\partial \widetilde{F}^*}{\partial x^h}\right), \tag{3.19}$$

and

$$\frac{\delta^*}{\delta^* x^i} = \frac{\partial}{\partial x^i} - \widetilde{G}^j_i \frac{\partial}{\partial y^j}, \tag{3.20}$$

respectively. We recall (see Section 1.7) that $\widetilde{\mathbf{FC}}^*$ is a metric Finsler connection and two of its torsion Finsler tensor fields vanish. More precisely, we have

$$(a)\ \ \tilde{g}_{ij|*k} = 0; \ \ (b)\ \ \tilde{g}_{ij\|*k} = 0, \tag{3.21}$$

and

$$(a)\ \ \widetilde{T}^{*k}_{i\ j} = 0; \ \ (b)\ \ \widetilde{S}^{*k}_{ij} = 0. \tag{3.22}$$

Moreover, $\widetilde{G}^k_j$ and $\widetilde{F}^{*k}_{i\ j}$ are related by (cf., (1.7.29))

$$y^i \widetilde{F}^{*k}_{i\ j} = \widetilde{G}^k_j. \tag{3.23}$$

Remark 3.2. Throughout the book we use c, b and r on all geometric objects induced by the Cartan, Berwald, and Rund connections, respectively. However, we omit these letters on N^γ_α and H^a_α, because according to (1.21) and (1.27) we have

$$(a)\ N^\gamma_\alpha = \widetilde{B}^\gamma_i(B^i_{\alpha 0} + B^j_\alpha \widetilde{G}^i_j) \text{ and } (b)\ H^a_\alpha = \widetilde{B}^a_i(B^i_{\alpha 0} + B^j_\alpha \widetilde{G}^i_j), \qquad (3.24)$$

that is, these geometric objects depend only on the cononical non-linear connection of $\widetilde{\mathbb{F}}^{m+n}$, which is the same for all special Finsler connections we are dealing with in the book. ■

THEOREM 3.1 *Let $\mathbb{F}^m$ be a pseudo-Finsler submanifold of a pseudo-Finsler manifold $\widetilde{\mathbb{F}}^{m+n}$ endowed with the Cartan connection* $\widetilde{\mathbf{FC}}^* = (\widetilde{G}^k_i, \widetilde{F}^{*k}_{i\ j}, \tilde{g}_i{}^k{}_j)$.
Then we have the assertions:
(i) The local components of the h-second fundamental form $\overset{c}{H} = (\overset{c}{H}{}^a{}_{\alpha\beta})$ and of the v-second fundamental form $\overset{c}{V} = (\overset{c}{V}{}^a{}_{\alpha\beta})$ are given by

$$\overset{c}{H}{}^a{}_{\alpha\beta} = \widetilde{B}^a_k(B^k_{\alpha\beta} + \widetilde{F}^{*k}_{i\ j} B^{ij}_{\alpha\beta}) + g_b{}^a{}_\alpha H^b_\beta, \qquad (3.25)$$

and

$$\overset{c}{V}{}^a{}_{\alpha\beta} = g^a{}_{\alpha\beta} = \tilde{g}_i{}^k{}_j B^{ij}_{\alpha\beta} \widetilde{B}^a_k, \qquad (3.26)$$

respectively.
(ii) The local components of the shape operators $\overset{c}{A}_a$ are given by

$$(a)\ \overset{c}{H}{}'^{\alpha}_{a\ \beta} = \overset{c}{H}{}_a{}^\alpha{}_\beta \text{ and } (b)\ \overset{c}{V}{}'^{\alpha}_{a\ \beta} = \overset{c}{V}{}_a{}^\alpha{}_\beta = g_a{}^\alpha{}_\beta. \qquad (3.27)$$

PROOF. The assertion (i) follows from the assertion (ii) of Theorem 1.1 and (3.3a) and (3.3b), taking into account Remark 3.1. Next, using (1.33) and (1.35) and taking into account that $\widetilde{\nabla}^*$ is a metric linear connection on $V\widetilde{M}'$, we deduce that

$$\tilde{g}(\overset{c}{H}(hX, hY), W) = g(\overset{c}{A}_W hX, QhY), \qquad (3.28)$$

for any $X, Y \in \Gamma(TM')$ and $W \in \Gamma(VM'^\perp)$. We replace hX, hY and W from (3.28) by $\delta/\delta u^\beta$, $\delta/\delta u^\alpha$ and B_a, respectively, and using (1.4.8), (1.49a), (1.62a) and Remark 3.1 we obtain (3.27a). Similarly, from (1.34) and (1.35) we infer that

$$\tilde{g}(\overset{c}{V}(vX, vY), W) = g(\overset{c}{A}_W vX, vY), \qquad (3.29)$$

which implies (3.27b) via (3.26). ■

COROLLARY 3.1 *Let $\mathbb{F}^m$ and $\widetilde{\mathbb{F}}^{m+n}$ be as in Theorem 3.1. Then we have*

$$(a)\quad v^\alpha \overset{c}{H}{}^a{}_{\alpha\beta} = H^a_\beta; \qquad (b)\quad \overset{c}{H}{}^a{}_{\alpha\beta}\, v^\beta = H^a_\alpha + g_b{}^a{}_\alpha H^b_\beta v^\beta, \tag{3.30}$$

and

$$(a)\quad v^\alpha \overset{c}{V}{}^a{}_{\alpha\beta} = 0; \quad (b)\quad \overset{c}{V}{}^a{}_{\alpha\beta}\, v^\beta = 0; \quad (c)\quad \overset{c}{V}{}^a{}_{\alpha\beta} = \overset{c}{V}{}^a{}_{\beta\alpha}\,. \tag{3.31}$$

PROOF. Using (1.6.22) and (3.3b) we derive that

$$g_b{}^a{}_\alpha v^\alpha = 0. \tag{3.32}$$

Then contracting (3.25) by v^α and taking into account (1.2a), (3.23), (3.32) and (3.24b) we obtain (3.30a). Similarly, we deduce (3.30b), (3.31a) and (3.31b) from (3.25) and (3.26) respectively. Finally, (3.31c) is a direct consequence of (3.26). ■

Remark 3.3. As $\overset{c}{V}$ is a $\Gamma(VM'^\perp)$–valued symmetric bilinear form on $\Gamma(VM')$, from (3.29) and (1.38) it follows that the v–shape operators $\overset{c}{A}{}^v_a$ are self–adjoint on $\Gamma(VM')$. On the contrary, as $\overset{c}{H}$ in general is not symmetric (see (3.25)), the above assertion does not hold for the h–shape operators defined by (1.37). This might be considered a weak point of the Cartan connection. ■

To state the next theorem we need the notations:

$$(a)\quad L^\gamma{}_{\alpha\beta} = \widetilde{P}^{*k}{}_{ij} B^{ij}_{\alpha\beta} \widetilde{B}^\gamma_k; \quad (b)\quad L_a{}^\gamma{}_\alpha = \widetilde{P}^{*k}{}_{ij} B^i_\alpha B^j_a \widetilde{B}^\gamma_k, \tag{3.33}$$

where $\widetilde{P}^{*k}{}_{ij}$ are the local components of one of the torsion Finsler tensor fields of $\widetilde{\mathbf{FC}}^*$ given (see (1.5.42b)) by

$$\widetilde{P}^{*k}{}_{ij} = \frac{\partial \widetilde{G}^k_i}{\partial y^j} - \widetilde{F}^{*k}_j{}_i. \tag{3.34}$$

THEOREM 3.2 *Let $\mathbb{F}^m$ and $\widetilde{\mathbb{F}}^{m+n}$ be as in Theorem 3.1. Then we have the assertions:*

(i) The local coefficients of the induced Finsler connection $I\overset{c}{F}C = (HM', \overset{c}{\nabla}) = (N^\gamma_\alpha, \overset{c}{F}_\alpha{}^\gamma{}_\beta, \overset{c}{C}_\alpha{}^\gamma{}_\beta)$ are given by (3.24a) and

$$\overset{c}{F}_\alpha{}^\gamma{}_\beta = \widetilde{B}^\gamma_k (B^k_{\alpha\beta} + \widetilde{F}^{*k}_i{}_j B^{ij}_{\alpha\beta}) + g_a{}^\gamma{}_\alpha H^a_\beta, \tag{3.35}$$

$$\overset{c}{C}_{\alpha}{}^{\gamma}{}_{\beta} = g_{\alpha}{}^{\gamma}{}_{\beta} = \widetilde{g}_{i}{}^{k}{}_{j} B^{ij}_{\alpha\beta} \widetilde{B}^{\gamma}_{k}. \tag{3.36}$$

(ii) The local coefficients N^{γ}_{β} *and* $\overset{c}{F}_{\alpha}{}^{\gamma}{}_{\beta}$ *are related by*

$$N^{\gamma}_{\beta} = v^{\alpha} \overset{c}{F}_{\alpha}{}^{\gamma}{}_{\beta}\,. \tag{3.37}$$

(iii) $I\overset{c}{F}C$ *is a metric Finsler connection, i.e., we have*

$$g_{\alpha\beta|\gamma} = g_{\alpha\beta||\gamma} = 0. \tag{3.38}$$

(iv) The torsion Finsler tensor fields of $I\overset{c}{F}C$ *are given by (3.36) and the following formulas:*

$$\overset{c}{R}{}^{\gamma}{}_{\alpha\beta} = \widetilde{R}^{*k}{}_{ij} B^{ij}_{\alpha\beta} \widetilde{B}^{\gamma}_{k} + \mathcal{A}_{(\alpha\beta)}\{(L_{a}{}^{\gamma}{}_{\alpha} - \overset{c}{H}_{a}{}^{\gamma}{}_{\alpha}) H^{a}_{\beta}\}, \tag{3.39}$$

$$\overset{c}{T}_{\alpha}{}^{\gamma}{}_{\beta} = \mathcal{A}_{(\alpha\beta)}\{g_{a}{}^{\gamma}{}_{\alpha} H^{a}_{\beta}\}, \tag{3.40}$$

$$\overset{c}{P}{}^{\gamma}{}_{\alpha\beta} = L^{\gamma}{}_{\alpha\beta} + H^{a}_{\alpha} g_{a}{}^{\gamma}{}_{\beta}, \tag{3.41}$$

$$\overset{c}{S}{}^{\gamma}{}_{\alpha\beta} = 0. \tag{3.42}$$

(v) The normal Finsler connection

$$N\overset{c}{F}C = (HM', \overset{c}{\nabla}{}^{\perp}) = (N^{\gamma}_{\alpha}, \overset{c}{F}_{a}{}^{b}{}_{\alpha}, \overset{c}{C}_{a}{}^{b}{}_{\alpha})$$

is a metric vectorial Finsler connection, i.e., we have

$$(\overset{c}{\nabla}{}^{\perp}_{X}\, \tilde{g})(W_1, W_2) = 0, \quad \forall X \in \Gamma(TM'), W_1, W_2 \in \Gamma(VM'^{\perp}), \tag{3.43}$$

or equivalently

$$(a) \quad \delta_{ab|\alpha} = 0 \quad \text{and} \quad \delta_{ab||\alpha} = 0 \tag{3.44}$$

where the covariant derivatives are considered with respect to the pair $(I\overset{c}{F}C, N\overset{c}{F}C)$ *on the pseudo–Finsler normal bundle* $VM'^{\perp}$*.*

PROOF. The assertion (i) is a direct consequence of assertion (i) in Theorem 1.1 via (3.2) and (3.3a). Contracting (3.3a) by v^{β} and using (1.2a) and (1.6.22) we deduce that

$$g_{a}{}^{\alpha}{}_{\beta} v^{\beta} = 0. \tag{3.45}$$

Then, contracting (3.35) by v^{α} and taking into account (3.45), (1.2a), (3.23) and (3.24a), we obtain (3.37). Next, using (3.21) in (1.77) and (1.78) we derive (3.38). Using (3.22b), (3.33b) and (3.27a) in (2.62) we obtain (3.39). Similarly, (3.40), (3.41) and (3.42) are obtained from (2.61), (2.70), and (2.71), respectively. Finally, taking into account that $\widetilde{\nabla}^*$ of $\widetilde{\mathbf{FC}}^*$ is a metric linear connection on $V\widetilde{M}'$ and using (1.35) we obtain that $\overset{c}{\nabla}{}^{\perp}$ of $N\overset{c}{F}C$ is a metric linear connection on $VM'^{\perp}$. Hence $N\overset{c}{F}C$ is a metric vectorial Finsler connection. ■

Taking into account (3.44), (1.5.20) and (1.5.21) we may state the following corollary.

COROLLARY 3.2 *The local coefficients of the normal Finsler connection induced by the Cartan connection satisfy*

$$(a) \quad \overset{c}{F}{}_{a\ \alpha}^{\ b} + \overset{c}{F}{}_{b\ \alpha}^{\ a} = 0 \quad \text{and} \quad (b) \quad \overset{c}{C}{}_{a\ \alpha}^{\ b} + \overset{c}{C}{}_{b\ \alpha}^{\ a} = 0. \tag{3.46}$$

As the Cartan connection is a metric Finsler connection, the curvature tensor $\widetilde{R}^*$ of $\widetilde{\nabla}^*$ of type (0, 4) satisfies (see (1.7.50b))

$$\widetilde{R}^*(X, Y, W_1, W_2) + \widetilde{R}^*(X, Y, W_2, W_1) = 0, \tag{3.47}$$

for any $X, Y \in \Gamma(T\widetilde{M}'_{|M'})$ and $W_1, W_2 \in \Gamma(VM'^{\perp})$. Using (3.28), (3.29) and (3.47) we easily obtain that the A–and B–Codazzi equations coincide. Thus our general structure equations (2.5), (2.6), (2.10), and (2.11) enable us to state the following result:

THEOREM 3.3 *(Bejancu [6]). Let $\widetilde{\mathbb{F}}^{m+n}$ be a pseudo-Finsler manifold endowed with the Cartan connection $\widetilde{\mathbf{FC}}^* = (G\widetilde{M}', \widetilde{\nabla}^*) = (\widetilde{G}_i^k, \widetilde{F}{}_{i\ j}^{*k}, \widetilde{g}_{i\ j}^{\ k})$ and $\mathbb{F}^m$ be a pseudo-Finsler submanifold of $\widetilde{F}^{m+n}$. Then the Gauss, Codazzi and Ricci equations for $\mathbb{F}^m$ are given by*

$$\begin{aligned}
&\widetilde{R}^*(X,Y, vZ, vU) \\
&\quad = \overset{c}{R}(X, Y, vZ, vU) \qquad\qquad (3.48)\\
&\qquad + \widetilde{g}(\overset{c}{B}(X, vZ), \overset{c}{B}(Y, vU)) - \widetilde{g}(\overset{c}{B}(Y, vZ), \overset{c}{B}(X, vU)),
\end{aligned}$$

$$\begin{aligned}
\widetilde{R}^*(X, Y, vZ, W) &= \widetilde{g}((\overset{c}{\nabla}{}'_X \overset{c}{B})(Y, vZ) - (\overset{c}{\nabla}{}'_Y \overset{c}{B})(X, vZ), W) \\
&\quad + \widetilde{g}(\overset{c}{B}(\overset{c}{T}(X, Y), vZ), W),
\end{aligned} \tag{3.49}$$

and

$$\widetilde{R}^*(X,Y,W_1,W_2) = \overset{c}{R^{\perp}}(X,Y,W_1,W_2)$$

$$+g(\overset{c}{A}_{W_1}X, \overset{c}{A}_{W_2}Y) - g(\overset{c}{A}_{W_2}X, \overset{c}{A}_{W_1}Y), \quad (3.50)$$

respectively, for any $X, Y, Z, U \in \Gamma(TM')$ *and* $W_1, W_2 \in \Gamma(VM'^{\perp})$, *where we set*

$$\overset{c}{R}(X,Y,vZ,vU) = g(\overset{c}{R}(X,Y)vZ,vU)$$

and

$$\overset{c}{R^{\perp}}(X,Y,W_1,W_2) = \tilde{g}(\overset{c}{R^{\perp}}(X,Y)W_1,W_2).$$

According to the Remark 3.3, from (3.50) we obtain

$$\widetilde{R}^*(vX, vY, W_1, W_2)$$

$$= \overset{c}{R^{\perp}}(vX,vY,W_1,W_2) + g(vX,[\overset{c}{A}{}^{v}_{W_1}, \overset{c}{A}{}^{v}_{W_2}]vY). \quad (3.51)$$

Finally, lowering and raising indices in the equations presented in Theorem 2.2 and taking into account Theorems 3.1–3.3 we state the following theorem.

THEOREM 3.4 *Let* $\mathbb{F}^m$ *and* $\widetilde{\mathbb{F}}^{m+n}$ *be as in Theorem 3.3. Then we have the assertions:*

(i) The local expressions of the Gauss equations are given by:

$$\begin{cases} \bar{R}^*_{ij\alpha\beta}B^{ij}_{\gamma\varepsilon} = \overset{c}{R}_{\gamma\varepsilon\alpha\beta} + \mathcal{A}_{(\alpha\beta)}\{\overset{c}{H}_{\gamma\beta}\overset{c}{H}_{a\varepsilon\alpha}\} & (a) \\ \bar{P}^*_{ij\alpha\beta}B^{ij}_{\gamma\varepsilon} = \overset{c}{P}_{\gamma\varepsilon\alpha\beta} + g^{a}{}_{\gamma\beta}\overset{c}{H}_{a\varepsilon\alpha} - \overset{c}{H}{}^{a}{}_{\gamma\alpha}\, g_{a\varepsilon\beta} & (b) \\ \bar{S}^*_{ij\alpha\beta}B^{ij}_{\gamma\varepsilon} = \overset{c}{S}_{\gamma\varepsilon\alpha\beta} + \mathcal{A}_{(\alpha\beta)}\{g^{a}{}_{\gamma\beta}g_{a\varepsilon\alpha}\} & (c). \end{cases} \quad (3.52)$$

(ii) The local expressions of the Codazzi equations are given by:

$$\begin{cases} \bar{R}^*_{ij\alpha\beta}B^i_{\gamma}B^j_a = \overset{c}{T}_{\alpha}{}^{\mu}{}_{\beta}\overset{c}{H}_{a\gamma\mu} + \overset{c}{R}{}^{\mu}{}_{\alpha\beta}\, g_{a\gamma\mu} + \mathcal{A}_{(\alpha\beta)}\{\overset{c}{H}_{a\gamma\alpha|\beta}\} & (a) \\ \bar{P}^*_{ij\alpha\beta}B^i_{\gamma}B^j_a = g_{\alpha}{}^{\mu}{}_{\beta}\,\overset{c}{H}_{a\gamma\mu} + \overset{c}{P}{}^{\mu}{}_{\alpha\beta}\, g_{a\gamma\mu} + \overset{c}{H}_{a\gamma\alpha\|\beta} - g_{a\gamma\beta|\alpha} & (b) \\ \bar{S}^*_{ij\alpha\beta}B^i_{\gamma}B^j_a = \mathcal{A}_{(\alpha\beta)}\{g_{a\gamma\alpha\|\beta}\} & (c). \end{cases} \quad (3.53)$$

(iii) The local expressions of the Ricci equations are given by:

$$\begin{cases} \bar{R}^*_{ij\alpha\beta}B^i_a B^j_b &= \overset{c}{R}{}^{\perp}{}_{ab\alpha\beta} + \mathcal{A}_{(\alpha\beta)}\{\overset{c}{H}{}_a{}^{\gamma}{}_{\beta}\overset{c}{H}_{b\gamma\alpha}\} \qquad (a) \\ \bar{P}^*_{ij\alpha\beta}B^i_a B^j_b &= \overset{c}{P}{}^{\perp}{}_{ab\alpha\beta} + g_a{}^{\gamma}{}_{\beta}\,\overset{c}{H}_{b\gamma\alpha} - \overset{c}{H}{}_a{}^{\gamma}{}_{\alpha}\, g_{b\gamma\beta} \quad (b) \\ \bar{S}^*_{ij\alpha\beta}B^i_a B^j_b &= \overset{c}{S}{}^{\perp}{}_{ab\alpha\beta} + \mathcal{A}_{(\alpha\beta)}\{g_a{}^{\gamma}{}_{\beta}g_{b\gamma\alpha}\} \qquad (c), \end{cases} \tag{3.54}$$

where $\bar{R}^*{}_{ij\alpha\beta}$, $\bar{P}^*{}_{ij\alpha\beta}$ and $\bar{S}^*_{ij\alpha\beta}$ are given by (see (2.16)–(2.18))

$$\bar{R}^*_{ij\alpha\beta} = \widetilde{R}^*_{ijkh}B^{kh}_{\alpha\beta} + \widetilde{P}^*_{ijkh}B^h_a(B^k_\alpha H^a_\beta - B^k_\beta H^a_\alpha) + \widetilde{S}^*_{ijkh}B^k_a B^h_b H^a_\alpha H^b_\beta, \tag{3.55}$$

$$\bar{P}^*_{ij\alpha\beta} = \widetilde{P}^*_{ijkh}B^{kh}_{\alpha\beta} + \widetilde{S}^*_{ijkh}B^k_a H^a_\alpha B^h_\beta, \tag{3.56}$$

and

$$\bar{S}^*_{ij\alpha\beta} = \widetilde{S}^*_{ijkh}B^{kh}_{\alpha\beta}, \tag{3.57}$$

respectively.

Contracting (3.52a) with v^γ and taking into account (3.30a), (3.55), (1.7.57a), (1.7.61a) and (3.33b) we deduce that

$$B^j_\varepsilon \widetilde{R}^*_{jkh}B^{kh}_{\alpha\beta} + \mathcal{A}_{(\alpha\beta)}\{L_{a\varepsilon\alpha}H^a_\beta\} = v^\gamma \overset{c}{R}_{\gamma\varepsilon\alpha\beta} + \mathcal{A}_{(\alpha\beta)}\{\overset{c}{H}_{a\varepsilon\alpha}\; H^a_\beta\}. \tag{3.58}$$

On the other hand, contracting (3.39) by $g_{\gamma\varepsilon}$ and using (1.13) we infer that

$$\overset{c}{R}_{\varepsilon\alpha\beta} = B^j_\varepsilon \widetilde{R}^*_{jkh}B^{kh}_{\alpha\beta} + \mathcal{A}_{(\alpha\beta)}\{(L_{a\varepsilon\alpha} - \overset{c}{H}_{a\varepsilon\alpha})H^a_\beta\}. \tag{3.59}$$

By comparing (3.58) and (3.59) we obtain

$$\overset{c}{R}_{\varepsilon\alpha\beta} = v^\gamma \overset{c}{R}_{\gamma\varepsilon\alpha\beta}\,. \tag{3.60}$$

We close the section with the following proposition.

PROPOSITION 3.2 *Let $\mathbb{F}^m$ and $\widetilde{\mathbb{F}}^{m+n}$ be as in Theorem 3.3. Then we have:*

$$\frac{\partial(H^a_\varepsilon v^\varepsilon)}{\partial v^\beta} = 2H^a_\beta + (g_b{}^a{}_\beta - \overset{c}{C}_b{}^a{}_\beta)H^b_\varepsilon v^\varepsilon, \tag{3.61}$$

and

$$\mathcal{A}_{(\alpha\beta)}\left\{\frac{\partial g_{a\gamma\alpha}}{\partial v^\beta} - g_{b\gamma\alpha}\,\overset{c}{C}_a{}^b{}_\beta - 2g_{a\alpha\varepsilon}g_\gamma{}^\varepsilon{}_\beta - g_{ab\alpha}g^b{}_{\gamma\beta}\right\} = 0. \tag{3.62}$$

PROOF. First, in this case (2.72b) becomes

$$\frac{\partial H^a_\varepsilon}{\partial v^\beta} + H^b_\varepsilon \overset{c}{C}{}_b{}^a{}_\beta - \overset{c}{H}{}^a{}_{\beta\varepsilon} = L^a{}_{\varepsilon\beta}, \tag{3.63}$$

since $\widetilde{S}^{*i}{}_{jk} = 0$. Contracting (3.63) by v^ε and using (3.30b) and taking into account that $L^b{}_{\varepsilon\beta} v^\varepsilon = 0$, we obtain

$$\frac{\partial H^a_\varepsilon}{\partial v^\beta} v^\varepsilon - H^a_\beta + (\overset{c}{C}{}_b{}^a{}_\beta - g_b{}^a{}_\beta) H^b_\varepsilon v^\varepsilon = 0,$$

which is equivalent to (3.61). Next, by using (3.57) and (1.7.75) in (3.53c) we deduce that

$$\mathcal{A}_{\alpha\beta}\left\{\frac{\partial g_{a\gamma\alpha}}{\partial v^\beta} - g_{b\gamma\alpha}\overset{c}{C}{}_a{}^b{}_\beta - g_{a\alpha\varepsilon} g_\gamma{}^\varepsilon{}_\beta\right\}$$

$$= (\tilde{g}_i{}^r{}_h \tilde{g}_{rjk} - \tilde{g}_i{}^r{}_k \tilde{g}_{rjh})\, B^{ikh}_{\gamma\alpha\beta} B^j_a. \tag{3.64}$$

On the other hand, by using (3.3a) and (3.3b) it is easy to check that

$$\tilde{g}_{rjk} B^j_a B^k_\alpha = g_{ab\alpha} \widetilde{B}^b_r + g_{a\alpha\varepsilon} \widetilde{B}^\varepsilon_r. \tag{3.65}$$

Then (3.62) follows from (3.64) by using (3.65). ■

4. Induced Berwald Connection

In the present section we consider a pseudo–Finlser manifold $\widetilde{\mathbb{F}}^{m+n}$ endowed with the Berwald connection $\widetilde{\mathbf{BFC}} = (G\widetilde{M}', \widetilde{\nabla}') = (\widetilde{G}^k_i, \widetilde{G}_i{}^k{}_j, 0)$, where $\widetilde{G}^k_i$ are the local coefficients of the canonical non–linear connection on $\widetilde{M}'$ (see (3.16)) and according to (1.6.27) and (1.6.17) we have

$$\widetilde{G}_i{}^k{}_j = \frac{\partial \widetilde{G}^k_j}{\partial y^i} = \frac{\partial^2 \widetilde{G}^k}{\partial y^i \partial y^j}. \tag{4.1}$$

As we have seen in Section 1.6 all torsions of $\widetilde{\mathbf{BFC}}$ vanish except $\widetilde{R}^k{}_{ij}$ which coincides with the corresponding one for the Cartan connection (see (1.6.29)). Thus we have (cf., (1.6.28))

$$(a)\ \ \widetilde{T}_i{}^k{}_j = 0; \quad (b)\ \ \widetilde{C}_i{}^k{}_j = 0; \quad (c)\ \ \widetilde{P}^k{}_{ij} = 0; \quad (d)\ \ \widetilde{S}^k{}_{ij} = 0. \tag{4.2}$$

Moreover, (1.7.57a) yields

$$\widetilde{P}^{*k}_{ij} = \widetilde{G}_i{}^k{}_j - \widetilde{F}^{*k}_{i\ j}, \tag{4.3}$$

where $\widetilde{F}^{*k}_{i\ j}$ are the local coefficients of the Cartan connection on $\widetilde{\mathbb{F}}^{m+n}$ (see(3.17)).

Taking into account (4.2b) in (1.54)–(1.57) we may state the following theorem.

THEOREM 4.1 *Let $\widetilde{\mathbb{F}}^{m+n}$ be a pseudo–Finsler manifold endowed with the Berwald connection $\widetilde{\mathbf{BFC}} = (\widetilde{G}^k_i, \widetilde{G}_i{}^k{}_j, 0)$ and $\mathbb{F}^m$ be a pseudo–Finsler submanifold of $\widetilde{F}^{m+n}$. Then we have the assertions:*
(i) The local coefficients of the induced Finsler connection

$$IFC = (HM', \overset{b}{\nabla}) = (\overset{b}{N}{}^{\gamma}_{\alpha}, \overset{b}{F}_{\alpha}{}^{\gamma}{}_{\beta}, \overset{b}{C}_{\alpha}{}^{\gamma}{}_{\beta}),$$

are given by (3.24a) and

$$(a)\quad \overset{b}{F}_{\alpha}{}^{\gamma}{}_{\beta} = \widetilde{B}^{\gamma}_k(B^k_{\alpha\beta} + \widetilde{G}_i{}^k{}_j B^{ij}_{\alpha\beta});\quad (b)\quad \overset{b}{C}_{\alpha}{}^{\gamma}{}_{\beta} = 0. \tag{4.4}$$

(ii) The local components of the h–and v–second fundamental forms are given by

$$(a)\quad \overset{b}{H}{}^a{}_{\alpha\beta} = \widetilde{B}^a_k(B^k_{\alpha\beta} + \widetilde{G}_i{}^k{}_j B^{ij}_{\alpha\beta}) \quad \text{and} \quad (b)\quad \overset{b}{V}{}^a{}_{\alpha\beta} = 0, \tag{4.5}$$

respectively.

Contracting (4.4a) and (4.5a) by v^{α} or v^{β}, and taking into account (1.2a), (1.6.30) and (3.24) we obtain:

$$(a)\quad \overset{b}{F}_{\alpha}{}^{\gamma}{}_{\varepsilon}\, v^{\alpha} = \overset{b}{F}_{\varepsilon}{}^{\gamma}{}_{\beta}\, v^{\beta} = N^{\gamma}_{\varepsilon};\quad (b)\quad \overset{b}{H}{}^a{}_{\alpha\varepsilon}\, v^{\alpha} = \overset{b}{H}{}^a{}_{\varepsilon\beta}\, v^{\beta} = H^a_{\varepsilon}. \tag{4.6}$$

The inter–relations between the shape operators and second fundamental forms are presented in the next proposition.

PROPOSITION 4.1 *Let $\mathbb{F}^m$ and $\widetilde{\mathbb{F}}^{m+n}$ be as in Theorem 4.1. Then the covariant local components of the shape operators $\overset{b}{A}_a$ are given by:*

$$(a)\quad \overset{b}{H}{}'_{a\alpha\beta} = \overset{b}{H}_{a\alpha\beta} + 2(g_{ab\alpha}H^b_{\beta} - L_{a\alpha\beta});\quad (b)\quad \overset{b}{V}{}'_{a\alpha\beta} = 2g_{a\alpha\beta}. \tag{4.7}$$

PROOF. Applying (1.46) and (1.47) for the Berwald connection we deduce that

$$(a)\ \bar{F}_{i\ \alpha}^{\ k} = \widetilde{G}_{i\ j}^{\ k} B_{\alpha}^{j} \quad \text{and} \quad (b)\ \bar{C}_{i\ \alpha}^{\ k} = 0. \tag{4.8}$$

Then, using (4.8a) and (4.3) in (1.67a) and, taking into account (3.33b) and Remark 3.1, we obtain

$$\overset{\mathsf{b}}{H'}_{a\alpha\beta} = -\widetilde{B}_{k}^{\gamma} \left(\frac{\delta B_{a}^{k}}{\delta u^{\beta}} + B_{a}^{i} \widetilde{F}_{i\ j}^{*k} B_{\beta}^{j} \right) g_{\gamma\alpha} - L_{a\alpha\beta}. \tag{4.9}$$

Taking into account (1.67a) for the local components $\overset{\mathsf{c}}{H'}_{a\alpha\beta}$ and using (1.46), (3.3b) and (3.27a) we see that (4.9) becomes

$$\overset{\mathsf{b}}{H'}_{a\alpha\beta} = \overset{\mathsf{c}}{H}_{a\alpha\beta} + g_{ab\alpha} H_{\beta}^{b} - L_{a\alpha\beta}. \tag{4.10}$$

Next, using (4.3) in (4.5a) and taking into account (3.25) we infer that

$$\overset{\mathsf{c}}{H}_{a\alpha\beta} = \overset{\mathsf{b}}{H}_{a\alpha\beta} + g_{ab\alpha} H_{\beta}^{b} - L_{a\alpha\beta}. \tag{4.11}$$

Thus (4.7a) follows from (4.10) and (4.11). Similarly, from (1.67b) using (4.8) and (3.13) we obtain (4.7b). ■

PROPOSITION 4.2 *Let $\mathbb{F}^m$ be a pseudo–Finsler submanifold of $\widetilde{\mathbb{F}}^{m+n}$. Then the local coefficients of the normal Finsler connections of $\mathbb{F}^m$ induced by the Cartan and Berwald connections of $\widetilde{\mathbb{F}}^{m+n}$ are related by*

$$\overset{\mathsf{c}}{F}_{a\ \alpha}^{\ b} = \overset{\mathsf{b}}{F}_{a\ \alpha}^{\ b} - L_{a\ \alpha}^{\ b} + g_{a\ c}^{\ b} H_{\alpha}^{c}, \tag{4.12}$$

and

$$\overset{\mathsf{c}}{C}_{a\ \alpha}^{\ b} = \overset{\mathsf{b}}{C}_{a\ \alpha}^{\ b} + g_{a\ \alpha}^{\ b}, \tag{4.13}$$

where $g_{a\ \alpha}^{\ b}$ and $g_{a\ c}^{\ b}$ are obtained by raising an index in (3.3b) and (3.3c) respectively, and we set

$$L_{a\ \alpha}^{\ b} = \widetilde{P}^{*k}_{\ \ ij} B_{a}^{i} \widetilde{B}_{k}^{b} B_{\alpha}^{j}. \tag{4.14}$$

PROOF. By using (1.68a) for the normal Finsler connections induced by the Cartan and Berwald connections, and taking into account (1.46), (4.3), and (4.14), we deduce (4.12). Similarly (1.68b) gives (4.13). ■

Combining Corollary 3.2 with Proposition 4.2 and taking into account that

$$(a)\ L_{a\ \alpha}^{\ b} = L_{b\ \alpha}^{\ a}; \quad (b)\ g_{a\ \alpha}^{\ b} = g_{b\ \alpha}^{\ a}; \quad (c)\ g_{a\ c}^{\ b} = g_{b\ c}^{\ a}, \tag{4.15}$$

we obtain the following corollary.

COROLLARY 4.1 *The local coefficients of the normal Finsler connection* $\overset{b}{NFC}$ *of* $\mathbb{F}^m$ *induced by the Berwald connection of* $\widetilde{\mathbb{F}}^{m+n}$ *satisfy*

$$(a)\quad \overset{b}{F}{}_a{}^b{}_\alpha + \overset{b}{F}{}_b{}^a{}_\alpha = 2(L_a{}^b{}_\alpha - g_a{}^b{}_c H^c_\alpha);$$

$$(b)\quad \overset{b}{C}{}_a{}^b{}_\alpha + \overset{b}{C}{}_b{}^a{}_\alpha = -2g_a{}^b{}_\alpha. \tag{4.16}$$

THEOREM 4.2 *Let* $\widetilde{F}^{m+n}$ *be a pseudo-Finsler manifold endowed with the Berwald connection and* $\mathbb{F}^m$ *be a pseudo-Finsler submanifold of* $\widetilde{\mathbb{F}}^{m+n}$. *Then the torsion Finsler tensor fields of the induced Finsler connection of* $\mathbb{F}^m$ *are given by (4.4b) and:*

$$\begin{cases} (a)\quad \overset{b}{T}{}_\alpha{}^\gamma{}_\beta = 0;\quad (b)\quad \overset{b}{P}{}^\gamma{}_{\alpha\beta} = 2H^a_\alpha g_a{}^\gamma{}_\beta;\quad (c)\quad \overset{b}{S}{}^\gamma{}_{\alpha\beta} = 0; \\ (d)\quad \overset{b}{R}{}^\gamma{}_{\alpha\beta} = \widetilde{R}^{*k}{}_{ij} B^{ij}_{\alpha\beta} \bar{B}^\gamma_k + \mathcal{A}_{(\alpha\beta)}\{H^a_\alpha\, \overset{b}{H'}{}_a{}^\gamma{}_\beta\}. \end{cases} \tag{4.17}$$

PROOF. Using (4.2) in (2.61), (2.62), and (2.71) we obtain (4.17a), (4.17d) and (4.17c) respectively. Finally, (4.17b) follows from (2.70) taking into account (4.2c), (4.2d) and (4.7b). ■

Next, using the local components of the curvature Finsler tensor fields of the Berwald connection (see Section 1.6), we see that (2.16), (2.17), and (2.18) become

$$\bar{R}_{ij\alpha\beta} = \widetilde{H}_{ijkh} B^{kh}_{\alpha\beta} + \widetilde{G}_{ijkh} B^h_b \mathcal{A}_{(\alpha\beta)}\{B^k_\alpha H^b_\beta\}, \tag{4.18}$$

$$\bar{P}_{ij\alpha\beta} = \widetilde{G}_{ijkh} B^{kh}_{\alpha\beta}, \tag{4.19}$$

and

$$\bar{S}_{ij\alpha\beta} = 0, \tag{4.20}$$

since $\widetilde{S}_{ijkh} = 0$. Finally, taking into account (4.5b), (4.7b) and Theorems 2.2 and 4.2 we may state the following theorem.

THEOREM 4.3 *Let* $\mathbb{F}^m$ *and* $\widetilde{\mathbb{F}}^{m+n}$ *be as in Theorem 4.2. Then we have the assertions:*

(i) The Gauss equations for $\mathbb{F}^m$ are given by

$$\left\{\begin{aligned} \bar{R}_{ij\alpha\beta}B^{ij}_{\gamma\varepsilon} &= \overset{b}{R}_{\gamma\varepsilon\alpha\beta} + \mathcal{A}_{(\alpha\beta)}\{\overset{b}{H}{}^a{}_{\gamma\beta}\overset{b}{H}{}'_{a\varepsilon\alpha}\} && (a)\\ \bar{P}_{ij\alpha\beta}B^{ij}_{\gamma\varepsilon} &= \overset{b}{P}_{\gamma\varepsilon\alpha\beta} - 2\,\overset{b}{H}{}^a{}_{\gamma\alpha}\, g_{a\varepsilon\beta}. && (b)\end{aligned}\right. \tag{4.21}$$

(ii) The A–Codazzi equations and the B–Codazzi equations for $\mathbb{F}^m$ are given by

$$\left\{\begin{aligned} \bar{R}_i{}^j{}_{\alpha\beta}B^i_a\widetilde{B}^\gamma_j &= \mathcal{A}_{(\alpha\beta)}\{\overset{b}{H}{}'_a{}^\gamma{}_{\beta|\alpha}\} - 2g_a{}^\gamma{}_\varepsilon\,\overset{b}{R}{}^\varepsilon{}_{\alpha\beta} && (a)\\ \bar{P}_i{}^j{}_{\alpha\beta}B^i_a\widetilde{B}^\gamma_j &= 2g_a{}^\gamma{}_{\beta|\alpha} - \overset{b}{H}{}'_a{}^\gamma{}_{\alpha||\beta} - 2g_a{}^\gamma{}_\varepsilon\,\overset{b}{P}{}^\varepsilon{}_{\alpha\beta} && (b)\\ g_a{}^\alpha{}_{\beta||\gamma} - g_a{}^\alpha{}_{\gamma||\beta} &= 0, && (c)\end{aligned}\right. \tag{4.22}$$

and

$$\left\{\begin{aligned} \bar{R}_i{}^j{}_{\alpha\beta}B^i_\gamma\widetilde{B}^a_j &= \mathcal{A}_{(\alpha\beta)}\{\overset{b}{H}{}^a{}_{\gamma\alpha|\beta}\} && (a)\\ \bar{P}_i{}^j{}_{\alpha\beta}B^i_\gamma\widetilde{B}^a_j &= \overset{b}{H}{}^a{}_{\gamma\alpha||\beta} && (b)\end{aligned}\right. \tag{4.23}$$

respectively.

(iii) The Ricci equations for $\mathbb{F}^m$ are given by

$$\left\{\begin{aligned} \bar{R}_{ij\alpha\beta}B^i_aB^j_b &= \overset{b}{R}{}^\perp_{ab\alpha\beta} + \mathcal{A}_{(\alpha\beta)}\{\overset{b}{H}{}'_a{}^\gamma{}_\beta\,\overset{b}{H}_{b\gamma\alpha}\} && (a)\\ \bar{P}_{ij\alpha\beta}B^i_aB^j_b &= \overset{b}{P}{}^\perp_{ab\alpha\beta} + 2g_a{}^\gamma{}_\beta\,\overset{b}{H}_{b\gamma\alpha} && (b)\\ 0 &= \overset{b}{S}{}^\perp_{ab\alpha\beta}\,. && (c)\end{aligned}\right. \tag{4.24}$$

We also note that following the same calculations as we performed for (3.60), but with respect to the Berwald connection, we obtain

$$\overset{b}{R}_{\varepsilon\alpha\beta} = v^\gamma\,\overset{b}{R}_{\gamma\varepsilon\alpha\beta}\,. \tag{4.25}$$

Finally, by using (1.5.20) and (1.5.21) with respect to the pair $(\overset{b}{IFC}, \overset{b}{NFC})$ and (4.16) we deduce that

$$\delta_{ab_{|\alpha}} = 2(g_{abc}H^c_\alpha - L_{ab\alpha}), \tag{4.26}$$

and

$$\delta_{ab\|\alpha} = 2g_{ab\alpha}. \tag{4.27}$$

Thus raising and lowering of Latin indices in (4.22) and (4.23) should be done taking into account (4.26) and (4.27).

5. Induced Rund Connection

Let $\widetilde{\mathbb{F}}^{m+n} = (\widetilde{M}, \widetilde{M}', \widetilde{F}^*)$ be a pseudo–Finsler manifold endowed with the Rund connection $\widetilde{\mathbf{RFC}} = (G\widetilde{M}', \widetilde{\nabla}'') = (\widetilde{G}_i^k, \widetilde{F}^{*k}_{i\ j}, 0)$ where $\widetilde{G}_i^k$ are the local coefficients of the canonical non–linear connection (see (3.16)) and $\widetilde{F}^{*k}_{i\ j}$ are the same as the coefficients of the Cartan connection (see (3.17)). Three of the torsion Finsler tensor fields of $\widetilde{\mathbf{RFC}}$ vanish, that is, we have:

$$(a)\ \ \widetilde{T}_i{}^k{}_j = 0; \quad (b)\ \ \widetilde{C}_i{}^k{}_j = 0; \quad (c)\ \ \widetilde{S}^k{}_{ij} = 0, \tag{5.1}$$

while $\widetilde{R}^k{}_{ij}$ and $\widetilde{P}^k{}_{ij}$ coincide with $\widetilde{R}^{*k}{}_{ij}$ and $\widetilde{P}^{*k}{}_{ij}$ of the Cartan connection, that is, they are given by (1.6.29) and (3.34) respectively.

In this case, using Theorem 1.1, we obtain the following theorem.

THEOREM 5.1 *Let $\widetilde{\mathbb{F}}^{m+n}$ be a pseudo–Finsler manifold endowed with the Rund connection and $\mathbb{F}^m = (M, M', F^*)$ be a pseudo–Finsler submanifold of $\widetilde{\mathbb{F}}^{m+n}$. Then we have the assertions:*

(i) The local coefficients of the induced Finsler connection

$$\overset{r}{IFC} = (HM', \overset{r}{\nabla}) = (N_\alpha^\gamma, \overset{r}{F}_\alpha{}^\gamma{}_\beta, \overset{r}{C}_\alpha{}^\gamma{}_\beta),$$

are given by (3.24a) and

$$(a)\ \ \overset{r}{F}_\alpha{}^\gamma{}_\beta = \widetilde{B}_k^\gamma(B_{\alpha\beta}^k + \widetilde{F}^{*k}_{i\ j} B_{\alpha\beta}^{ij}); \quad (b)\ \ \overset{r}{C}_\alpha{}^\gamma{}_\beta = 0. \tag{5.2}$$

(ii) The local components of the h–and v–second fundamental forms are given by

$$(a)\ \ \overset{r}{H}{}^a{}_{\alpha\beta} = \widetilde{B}_k^a(B_{\alpha\beta}^k + \widetilde{F}^{*k}_{i\ j} B_{\alpha\beta}^{ij}); \quad (b)\ \ \overset{r}{V}{}^a{}_{\alpha\beta} = 0. \tag{5.3}$$

Contracting (5.2a) and (5.3a) by v^α or v^β we obtain the local coefficients of the induced non–linear connection and of the normal curvature forms, i.e., we have

$$(a)\ \ \overset{r}{F}_\alpha{}^\gamma{}_\varepsilon\, v^\alpha = \overset{r}{F}_\varepsilon{}^\gamma{}_\beta\, v^\beta = N_\varepsilon^\gamma; \quad (b)\ \ \overset{r}{H}{}^a{}_{\alpha\varepsilon}\, v^\alpha = \overset{r}{H}{}^a{}_{\varepsilon\beta}\, v^\beta = H_\varepsilon^a. \tag{5.4}$$

As we have seen in Section 1.7 the Rund connection is an h–metric connection. Unfortunately, $\overset{r}{IFC}$ does not inherit this property. More precisely, using (1.7.76), (3.3a) and (3.1) in (1.77) and (1.78) we obtain

$$(a) \quad g_{\alpha\beta|\gamma} = 2g_{a\alpha\beta}H^{a}_{\gamma}; \quad (b) \quad g_{\alpha\beta||\gamma} = 2g_{\alpha\beta\gamma}. \tag{5.5}$$

PROPOSITION 5.1 *Let $\mathbb{F}^m$ and $\widetilde{\mathbb{F}}^{m+n}$ be as in Theorem 5.1. Then the local components of the shape operators $\overset{r}{A}_a$ are given by*

$$\overset{r}{H'}_{a\alpha\beta} = \overset{r}{H}_{a\alpha\beta} + 2g_{ab\alpha}H^{b}_{\beta}; \quad (b) \quad \overset{r}{V'}_{a\alpha\beta} = 2g_{a\alpha\beta}. \tag{5.6}$$

PROOF. Applying (1.67a) for both the Rund connection and the Cartan connection and using (1.46) and (3.3b) we derive

$$\overset{r}{H'}_{a\alpha\beta} = \overset{c}{H'}_{a\alpha\beta} + g_{ab\alpha}H^{b}_{\beta}. \tag{5.7}$$

On the other hand, comparing (5.3a) with (3.25) we deduce that

$$\overset{c}{H}_{a\alpha\beta} = \overset{r}{H}_{a\alpha\beta} + g_{ab\alpha}H^{b}_{\beta}. \tag{5.8}$$

Thus (5.6a) follows from (5.7) and (5.8) via (3.27a). Finally, using (1.67b) for Rund connection and taking into account (3.13) we obtain (5.6b). ■

The proof of the next proposition follows from (1.68) written for both the Cartan connection and the Rund connection.

PROPOSITION 5.2 *Let $\mathbb{F}^m$ be a pseudo–Finsler submanifold of $\widetilde{\mathbb{F}}^{m+n}$. Then the local coefficients of the normal Finsler connections of $\mathbb{F}^m$ induced by the Cartan and Rund connections of $\widetilde{\mathbb{F}}^{m+n}$ are related by:*

$$(a) \quad \overset{c}{F}_{a}{}^{b}{}_{\alpha} = \overset{r}{F}_{a}{}^{b}{}_{\alpha} + g_{a}{}^{b}{}_{c}H^{c}_{\alpha}; \quad (b) \quad \overset{c}{C}_{a}{}^{b}{}_{\alpha} = \overset{r}{C}_{a}{}^{b}{}_{\alpha} + g_{a}{}^{b}{}_{\alpha}. \tag{5.9}$$

Moreover, by using Corollary 3.2, (4.15b), (4.15c), and (5.9) we may state the following corollary.

COROLLARY 5.1 *The local coefficients of the normal Finsler connection of $\mathbb{F}^m$ induced by the Rund connection of $\widetilde{\mathbb{F}}^{m+n}$ satisfy*

$$(a) \quad \overset{r}{F}_{a}{}^{b}{}_{\alpha} + \overset{r}{F}_{b}{}^{a}{}_{\alpha} = -2g_{a}{}^{b}{}_{c}H^{c}_{\alpha}; \quad (b) \quad \overset{r}{C}_{a}{}^{b}{}_{\alpha} + \overset{r}{C}_{b}{}^{a}{}_{\alpha} = -2g_{a}{}^{b}{}_{\alpha}. \tag{5.10}$$

Applying (2.61), (2.62), (2.70), and (2.71) for the Rund connection we obtain the following theorem.

THEOREM 5.2 *Let $\widetilde{\mathbb{F}}^{m+n}$ be a pseudo–Finsler manifold endowed with the Rund connection and $\mathbb{F}^m$ be a pseudo–Finsler submanifold of $\widetilde{\mathbb{F}}^{m+n}$. Then the torsion Finsler tensor fields of $\overset{r}{IFC}$ are given by (5.2b) and*

$$\left\{\begin{array}{l} (a)\ \overset{r}{T}{}_{\alpha}{}^{\gamma}{}_{\beta}=0;\ (b)\ \overset{r}{P}{}^{\gamma}{}_{\alpha\beta}=2H^{a}_{\alpha}g_{a}{}^{\gamma}{}_{\beta}+L^{\gamma}{}_{\alpha\beta};\ (c)\ \overset{r}{S}{}^{\gamma}{}_{\alpha\beta}=0; \\ (d)\ \ \overset{r}{R}{}^{\gamma}{}_{\alpha\beta}=\ \widetilde{R}^{*k}{}_{ij}B^{ij}_{\alpha\beta}\widetilde{B}^{\gamma}_{k}+\mathcal{A}_{(\alpha\beta)}\{(L_{a}{}^{\gamma}{}_{\alpha}-\overset{r}{H}{}'^{\gamma}_{a}{}_{\alpha})H^{a}_{\beta}\}. \end{array}\right. \tag{5.11}$$

Taking into account that the v–curvature Finsler tensor field of $\widetilde{\mathbf{RFC}}$ vanishes on $\widetilde{M}'$, from (2.16)–(2.18) we obtain

$$\bar{R}_{ij\alpha\beta}\ =\ \widetilde{K}_{ijkh}B^{kh}_{\alpha\beta}+\widetilde{F}_{ijkh}B^{h}_{a}(B^{k}_{\alpha}H^{a}_{\beta}-B^{k}_{\beta}H^{a}_{\alpha}), \tag{5.12}$$

$$\bar{P}_{ij\alpha\beta}\ =\ \widetilde{F}_{ijkh}B^{kh}_{\alpha\beta}, \tag{5.13}$$

and

$$\bar{S}_{ij\alpha\beta}=0, \tag{5.14}$$

where $\widetilde{K}$ and $\widetilde{F}$ are the h–and hv–curvature Finsler tensor fields of the Rund connection (see (1.7.77a) and (1.7.77b)). Moreover, applying the general result from Theorem 2.2 we state the following theorem on the structure equations of $\mathbb{F}^m$.

THEOREM 5.3 *Let $\mathbb{F}^m$ and $\widetilde{\mathbb{F}}^{m+n}$ be as in Theorem 5.2. Then we have the following assertions:*
(i) The Gauss equations for $\mathbb{F}^m$ are given by

$$\left\{\begin{array}{lcll} \bar{R}_{ij\alpha\beta}B^{ij}_{\gamma\varepsilon} & = & \overset{r}{R}_{\gamma\varepsilon\alpha\beta}+\mathcal{A}_{(\alpha\beta)}\{\overset{r}{H}{}^{a}{}_{\gamma\beta}\overset{r}{H}{}'_{a\varepsilon\alpha}\} & (a) \\ \bar{P}_{ij\alpha\beta}B^{ij}_{\gamma\varepsilon} & = & \overset{r}{P}_{\gamma\varepsilon\alpha\beta}\ -\ 2\ \overset{r}{H}{}^{a}{}_{\gamma\alpha}\ g_{a\varepsilon\beta}. & (b) \end{array}\right. \tag{5.15}$$

(ii) The A–Codazzi equations and the B–Codazzi equations for $\mathbb{F}^m$ are given by

$$\left\{\begin{array}{lcll} \bar{R}_{i}{}^{j}{}_{\alpha\beta}B^{i}_{a}\widetilde{B}^{\gamma}_{j} & = & \mathcal{A}_{(\alpha\beta)}\{\overset{r}{H}{}'^{\gamma}_{a}{}_{\beta|\alpha}\}-\ 2g_{a}{}^{\gamma}{}_{\varepsilon}\overset{r}{R}{}^{\varepsilon}{}_{\alpha\beta} & (a) \\ \bar{P}_{i}{}^{j}{}_{\alpha\beta}B^{i}_{a}\widetilde{B}^{\gamma}_{j} & = & 2g_{a}{}^{\gamma}{}_{\beta|\alpha}-\ \overset{r}{H}{}'^{\gamma}_{a}{}_{\alpha\|\beta}\ -2g_{a}{}^{\gamma}{}_{\varepsilon}\ \overset{r}{P}{}^{\varepsilon}{}_{\alpha\beta} & (b) \\ g_{a}{}^{\alpha}{}_{\beta\|\gamma}-g_{a}{}^{\alpha}{}_{\gamma\|\beta} & = & 0, & (c) \end{array}\right. \tag{5.16}$$

and

$$\left\{\begin{array}{lcll} \bar{R}_{i}{}^{j}{}_{\alpha\beta}B^{i}_{\gamma}\widetilde{B}^{a}_{j} & = & \mathcal{A}_{(\alpha\beta)}\{\overset{r}{H}{}^{a}{}_{\gamma\alpha|\beta}\} & (a) \\ \bar{P}_{i}{}^{j}{}_{\alpha\beta}B^{i}_{\gamma}\widetilde{B}^{a}_{j} & = & \overset{r}{H}{}^{a}{}_{\gamma\alpha\|\beta}, & (b) \end{array}\right. \tag{5.17}$$

respectively.

(iii) The Ricci equations for $\mathbb{F}^m$ *are given by*

$$\left\{\begin{array}{rcll} \bar{R}_{ij\alpha\beta}B^i_a B^j_b & = & \overset{r}{R}{}^{\perp}{}_{ab\alpha\beta} + \mathcal{A}_{(\alpha\beta)}\{\overset{r}{H}{}'^{\gamma}_{a\ \beta}\overset{r}{H}_{b\gamma\alpha}\} & (a) \\ \bar{P}_{ij\alpha\beta}B^i_a B^j_b & = & \overset{r}{P}{}^{\perp}{}_{ab\alpha\beta} + 2g_{a\ \beta}^{\ \gamma}\ \overset{r}{H}_{b\gamma\alpha} & (b) \\ 0 & = & \overset{r}{S}{}^{\perp}{}_{ab\alpha\beta}\,. & (c) \end{array}\right. \tag{5.18}$$

Comparing the structure equations of $\mathbb{F}^m$ from Theorems 4.3 and 5.3 we see that they are formally the same. However, the geometric objects involved in these equations are different from each other. It is also interesting to note that, as in the case of the Cartan and Berwald connections, we have

$$\overset{r}{R}_{\varepsilon\alpha\beta} = v^{\gamma}\overset{r}{R}_{\gamma\varepsilon\alpha\beta}, \tag{5.19}$$

for the Rund connection.

Finally, as in the case of Berwald connection, by using (5.10) we deduce that

$$(a)\ \ \delta_{ab|\alpha} = 2g_{abc}H^c_{\alpha}\ \text{ and }\ (b)\ \ \delta_{ab||\alpha} = 2g_{ab\alpha}. \tag{5.20}$$

6. A Comparison Between the Induced and the Intrinsic Finsler Connections

Let $\widetilde{\mathbb{F}}^{m+n} = (\widetilde{M}, \widetilde{M}', \widetilde{F}^*)$ be a pseudo–Finsler manifold endowed with the Cartan connection $\widetilde{\mathbf{FC}}^* = (G\widetilde{M}', \widetilde{\nabla}^*) = (\widetilde{G}^k_i, \widetilde{F}^{*k}_{i\ j}, \widetilde{g}_i{}^k{}_j)$ and $\mathbb{F}^m$ be a pseudo–Finsler submanifold of $\widetilde{\mathbb{F}}^{m+n}$. In Section 3 we constructed the induced Finsler connection $I\overset{c}{F}C = (HM', \overset{c}{\nabla}) = (\overset{c}{N}{}^{\gamma}_{\alpha}, \overset{c}{F}_{\alpha\ \beta}^{\ \gamma}, \overset{c}{C}_{\alpha\ \beta}^{\ \gamma})$, whose local coefficients are given by (3.24a), (3.35), and (3.36). On the other hand, on $\mathbb{F}^m$ lives the intrinsic Cartan connection $\mathbf{FC}^* = (GM', \nabla^*) = (G^{\gamma}_{\alpha}, F^{*\gamma}_{\alpha\ \beta}, g_{\alpha\ \beta}^{\ \gamma})$. More precisely, the canonical horizontal distribution GM' is locally represented by the vector fields

$$\frac{\delta^*}{\delta^* u^{\alpha}} = \frac{\partial}{\partial u^{\alpha}} - G^{\gamma}_{\alpha}\frac{\partial}{\partial v^{\gamma}}, \tag{6.1}$$

with the local coefficients

$$(a)\ \ G^{\gamma}_{\alpha} = \frac{\partial G^{\gamma}}{\partial v^{\alpha}};\quad (b)\ \ G^{\gamma} = \frac{1}{4}g^{\gamma\beta}\left(\frac{\partial^2 F^*}{\partial v^{\beta}\partial u^{\varepsilon}}v^{\varepsilon} - \frac{\partial F^*}{\partial u^{\beta}}\right). \tag{6.2}$$

Also, $(F^{*\gamma}_{\alpha\ \beta}, g_{\alpha\ \beta}^{\ \gamma})$ are the local coefficients of the linear connection ∇^* on VM' with respect to the local field of frames $\{\delta^*/\delta^* u^\alpha, \partial/\partial v^\alpha\}$ on M', that is, we set

$$\nabla^*_{\delta^*/\delta^* u^\beta}\frac{\partial}{\partial v^\alpha} = F^{*\gamma}_{\alpha\ \beta}\frac{\partial}{\partial v^\gamma}; \quad (b)\ \nabla^*_{\partial/\partial v^\beta}\frac{\partial}{\partial v^\alpha} = g_{\alpha\ \beta}^{\ \gamma}\frac{\partial}{\partial v^\gamma}, \tag{6.3}$$

where we have:

$$\begin{aligned} F^{*\gamma}_{\alpha\ \beta} &= \frac{1}{2}g^{\gamma\varepsilon}\left(\frac{\delta^* g_{\varepsilon\alpha}}{\delta^* u^\beta} + \frac{\delta^* g_{\varepsilon\beta}}{\delta^* u^\alpha} - \frac{\delta^* g_{\alpha\beta}}{\delta^* u^\varepsilon}\right) \quad (a)\\ g_{\alpha\ \beta}^{\ \gamma} &= \frac{1}{2}g^{\gamma\varepsilon}\frac{\partial g_{\varepsilon\alpha}}{\partial v^\beta}. \quad (b) \end{aligned} \tag{6.4}$$

Now, we recall from the theory of Riemannian (pseudo–Riemannian) submanifolds (cf., Chen [1], Dajczer et al. [1], O'Neill [1]) that the induced linear connection by the Levi–Civita connection is just the intrinsic Levi–Civita connection of the submanifold. Hence an important question for Finslerian geometry arises: does the induced Finsler connection $I\overset{c}{F}C$ coincide with the intrinsic Cartan connection $\mathbf{FC}^*$ of $\mathbb{F}^m$? Taking into account (3.40) which shows that, in general, $I\overset{c}{F}C$ has a non–zero torsion Finsler tensor field $\overset{c}{T}_{\alpha\ \beta}^{\ \gamma}$, the answer to the above question is in the negative. Thus a comparison between $I\overset{c}{F}C$ and $\mathbf{FC}^*$ shows an imperious need for better understanding of the geometry of $\mathbb{F}^m$ in $\widetilde{\mathbb{F}}^{m+n}$.

We start with the comparison of the non–linear connections of $I\overset{c}{F}C$ and $\mathbf{FC}^*$.

THEOREM 6.1 *(Bejancu–Farran [2]) Let $\widetilde{\mathbb{F}}^{m+n}$ be a pseudo–Finsler manifold endowed with the canonical non–linear connection $G\widetilde{M}' = (\widetilde{G}_i^k)$ and $\mathbb{F}^m$ be a pseudo–Finsler submanifold of $\widetilde{\mathbb{F}}^{m+n}$. Then the canonical non–linear connection $GM' = (G_\alpha^\gamma)$ and the induced non–linear connection $HM' = (N_\alpha^\gamma)$ of $\mathbb{F}^m$ are related by*

$$G_\alpha^\gamma = N_\alpha^\gamma + D_\alpha^\gamma, \tag{6.5}$$

where we set

$$D_\alpha^\gamma = g_{a\ \alpha}^{\ \gamma} H_\beta^a v^\beta. \tag{6.6}$$

PROOF. By straightforward calculations using (6.2b), (1.5), (1.6) and (3.19) we obtain

$$2G_\alpha = 2g_{\alpha\beta}G^\beta = B_\alpha^i \tilde{g}_{ij}(2\widetilde{G}^j + B_{\varepsilon\mu}^j v^\varepsilon v^\mu). \tag{6.7}$$

Contracting (6.7) by $g^{\alpha\gamma}$ and taking into account (1.13) we infer that

$$2G^{\gamma} = \widetilde{B}^{\gamma}_{j}(2\widetilde{G}^{j} + B^{j}_{\varepsilon\mu}v^{\varepsilon}v^{\mu}). \tag{6.8}$$

Differentiating (6.8) with respect to v^{α} and using (6.2a), (3.5), (3.16), and (3.24a) we deduce that

$$G^{\gamma}_{\alpha} = N^{\gamma}_{\alpha} + g_{a}{}^{\gamma}{}_{\alpha}\widetilde{B}^{a}_{j}(2\widetilde{G}^{j} + B^{j}_{\varepsilon\mu}v^{\varepsilon}v^{\mu}). \tag{6.9}$$

As $\widetilde{G}^{j}$ are positive homogeneous of degree two with respect to (y^{i}) we have

$$2\widetilde{G}^{j} = \widetilde{G}^{j}_{i}y^{i} = \widetilde{G}^{j}_{i}B^{i}_{\varepsilon}v^{\varepsilon}. \tag{6.10}$$

Finally, we replace $\widetilde{G}^{j}$ from (6.10) into (6.9) and use (3.24b) and (6.6) to obtain (6.5). ■

The induced horizontal distribution HM' is locally represented by the vector fields

$$\frac{\delta}{\delta u^{\alpha}} = \frac{\partial}{\partial u^{\alpha}} - N^{\gamma}_{\alpha}\frac{\partial}{\partial v^{\gamma}}, \tag{6.11}$$

where N^{γ}_{α} are given by (3.24a). Then using (6.5) in (6.1) and taking into account (6.11) we derive that

$$\frac{\delta^{*}}{\delta^{*}u^{\alpha}} = \frac{\delta}{\delta u^{\alpha}} - D^{\gamma}_{\alpha}\frac{\partial}{\partial v^{\gamma}}. \tag{6.12}$$

Next, contracting (6.6) by v^{α} and using (3.45) we deduce that

$$D^{\gamma}_{\alpha}v^{\alpha} = 0. \tag{6.13}$$

Finally, contracting (6.5) by v^{α} and using (6.13) we obtain that

$$G^{\gamma}_{\alpha}v^{\alpha} = N^{\gamma}_{\alpha}v^{\alpha}. \tag{6.14}$$

It is easy to see that D^{γ}_{α} given by (6.6) are the local components of a Finsler tensor field **D** of type (1,1) on $\mathbb{F}^{m}$. As $\mathbf{D} = 0$ if and only if $G^{\gamma}_{\alpha} = N^{\gamma}_{\alpha}$, we call **D** the **deformation Finsler tensor field** with respect to the pair of non–linear connections (GM', HM') of $\mathbb{F}^{m}$.

We now recall from Section 3 that the induced Finsler connection

$$I\overset{c}{F}C = (HM', \overset{c}{\nabla})$$

is a metric Finsler connection (cf., (3.38)). Then, according to the general result with respect to the existence and uniqueness of a metric linear connection on the vertical vector bundle with some prescribed torsion Finsler tensor fields (see Theorem 1.7.1) we may state the following theorem.

THEOREM 6.2 *Let $HM' = (N^{\gamma}_{\alpha})$ be the induced non–linear connection on M' by the canonical non–linear connection $G\widetilde{M}' = (\widetilde{G}^k_i)$ on $\widetilde{M}'$. Then the linear connection $\overset{c}{\nabla}$ from $I\overset{c}{F}C$ with local coefficients given by (3.35) and (3.36) is the unique metric linear connection on VM' with torsion Finsler tensor fields $\overset{c}{T}$ and $\overset{c}{S}$ given by (3.40) and (3.42).*

More precisely, taking into account that $\overset{c}{S}= 0$, from (1.7.3) and (1.7.4) we deduce that the v–covariant derivative and h–covariant derivative determined by $\overset{c}{\nabla}$ are given by

$$\begin{aligned} 2g(\overset{c}{\nabla}_{vX}\, vY, vZ) &= vX\,(g(vY,vZ)) + vY\,(g(vZ,vX)) \\ &\quad -vZ\,(g(vX,vY)) - g(vX,[vY,vZ]) \\ &\quad +g(vY,[vZ,vX]) + g(vZ,[vX,vY]), \qquad (6.15) \end{aligned}$$

and

$$\begin{aligned} &2g(\overset{c}{\nabla}_{hX}QhY,QhZ) \\ &\quad = hX(g(QhY,QhZ)) + hY(g(QhZ,QhX)) \\ &\qquad -hZ(g(QhX,QhY)) - g(QhX,Qh[hY,hZ]) + g(QhY,Qh[hZ,hX]) \\ &\qquad +g(QhZ,Qh[hX,hY]) - g(QhX,\overset{c}{T}\,(QhY,QhZ)) \\ &\qquad +g(QhY,\overset{c}{T}\,(QhZ,QhX)) + g(QhZ,\overset{c}{T}\,(QhX,QhY)), \qquad (6.16) \end{aligned}$$

for any $X,Y,Z \in \Gamma(TM')$, where h and v are the projection morphisms of TM' on HM' and VM', respectively, and Q is the associate almost product structure to HM' (see (1.4.8)). For further calculations we need the following (cf., (1.4.11) and (1.4.26))

$$(a)\ \left[\frac{\delta}{\delta u^{\alpha}},\frac{\delta}{\delta u^{\beta}}\right] = \overset{c}{R}{}^{\gamma}{}_{\alpha\beta}\,\frac{\partial}{\partial v^{\gamma}} \ \text{ and } \ (b)\ \left[\frac{\delta}{\delta u^{\alpha}},\frac{\partial}{\partial v^{\beta}}\right] = \frac{\partial N^{\gamma}_{\alpha}}{\partial v^{\beta}}\frac{\partial}{\partial v^{\gamma}}. \qquad (6.17)$$

Also we have

$$(a)\ \ \overset{c}{T}\left(\frac{\partial}{\partial v^{\beta}},\frac{\partial}{\partial v^{\alpha}}\right) = \overset{c}{T}_{\alpha}{}^{\gamma}{}_{\beta}\,\frac{\partial}{\partial v^{\gamma}};\quad (b)\ \ \overset{c}{T}_{\alpha\varepsilon\beta}= g_{\varepsilon\gamma}\,\overset{c}{T}_{\alpha}{}^{\gamma}{}_{\beta}\,. \qquad (6.18)$$

Finally, from (1.4.8) we deduce that

$$(a)\ Q\left(\frac{\delta}{\delta u^{\alpha}}\right) = \frac{\partial}{\partial v^{\alpha}} \ \text{ and } \ (b)\ Q\left(\frac{\partial}{\partial v^{\alpha}}\right) = \frac{\delta}{\delta u^{\alpha}}. \qquad (6.19)$$

Next, we replace hX, hY and hZ from (6.16) by $\delta/\delta u^\beta$, $\delta/\delta u^\alpha$ and $\delta/\delta u^\varepsilon$ respectively, and using (6.19a), (1.48a), (6.17a) and (6.18) we obtain

$$2g_{\varepsilon\mu}\overset{c}{F}{}_{\alpha}{}^{\mu}{}_{\beta} = \frac{\delta g_{\varepsilon\alpha}}{\delta u^\beta} + \frac{\delta g_{\varepsilon\beta}}{\delta u^\alpha} - \frac{\delta g_{\alpha\beta}}{\delta u^\varepsilon} + \overset{c}{T}_{\alpha\varepsilon\beta} + \overset{c}{T}_{\beta\alpha\varepsilon} - \overset{c}{T}_{\varepsilon\beta\alpha}\,. \quad (6.20)$$

Contracting (6.20) by $g^{\varepsilon\gamma}$ and using (6.12), (6.4b) and (3.40) we infer that

$$\overset{c}{F}{}_{\alpha}{}^{\gamma}{}_{\beta} = F^{*\gamma}_{\alpha\ \beta} + D^{\mu}_{\alpha}g_{\mu}{}^{\gamma}{}_{\beta} + D^{\mu}_{\beta}g_{\mu}{}^{\gamma}{}_{\alpha} - D^{\gamma\mu}g_{\mu\alpha\beta} + g_{a\alpha\beta}H^{a\gamma} - g_{a}{}^{\gamma}{}_{\beta}H^{a}_{\alpha}, \quad (6.21)$$

where we set

$$(a)\ \ D^{\gamma\mu} = g^{\gamma\varepsilon}D^{\mu}_{\varepsilon} \ \ \text{and} \ \ (b)\ \ H^{a\gamma} = g^{\gamma\varepsilon}H^{a}_{\varepsilon}. \quad (6.22)$$

Thus the above calculations enable us to state the following result.

THEOREM 6.3 *(Bejancu–Farran [2]). Let $\widetilde{\mathbb{F}}^{m+n}$ be a pseudo–Finsler manifold endowed with the Cartan connection $\widetilde{\mathbf{FC}}^*$, and $\mathbb{F}^m$ be a pseudo–Finsler submanifold of $\widetilde{\mathbb{F}}^{m+n}$. Then the local coefficients of the induced Finsler connection $I\overset{c}{F}C = (N^{\gamma}_{\alpha}, \overset{c}{F}{}_{\alpha}{}^{\gamma}{}_{\beta}, \overset{c}{C}{}_{\alpha}{}^{\gamma}{}_{\beta})$ and of the intrinsic Cartan connection $\mathbf{FC}^* = (G^{\gamma}_{\alpha}, F^{*\gamma}_{\alpha\ \beta}, g_{\alpha}{}^{\gamma}{}_{\beta})$ are related by (6.5), (6.21) and (3.36).*

Moreover, we prove the following theorem.

THEOREM 6.4 *(Bejancu–Farran [2]). Let $\mathbb{F}^m$ and $\widetilde{\mathbb{F}}^{m+n}$ be as in Theorem 6.3. Then the following assertions are equivalent:*

(i) The induced Finsler connection $I\overset{c}{F}C$ coincides with the intrinsic Cartan connection $\mathbf{FC}^*$ of $\mathbb{F}^m$.

(ii) The normal curvature forms of $\mathbb{F}^m$ satisfy

$$g_{a\alpha\beta}H^{a\gamma} - g_{a}{}^{\gamma}{}_{\beta}H^{a}_{\alpha} = 0. \quad (6.23)$$

(iii) The deformation Finsler tensor field $\mathbf{D}$ vanishes on M'.

(iv) The torsion Finsler tensor field $\overset{c}{T}{}_{\alpha}{}^{\gamma}{}_{\beta}$ of $I\overset{c}{F}C$ given by (3.40) vanishes on M'.

(v) The local coefficients $\overset{c}{F}{}_{\alpha}{}^{\gamma}{}_{\beta}$ of $I\overset{c}{F}C$ are given by

$$\overset{c}{F}{}_{\alpha}{}^{\gamma}{}_{\beta} = \frac{1}{2}g^{\gamma\varepsilon}\left(\frac{\delta g_{\varepsilon\alpha}}{\delta u^\beta} + \frac{\delta g_{\varepsilon\beta}}{\delta u^\alpha} - \frac{\delta g_{\alpha\beta}}{\delta u^\varepsilon}\right). \quad (6.24)$$

PROOF. (*i*) $\Longrightarrow$ (*ii*). As $N^{\gamma}_{\alpha} = G^{\gamma}_{\alpha}$, from (6.5) we infer that $D^{\gamma}_{\alpha} = 0$. Thus (6.23) follows from (6.21) since $\overset{c}{F}{}_{\alpha}{}^{\gamma}{}_{\beta} = F^{*\gamma}_{\alpha\ \beta}$.

(*ii*) $\Longrightarrow$ (*iii*). Contracting (6.23) by v^{α} and using (3.45) and (6.6) we obtain $D^{\beta}_{\alpha} = 0$ for any $\alpha, \beta \in \{1, \cdots, m\}$.

(*iii*) $\Longrightarrow$ (*iv*). Suppose $D^{\beta}_{\alpha} = 0$. Then by using (6.6) and (3.61) we deduce that

$$\frac{\partial g_{a\gamma\alpha}}{\partial v^{\beta}} H^{a}_{\varepsilon} v^{\varepsilon} + g_{a\gamma\alpha}(2H^{a}_{\beta} + (g_{b}{}^{a}{}_{\beta} - \overset{c}{C}_{b}{}^{a}{}_{\beta}) H^{b}_{\varepsilon} v^{\varepsilon}) = 0. \tag{6.25}$$

Contracting (3.62) by $H^{a}_{\varepsilon} v^{\varepsilon}$ and using (6.25) we derive

$$H^{a}_{\alpha} g_{a\gamma\beta} - H^{a}_{\beta} g_{a\gamma\alpha} = 0,$$

which is equivalent to $\overset{c}{T}{}_{\alpha}{}^{\gamma}{}_{\beta} = 0$.

(*iv*) $\Longrightarrow$ (*v*). As $\overset{c}{T}{}_{\alpha}{}^{\gamma}{}_{\beta} = 0$, from (6.20) we obtain (6.24).

(*v*) $\Longrightarrow$ (*i*). Using (6.24) in (6.20) we derive

$$\overset{c}{T}_{\varepsilon\beta\alpha} = \overset{c}{T}_{\alpha\varepsilon\beta} + \overset{c}{T}_{\beta\alpha\varepsilon}. \tag{6.26}$$

Taking into account (3.40) we easily see that (6.26) implies (6.23). Contracting (6.23) by v^{α} and using (3.45) and (6.6) we infer that $D^{\gamma}_{\beta} = 0$. Finally, from (6.5) and (6.21) we obtain $G^{\gamma}_{\alpha} = N^{\gamma}_{\alpha}$ and $\overset{c}{F}{}_{\alpha}{}^{\gamma}{}_{\beta} = F^{*\gamma}_{\alpha\ \beta}$ respectively. Thus the proof of the theorem is complete. ■

Now we are concerned with a comparison between the torsion and curvature Finsler tensor fields of $\overset{c}{IFC}$ and $\mathbf{FC}^*$ on $\mathbb{F}^m$. First we note that $\overset{c}{IFC}$ and $\mathbf{FC}^*$ have the torsion Finsler tensor fields: $\overset{c}{S} = 0$, $\overset{c}{T}$ given by (3.40), $\overset{c}{C}{}_{\alpha}{}^{\gamma}{}_{\beta} = g_{\alpha}{}^{\gamma}{}_{\beta}$ and $S^* = T^* = 0$, $C^{*\gamma}_{\alpha\ \beta} = g_{\alpha}{}^{\gamma}{}_{\beta}$, respectively. Thus it remains to find the relations between the torsion Finsler tensor fields $\{\overset{c}{R}{}^{\gamma}{}_{\alpha\beta}, \overset{c}{P}{}^{\gamma}{}_{\alpha\beta}\}$ and $\{R^{*\gamma}{}_{\alpha\beta}, P^{*\gamma}{}_{\alpha\beta}\}$.

By direct calculations using (1.4.12), (6.5), (6.12), and (1.5.42b) we obtain

$$\overset{c}{R}{}^{\gamma}{}_{\alpha\beta} = R^{*\gamma}{}_{\alpha\beta} + \mathcal{A}_{(\alpha\beta)} \left\{ D^{\gamma}_{\beta|*\alpha} + D^{\mu}_{\beta} \left(P^{*\gamma}{}_{\mu\alpha} - \frac{\partial D^{\gamma}_{\alpha}}{\partial v^{\mu}} \right) \right\}, \tag{6.27}$$

where according to the notation we introduced in Section 7, $D^{\gamma}_{\beta|*\alpha}$ is the horizontal covariant derivative of D^{γ}_{β} with respect to the Cartan connection $\mathbf{FC}^*$ of $\mathbb{F}^m$. Further, taking into account (6.22a), (6.6), and (3.3a) we deduce that

$$D^{\alpha\beta} = D^{\beta\alpha}, \tag{6.28}$$

which is equivalent to the fact that $D^{\gamma}_{\alpha}(u,v)$ are the components of a self-adjoint operator with respect to $g(u,v)$ on $VM'_{(u,v)}$. Then using (1.5.42b), (6.5), (6.21), and (6.22a) we infer that

$$\overset{c}{P}{}^{\gamma}{}_{\alpha\beta} = P^{*\gamma}{}_{\alpha\beta} - D^{\gamma}_{\alpha\|*\beta} - D^{\mu}_{\beta} g_{\mu}{}^{\gamma}{}_{\alpha} - g_{a\alpha\beta} H^{a\gamma} + g_{a}{}^{\gamma}{}_{\alpha} H^{a}_{\beta}. \tag{6.29}$$

Summing up, we may state the following theorem:

THEOREM 6.5 *The torsion Finsler tensor fields* $(\overset{c}{R}{}^{\gamma}{}_{\alpha\beta}, \overset{c}{P}{}^{\gamma}{}_{\alpha\beta}, \overset{c}{S}{}^{\gamma}{}_{\alpha\beta}, \overset{c}{T}{}_{\alpha}{}^{\gamma}{}_{\beta}, \overset{c}{C}{}_{\alpha}{}^{\gamma}{}_{\beta})$ *and* $(R^{*\gamma}{}_{\alpha\beta}, P^{*\gamma}{}_{\alpha\beta}, S^{*\gamma}{}_{\alpha\beta}, T^{*\gamma}_{\alpha\ \beta}, C^{*\gamma}_{\alpha\ \beta})$ *of* $I\overset{c}{F}C$ *and* **FC*** *of* $\mathbb{F}^m$ *are related by (6.27), (6.29), (3.40) and:*

$$(a)\ \ \overset{c}{S}{}^{\gamma}{}_{\alpha\beta} = S^{*\gamma}{}_{\alpha\beta} = 0; \ \ (b)\ \ T^{*\gamma}_{\alpha\ \beta} = 0; \ \ (c)\ \ \overset{c}{C}{}_{\alpha}{}^{\gamma}{}_{\beta} = C^{*\gamma}_{\alpha\ \beta} = g_{\alpha}{}^{\gamma}{}_{\beta}. \tag{6.30}$$

In order to obtain the inter-relations between the curvature Finsler tensor fields of $I\overset{c}{F}C$ and **FC*** we denote by h^* and v^* the projection morphisms of TM' on GM' and VM', respectively, and by Q^* the associate almost product structure to GM'. Then owing to (6.30) we have

$$\nabla^*_{v^*X} v^*Y = \overset{c}{\nabla}_{v^*X} v^*Y, \quad \forall X, Y \in \Gamma(TM'). \tag{6.31}$$

As in general, the h-covariant derivatives with respect to $I\overset{c}{F}C$ and **FC*** do not coincide, we set

$$\nabla^*_{h^*X} v^*Y = \overset{c}{\nabla}_{h^*X} v^*Y + \overset{c}{D}(Q^*h^*X, v^*Y), \tag{6.32}$$

for any $X, Y \in \Gamma(TM')$, where $\overset{c}{D}$ is a Finsler tensor field of type (1,2), which we call the **deformation Finsler tensor field** of the pair $(\nabla^*, \overset{c}{\nabla})$.

Next, we denote by R^* and $\overset{c}{R}$ the curvature forms (see (1.3.10)) of the linear connections ∇^* and $\overset{c}{\nabla}$ respectively. To be able to relate these two forms, we need a coordinate-free expression for the Cartan tensor field. To obtain this expression we first note that $g_{\alpha}{}^{\gamma}{}_{\beta}$ are the local components of the torsion Finsler tensor field T_4 (see.(1.5.41c) and (1.5.42c)). Then by using (1.5.36) we deduce that for the Cartan tensor field $C^* = (g_{\alpha}{}^{\gamma}{}_{\beta})$, we have the following coordinate-free expression

$$C^*(v^*X, Q^*h^*Y) = \nabla^*_{v^*X} Q^*h^*Y - Q^*h^*[v^*X, h^*Y], \tag{6.33}$$

for any $X, Y \in \Gamma(TM')$. Similarly, using (6.30b) and taking into account (1.5.41a) and (1.5.34) we obtain

$$\nabla^*_{h^*X} Q^*h^*Y - \nabla^*_{h^*Y} Q^*h^*X = Q^*h^*([h^*X, h^*Y]). \tag{6.34}$$

We are now in a position to state the following important theorem.

THEOREM 6.6 *(Bejancu–Farran [2]). The curvature forms* R^* *and* $\overset{c}{R}$ *of the linear connections* ∇^* *and* $\overset{c}{\nabla}$ *from* $\mathbf{FC}^*$ *and* $I\overset{c}{F}C$ *of* $\mathbb{F}^m$ *respectively, are related by*

$$\begin{aligned} R^*(h^*X\,,\ h^*Y)v^*Z &= \overset{c}{R}(h^*X, h^*Y)v^*Z + (\nabla^*_{h^*X}\overset{c}{D})(Q^*h^*Y, v^*Z) \\ &\quad -(\nabla^*_{h^*Y}\overset{c}{D})(Q^*h^*X, v^*Z) + \overset{c}{D}(Q^*h^*Y, \overset{c}{D}(Q^*h^*X, v^*Z)) \\ &\quad - \overset{c}{D}(Q^*h^*X, \overset{c}{D}(Q^*h^*Y, v^*Z)), \end{aligned} \tag{6.35}$$

$$\begin{aligned} R^*(v^*X, h^*Y)v^*Z &= \overset{c}{R}(v^*X, h^*Y)v^*Z + (\nabla^*_{v^*X}\overset{c}{D})(Q^*h^*Y, v^*Z) \\ &\quad + \overset{c}{D}(C^*(v^*X, Q^*h^*Y), v^*Z), \end{aligned} \tag{6.36}$$

$$R^*(v^*X, v^*Y)v^*Z = \overset{c}{R}(v^*X, v^*Y)v^*Z, \tag{6.37}$$

for any $X, Y, Z \in \Gamma(TM')$.

PROOF. Using (1.3.10) for both R^* and $\overset{c}{R}$, (6.31) and (6.32) we obtain

$$\begin{aligned} R^*(h^*X, h^*Y)v^*Z &= \overset{c}{R}(h^*X, h^*Y)v^*Z + \{\nabla^*_{h^*X}(\overset{c}{D}(Q^*h^*Y, v^*Z)) \\ &\quad - \overset{c}{D}(Q^*h^*Y, \nabla^*_{h^*X}v^*Z)\} - \{\nabla^*_{h^*Y}(\overset{c}{D}(Q^*h^*X, v^*Z)) \\ &\quad - \overset{c}{D}(Q^*h^*X, \nabla^*_{h^*Y}v^*Z)\} + \overset{c}{D}(Q^*h^*Y, \overset{c}{D}(Q^*h^*X, v^*Z)) \\ &\quad - \overset{c}{D}(Q^*h^*X, \overset{c}{D}(Q^*h^*Y, v^*Z)) - \overset{c}{D}(Q^*h^*[h^*X, h^*Y], v^*Z). \end{aligned} \tag{6.38}$$

Then substituting $Q^*h^*[h^*X, h^*Y]$ from (6.34) into (6.38) we deduce (6.35). Similarly, we infer that

$$\begin{aligned} R^*(v^*X, h^*Y)v^*Z &= \overset{c}{R}(v^*X, h^*Y)v^*Z \\ &\quad + \{\nabla^*_{v^*X}(\overset{c}{D}(Q^*h^*Y, v^*Z)) - \overset{c}{D}(Q^*h^*Y, \nabla^*_{v^*X}v^*Z)\} \\ &\quad - \overset{c}{D}(Q^*h^*[v^*X, h^*Y], v^*Z). \end{aligned} \tag{6.39}$$

Thus (6.36) follows from (6.39) via (6.33). Finally, (6.37) is a direct consequence of (6.31). ■

For applications of Finslerian geometry, it is important to express (6.35)–(6.37) by using the local components of Finsler tensor fields involved therein. First we set

$$\overset{c}{D}\left(\frac{\partial}{\partial v^\beta}, \frac{\partial}{\partial v^\alpha}\right) = \overset{c}{D}{}_\alpha{}^\gamma{}_\beta \frac{\partial}{\partial v^\gamma}. \tag{6.40}$$

Then replace h^*X and v^*Y from (6.32) by $\delta^*/\delta^* u^\beta$ and $\partial/\partial v^\alpha$ respectively, and use (6.3a), (1.48), (6.30) and (6.40) to deduce

$$\overset{c}{F}{}_\alpha{}^\gamma{}_\beta = F^{*\gamma}_{\alpha\ \beta} + D^\varepsilon_\beta g_\varepsilon{}^\gamma{}_\alpha - \overset{c}{D}{}_\alpha{}^\gamma{}_\beta\,. \tag{6.41}$$

Comparing (6.41) with (6.21), we infer that the local components of the deformation Finsler tensor field of the pair $(\nabla^*, \overset{c}{\nabla})$ are given by

$$\overset{c}{D}{}_\alpha{}^\gamma{}_\beta = D^\gamma_\mu g_\alpha{}^\mu{}_\beta - D^\mu_\alpha g_\mu{}^\gamma{}_\beta - g_{a\alpha\beta} H^{a\gamma} + g_a{}^\gamma{}_\beta H^a_\alpha. \tag{6.42}$$

Contracting (6.42), in turn, by v^α and v^β we derive

$$(a)\ \ v^\alpha\, \overset{c}{D}{}_\alpha{}^\gamma{}_\beta = D^\gamma_\beta \ \text{ and } \ (b)\ \ \overset{c}{D}{}_\alpha{}^\gamma{}_\beta\, v^\beta = 0, \tag{6.43}$$

that is, the deformation Finsler tensor field of the pair (GM', HM') is obtained from the deformation Finsler tensor field of the pair $(FC^*, I\overset{c}{F}C)$ by a contraction with the supporting element of $\mathbb{F}^m$.

Now, we consider the local components $(\overset{c}{R}{}_\alpha{}^\mu{}_{\beta\gamma}, \overset{c}{P}{}_\alpha{}^\mu{}_{\beta\gamma}, \overset{c}{S}{}_\alpha{}^\mu{}_{\beta\gamma})$ of $\overset{c}{R}$ (cf., (2.13)) and the local components $(R^{*\mu}_{\alpha\ \beta\gamma}, P^{*\mu}_{\alpha\ \beta\gamma}, S^{*\mu}_{\alpha\ \beta\gamma})$ of R^* (cf., (1.7.51)). Then we replace h^*X, h^*Y and v^*Z from (6.35) by $\delta^*/\delta^* u^\gamma$, $\delta^*/\delta^* u^\beta$ and $\partial/\partial v^\alpha$ respectively. Thus using (6.12), we obtain

$$\begin{aligned} R^*{}_{\alpha\varepsilon\beta\gamma} &= \overset{c}{R}_{\alpha\varepsilon\beta\gamma} + \overset{c}{S}_{\alpha\varepsilon\nu\mu}\, D^\nu_\beta D^\mu_\gamma \\ &\quad + \mathcal{A}_{(\beta\gamma)}\{\overset{c}{P}_{\alpha\varepsilon\gamma\mu}\, D^\mu_\beta + \overset{c}{D}_{\alpha\varepsilon\beta|*\gamma} + \overset{c}{D}{}_\alpha{}^\mu{}_\gamma \overset{c}{D}_{\mu\varepsilon\beta}\}. \end{aligned} \tag{6.44}$$

Similarly, (6.36) and (6.37) are locally expressed by

$$P^*_{\alpha\varepsilon\beta\gamma} = \overset{c}{P}_{\alpha\varepsilon\beta\gamma} + \overset{c}{S}_{\alpha\varepsilon\gamma\mu}\, D^\mu_\beta + \overset{c}{D}_{\alpha\varepsilon\beta\|*\gamma} + \overset{c}{D}_{\alpha\varepsilon\mu}\, g_\beta{}^\mu{}_\gamma, \tag{6.45}$$

and

$$S^*_{\alpha\varepsilon\beta\gamma} = \overset{c}{S}_{\alpha\varepsilon\beta\gamma}, \tag{6.46}$$

respectively. Therefore, we may state the following corollary.

COROLLARY 6.2 *The local components* $(R^{*\mu}_{\alpha\ \beta\gamma}, P^{*\mu}_{\alpha\ \beta\gamma}, S^{*\mu}_{\alpha\ \beta\gamma})$ *and* $(\overset{c}{R}_{\alpha}{}^{\mu}{}_{\beta\gamma}, \overset{c}{P}_{\alpha}{}^{\mu}{}_{\beta\gamma}, \overset{c}{S}_{\alpha}{}^{\mu}{}_{\beta\gamma})$ *of the curvature forms* R^* *and* $\overset{c}{R}$ *are related by* (6.44)–(6.46).

Finally, using (6.44)–(6.46) in (3.52) we obtain the Gauss equations for the pseudo–Finsler submanifold $\mathbb{F}^m$ in $\widetilde{\mathbb{F}}^{m+n}$ with respect to the Cartan connections $\mathbf{FC}^*$ and $\widetilde{\mathbf{FC}}^*$ respectively, as follows

$$\begin{aligned}\bar{R}^*_{ij\alpha\beta}B^{ij}_{\gamma\varepsilon} &= R^*_{\gamma\varepsilon\alpha\beta} + S^*_{\gamma\varepsilon\mu\nu}D^{\mu}_{\alpha}D^{\nu}_{\beta} \\ &+\mathcal{A}_{(\alpha\beta)}\{(P^*_{\gamma\varepsilon\alpha\nu} - \overset{c}{D}_{\gamma\varepsilon\alpha\|*\nu} - \overset{c}{D}_{\gamma\varepsilon\mu}\, g_{\alpha}{}^{\mu}{}_{\nu})D^{\nu}_{\beta} + \overset{c}{D}_{\gamma\varepsilon\beta|*\alpha} \\ &+ \overset{c}{D}_{\gamma}{}^{\nu}{}_{\alpha}\overset{c}{D}_{\nu\varepsilon\beta} + \overset{c}{H}{}^{a}{}_{\gamma\beta}\overset{c}{H}_{a\varepsilon\alpha}\},\end{aligned} \tag{6.47}$$

$$\begin{aligned}\bar{P}^*_{ij\alpha\beta}B^{ij}_{\gamma\varepsilon} &= P^*_{\gamma\varepsilon\alpha\beta} + S^*_{\gamma\varepsilon\mu\beta}D^{\mu}_{\alpha} - \overset{c}{D}_{\gamma\varepsilon\alpha\|*\beta} - \overset{c}{D}_{\gamma\varepsilon\mu}\, g_{\alpha}{}^{\mu}{}_{\beta} \\ &+g^{a}{}_{\gamma\beta}\,\overset{c}{H}_{a\varepsilon\alpha} - \overset{c}{H}{}^{a}{}_{\gamma\alpha}\, g_{a\varepsilon\beta},\end{aligned} \tag{6.48}$$

$$\bar{S}^*_{ij\alpha\beta}B^{ij}_{\gamma\varepsilon} = S^*_{\gamma\varepsilon\alpha\beta} + \mathcal{A}_{(\alpha\beta)}\{g^{a}{}_{\gamma\beta}g_{a\varepsilon\alpha}\}. \tag{6.49}$$

We call (6.47)–(6.49) the $(\mathbf{FC}^*, \widetilde{\mathbf{FC}}^*)$**–Gauss equations** for $\mathbb{F}^m$ in $\widetilde{\mathbb{F}}^{m+n}$.

Now, we consider on $\widetilde{\mathbb{F}}^{m+n}$ the Berwald connection $\widetilde{\mathbf{BFC}} = (G\widetilde{M}', \widetilde{\nabla}') = (\widetilde{G}^k_i, \widetilde{G}_i{}^k{}_j, 0)$ (see (4.1)) and look for a comparison between the induced Finsler connection $\overset{b}{IFC} = (HM', \overset{b}{\nabla}) = (N^{\gamma}_{\alpha}, \overset{b}{F}_{\alpha}{}^{\gamma}{}_{\beta}, 0)$ (see (4.4)) and the intrinsic Berwald connection $\mathbf{BFC} = (GM', \nabla') = (G^{\gamma}_{\alpha}, G_{\alpha}{}^{\gamma}{}_{\beta}, 0)$ of $\mathbb{F}^m$, where we set (cf., (4.1))

$$G_{\alpha}{}^{\gamma}{}_{\beta} = \frac{\partial G^{\gamma}_{\beta}}{\partial v^{\alpha}} = \frac{\partial^2 G^{\gamma}}{\partial v^{\alpha}\partial v^{\beta}}. \tag{6.50}$$

First we remark that the v–covariant derivatives with respect to $\overset{b}{IFC}$ and **BFC** coincide (cf., (4.4b)). Then we differentiate (6.5) with respect to v^{β} and using (1.5.42b) with respect to $\mathbb{F}^m$ and (6.50), we obtain

$$G_{\alpha}{}^{\gamma}{}_{\beta} = \overset{b}{F}_{\alpha}{}^{\gamma}{}_{\beta} + \overset{b}{P}{}^{\gamma}{}_{\alpha\beta} + D^{\gamma}_{\alpha\|_{b}\beta}. \tag{6.51}$$

Thus we may state the following theorem.

THEOREM 6.7 *Let $\widetilde{\mathbb{F}}^{m+n}$ be a pseudo–Finsler manifold endowed with the Berwald connection and $\mathbb{F}^m$ be a pseudo–Finsler submanifold of $\widetilde{\mathbb{F}}^{m+n}$. Then the local coefficients of the induced Finsler connection $I\overset{b}{F}C$ and the intrinsic Berwald connection* **BFC** *of $\mathbb{F}^m$ are related by (6.5) and (6.51).*

Moreover we prove the following result.

THEOREM 6.8 *The induced Finsler connection $I\overset{b}{F}C$ coincides with the intrinsic Berwald connection* **BFC** *of $\mathbb{F}^m$ if and only if*

$$\overset{b}{P}{}^{\gamma}{}_{\alpha\beta} = 0, \ \forall \alpha, \beta, \gamma \in \{1, \cdots, m\}. \tag{6.52}$$

PROOF. Suppose $I\overset{b}{F}C=$ **BFC**. Then from (6.5) we obtain $D^{\gamma}_{\alpha} = 0$. Thus (6.52) follows from (6.51) since $G_{\alpha}{}^{\gamma}{}_{\beta} = \overset{b}{F}_{\alpha}{}^{\gamma}{}_{\beta}$. Conversely, suppose (6.52) is satisfied. Contracting (6.52) by v^{α} and using (4.17b) and (6.6) we infer that $D^{\gamma}_{\beta} = 0$. Hence (6.5) and (6.51) yield $I\overset{b}{F}C=$ **BFC**. ■

Taking into account that the induced non–linear connection on $\mathbb{F}^m$ is the same for both the Cartan connection and the Berwald connection, we conclude that $\overset{b}{R}{}^{\gamma}{}_{\alpha\beta} = \overset{c}{R}{}^{\gamma}{}_{\alpha\beta}$ (see (6.27)). On the other hand, by straightforward calculations, using (6.5) and (6.12) we obtain

$$\overset{b}{R}{}^{\gamma}{}_{\alpha\beta} = R^{*\gamma}{}_{\alpha\beta} + \mathcal{A}_{(\alpha\beta)}\{D^{\gamma}_{\beta|_{\mathbf{b}}\alpha} + D^{\mu}_{\alpha}D^{\gamma}_{\beta||_{\mathbf{b}}\mu}\}. \tag{6.53}$$

The linear connections ∇' and $\overset{b}{\nabla}$ from **BFC** and $I\overset{b}{F}C$ of $\mathbb{F}^m$ are related by

$$\nabla'_{v^*X}v^*Y = \overset{b}{\nabla}_{v^*X}\, v^*Y, \tag{6.54}$$

and

$$\nabla'_{h^*X}v^*Y = \overset{b}{\nabla}_{h^*X}\, v^*Y + \overset{b}{D}(Q^*h^*X, v^*Y), \tag{6.55}$$

for any $X, Y \in \Gamma(TM')$. As in the case of the Cartan connection, we call $\overset{b}{D}$ the **deformation Finsler tensor field** of the pair $(\nabla', \overset{b}{\nabla})$. It follows from (1.6.28) that the torsions $T'_{\alpha}{}^{\gamma}{}_{\beta}$ and $L'_{\alpha}{}^{\gamma}{}_{\beta}$ of **BFC** both vanish. Thus by using (1.5.41a), (1.5.41c), (1.5.34) and (1.5.36) we deduce that

$$\nabla'_{h^*X}Q^*h^*Y - \nabla'_{h^*Y}Q^*h^*X = Q^*h^*[h^*X, h^*Y], \tag{6.56}$$

and

$$\nabla'_{v^*X}Q^*h^*Y = Q^*h^*[v^*X, h^*Y], \tag{6.57}$$

for any $X, Y \in \Gamma(TM')$. Then a proof similar to that of Theorem 6.6 enables us to state the following theorem.

THEOREM 6.9 *The curvature forms R' and $\overset{b}{R}$ of the linear connections ∇' and $\overset{b}{\nabla}$ from* **BFC** *and $\overset{b}{IFC}$ of $\mathbb{F}^m$ respectively, are related by*

$$\begin{aligned} & R'(h^* X, h^*Y)v^*Z \\ = \;& \overset{b}{R}(h^*X, h^*Y)v^*Z + (\nabla'_{h^*X}\overset{b}{D})(Q^*h^*Y, v^*Z) \\ & -(\nabla'_{h^*Y}\overset{b}{D})(Q^*h^*X, v^*Z) + \overset{b}{D}(Q^*h^*Y, \overset{b}{D}(Q^*h^*X, v^*Z)) \\ & - \overset{b}{D}(Q^*h^*X, \overset{b}{D}(Q^*h^*Y, v^*Z)), \end{aligned} \tag{6.58}$$

$$R'(v^*X, h^*Y)v^*Z = \overset{b}{R}(v^*X, h^*Y)v^*Z + (\nabla'_{v^*X}\overset{b}{D})(Q^*h^*Y, v^*Z), \tag{6.59}$$

$$R'(v^*X, v^*Y)v^*Z = \overset{b}{R}(v^*X, v^*Y)v^*Z = 0, \tag{6.60}$$

for $X, Y, Z \in \Gamma(TM')$.

In order to get the local expressions of (6.58)–(6.60) we denote by $\overset{b}{D}{}_{\alpha}{}^{\gamma}{}_{\beta}$ the local components of $\overset{b}{D}$. Then, using (6.55), (6.12) and (6.51), we obtain

$$\overset{b}{D}{}_{\alpha}{}^{\gamma}{}_{\beta} = \overset{b}{P}{}^{\gamma}{}_{\alpha\beta} + D^{\gamma}_{\alpha\|_{\mathbf{b}}\beta}. \tag{6.61}$$

Taking into account that $S^*_{\alpha\beta\gamma\varepsilon} = \overset{b}{S}_{\alpha\beta\gamma\varepsilon} = 0$, (6.58) and (6.59) have the local expressions:

$$H_{\alpha\varepsilon\beta\gamma} = \overset{b}{R}_{\alpha\varepsilon\beta\gamma} + \mathcal{A}_{(\beta\gamma)}\{\overset{b}{P}_{\alpha\varepsilon\gamma\mu}\, D^{\mu}_{\beta} + \overset{b}{D}_{\alpha\varepsilon\beta|_{\mathbf{b}}\gamma} + \overset{b}{D}{}_{\alpha}{}^{\mu}{}_{\gamma}\overset{b}{D}_{\mu\varepsilon\beta}\}, \tag{6.62}$$

and

$$G_{\alpha\varepsilon\beta\gamma} = \overset{b}{P}_{\alpha\varepsilon\beta\gamma} + \overset{b}{D}_{\alpha\varepsilon\beta\|_{\mathbf{b}}\gamma}, \tag{6.63}$$

respectively. Thus the Gauss equations (4.21) can be rewritten using the local components of the curvature forms of **BFC** and $\widetilde{\mathbf{BFC}}$ as follows:

$$\left\{ \begin{aligned} \bar{R}_{ij\alpha\beta}B^{ij}_{\gamma\varepsilon} \;&=\; H_{\gamma\varepsilon\alpha\beta} + \mathcal{A}_{(\alpha\beta)}\{(G_{\gamma\varepsilon\alpha\mu} - \overset{b}{D}_{\gamma\varepsilon\alpha\|_{\mathbf{b}}\mu})D^{\mu}_{\beta} \\ & \quad + \overset{b}{D}_{\gamma\varepsilon\beta|_{\mathbf{b}}\alpha} + \overset{b}{D}{}_{\gamma}{}^{\mu}{}_{\alpha}\overset{b}{D}_{\mu\varepsilon\beta} + \overset{b}{H}{}^{a}{}_{\gamma\beta}H'_{a\varepsilon\alpha}\} \quad (a) \\ \bar{P}_{ij\alpha\beta}B^{ij}_{\gamma\varepsilon} \;&=\; G_{\gamma\varepsilon\alpha\beta} - \overset{b}{D}_{\gamma\varepsilon\alpha\|_{\mathbf{b}}\beta} - 2\,\overset{b}{H}{}^{\alpha}{}_{\gamma\alpha}\, g_{a\varepsilon\beta}. \quad (b) \end{aligned} \right. \tag{6.64}$$

Finally, we consider on $\widetilde{\mathbb{F}}^{m+n}$, the Rund connection

$$\widetilde{\mathbf{RFC}} = (G\widetilde{M}', \widetilde{\nabla}'') = (\widetilde{G}_i^k, \widetilde{F}_{i\ j}^{*k}, 0),$$

where $\widetilde{F}_{i\ j}^{*k}$ are given by (3.17). It is our purpose to compare the induced Finsler connection

$$I\overset{\mathrm{r}}{F}C = (HM', \overset{\mathrm{r}}{\nabla}) = (N_\alpha^\gamma, \overset{\mathrm{r}}{F}_{\alpha\ \beta}^{\ \gamma}, 0)$$

(see (5.2)) and the intrinsic Rund connection

$$\mathbf{RFC} = (GM, \nabla'') = (G_\alpha^\gamma, F_{\alpha\ \beta}^{*\gamma}, 0)$$

(see (6.4a)).

Clearly the v–covariant derivatives with respect to $I\overset{\mathrm{r}}{F}C$ and $\mathbf{RFC}$ are the same. Comparing (3.35) with (5.2a) we deduce that

$$\overset{\mathrm{c}}{F}_{\alpha\ \beta}^{\ \gamma} = \overset{\mathrm{r}}{F}_{\alpha\ \beta}^{\ \gamma} + g_{a\ \alpha}^{\ \gamma} H_\beta^a. \tag{6.65}$$

Then, using (6.65) in (6.21), we obtain

$$\begin{aligned} \overset{\mathrm{r}}{F}_{\alpha\ \beta}^{\ \gamma} &= F_{\alpha\ \beta}^{*\gamma} + D_\alpha^\mu g_{\mu\ \beta}^{\ \gamma} + D_\beta^\mu g_{\mu\ \alpha}^{\ \gamma} - D^{\gamma\mu} g_{\mu\alpha\beta} \\ &\quad + g_{a\alpha\beta} H^{a\gamma} - g_{a\ \beta}^{\ \gamma} H_\alpha^a - g_{a\ \alpha}^{\ \gamma} H_\beta^a. \end{aligned} \tag{6.66}$$

Now we can prove the following theorem.

THEOREM 6.10 *The induced Finsler connection $I\overset{\mathrm{r}}{F}C$ coincides with the intrinsic Rund connection* $\mathbf{RFC}$ *of $\mathbb{F}^m$ if and only if*

$$g_{a\ \alpha}^{\ \gamma} H_\beta^a = 0, \quad \forall \alpha, \beta, \gamma \in \{1, \cdots, m\}. \tag{6.67}$$

PROOF. Suppose $I\overset{\mathrm{r}}{F}C = \mathbf{RFC}$. Hence $\delta^*/\delta^* u^\alpha = \delta/\delta u^\alpha$ and $\overset{\mathrm{r}}{F}_{\alpha\ \beta}^{\ \gamma} = F_{\alpha\ \beta}^{*\gamma}$, which yield $g_{\alpha\beta|_\gamma} = g_{\alpha\beta|\mathbf{r}\gamma} = 0$. Thus from (5.5a) we get (6.67). Conversely, suppose (6.67) is satisfied. Then contracting it by v^β we infer that $D_\alpha^\gamma = 0$. Thus (6.66) and (6.5) imply $I\overset{\mathrm{r}}{F}C = \mathbf{RFC}$. ∎

As in the case of the Cartan and Berwald connections, the linear connections ∇'' and $\overset{\mathrm{r}}{\nabla}$ from $\mathbf{RFC}$ and $I\overset{\mathrm{r}}{F}C$ of $\mathbb{F}^m$ are related by

$$\nabla''_{v^*X} v^*Y = \overset{\mathrm{r}}{\nabla}_{v^*X}\, v^*Y, \tag{6.68}$$

and

$$\nabla''_{h^*X} v^*Y = \overset{\mathrm{r}}{\nabla}_{h^*X}\, v^*Y + \overset{\mathrm{r}}{D}\,(Q^*h^*X, v^*Y), \tag{6.69}$$

for any $X, Y \in \Gamma(TM')$, where we call $\overset{r}{D}$ the **deformation Finsler tensor field** of the pair $(\nabla'', \overset{r}{\nabla})$. We also write the following relations

$$\begin{cases} \nabla''_{h^*X} Q^*h^*Y - \nabla''_{h^*Y} Q^*h^*X = Q^*h^*[h^*X, h^*Y] \\ \nabla''_{v^*X} Q^*h^*Y = Q^*h^*[v^*X, h^*Y]. \end{cases}$$

Finally, we state the following theorem:

THEOREM 6.11 *The curvature forms R'' and $\overset{r}{R}$ of the linear connections ∇'' and $\overset{r}{\nabla}$ from* **RFC** *and* $I\overset{r}{F}C$ *of $\mathbb{F}^m$ respectively, are related by*

$$\begin{aligned} R''(h^*X, h^*Y)v^*Z &= \overset{r}{R}(h^*X, h^*Y)\, v^*Z + (\nabla''_{h^*X} \overset{r}{D})(Q^*h^*Y, v^*Z) \\ &\quad -(\nabla''_{h^*Y} \overset{r}{D})(Q^*h^*X, v^*Z) + \overset{r}{D}(Q^*h^*Y, \overset{r}{D}(Q^*h^*X, v^*Z)) \\ &\quad - \overset{r}{D}(Q^*h^*X, \overset{r}{D}(Q^*h^*Y, v^*Z)), \end{aligned} \tag{6.70}$$

$$R''(v^*X, h^*Y)v^*Z = \overset{r}{R}(v^*X, h^*Y)v^*Z + (\nabla''_{v^*X} \overset{r}{D})(Q^*h^*Y, v^*Z), \tag{6.71}$$

$$R''(v^*X, v^*Y)v^*Z = \overset{r}{R}(v^*X, v^*Y)v^*Z = 0, \tag{6.72}$$

for any $X, Y, Z \in \Gamma(TM')$.

We denote by $\overset{r}{D}_{\alpha}{}^{\gamma}{}_{\beta}$ the local components of $\overset{r}{D}$, and obtain

$$\begin{aligned} \overset{r}{D}_{\alpha}{}^{\gamma}{}_{\beta} &= D^{\gamma\mu} g_{\mu\alpha\beta} - D^{\mu}_{\alpha} g_{\mu}{}^{\gamma}{}_{\beta} - D^{\mu}_{\beta} g_{\mu}{}^{\gamma}{}_{\alpha} - g_{a\alpha\beta} H^{a\gamma} \\ &\quad + g_{a}{}^{\gamma}{}_{\beta} H^{a}_{\alpha} + g_{a}{}^{\gamma}{}_{\alpha} H^{a}_{\beta}. \end{aligned} \tag{6.73}$$

As in the case of the Berwald connection, we have $S^*_{\alpha\beta\gamma\varepsilon} = \overset{r}{S}_{\alpha\beta\gamma\varepsilon} = 0$, and (6.70) and (6.71) have the local expressions:

$$K_{\alpha\varepsilon\beta\gamma} = \overset{r}{R}_{\alpha\varepsilon\beta\gamma} + \mathcal{A}_{(\beta\gamma)}\{\overset{r}{P}_{\alpha\varepsilon\gamma\mu}\, D^{\mu}_{\beta} + \overset{r}{D}_{\alpha\varepsilon\beta|\mathbf{r}\gamma} + \overset{r}{D}_{\alpha}{}^{\mu}{}_{\gamma} \overset{r}{D}_{\mu\varepsilon\beta}\}, \tag{6.74}$$

and

$$F_{\alpha\varepsilon\beta\gamma} = \overset{r}{P}_{\alpha\varepsilon\beta\gamma} + \overset{r}{D}_{\alpha\varepsilon\beta\|\mathbf{r}\gamma}, \tag{6.75}$$

respectively. Thus the Gauss equations (5.15) can be expressed by using the curvature Finsler tensor fields of the Rund connections of $\widetilde{\mathbb{F}}^{m+n}$ and $\mathbb{F}^m$ as follows

$$\begin{cases} \bar{R}_{ij\alpha\beta}B^{ij}_{\gamma\varepsilon} = K_{\gamma\varepsilon\alpha\beta} + \mathcal{A}_{(\alpha\beta)}\{(F_{\gamma\varepsilon\alpha\mu} - \overset{r}{D}_{\gamma\varepsilon\alpha\|\mathbf{r}\mu})D^{\mu}_{\beta} \\ \qquad + \overset{r}{D}_{\gamma\varepsilon\beta|\mathbf{r}\alpha} + \overset{r}{D}_{\gamma}{}^{\mu}{}_{\alpha}\overset{r}{D}_{\mu\varepsilon\beta} + \overset{r}{H}{}^{a}{}_{\gamma\beta}\overset{r}{H}{}'_{a\varepsilon\alpha}\} \quad (a) \\ \bar{P}_{ij\alpha\beta}B^{ij}_{\gamma\varepsilon} = F_{\gamma\varepsilon\alpha\beta} - \overset{r}{D}_{\gamma\varepsilon\alpha\|\mathbf{r}\beta} - 2\,\overset{r}{H}{}^{a}{}_{\gamma\alpha}\, g_{a\varepsilon\beta}. \quad (b) \end{cases} \tag{6.76}$$

The above study shows a striking similarity between the equations we obtained from $\widetilde{\mathbf{BFC}}$ on the one hand, and those from $\widetilde{\mathbf{RFC}}$ on the other hand (compare (6.62)–(6.64) with (6.74)–(6.76)). However, this similarity is only formal. This is because the geometric objects involved in these equations are quite different from each other (compare $\overset{b}{D}_{\alpha}{}^{\gamma}{}_{\beta}$ with $\overset{r}{D}_{\alpha}{}^{\gamma}{}_{\beta}$).

We close the Chapter with a short history of Finsler submanifolds. In 1918 Finsler [1] initiated a study of curves and surfaces in a space (manifold) endowed with a generalized metric. Thus we may claim that the theory of Finsler submanifolds is as old as Finsler geometry itself. However, a systematic study of the geometry of Finsler submanifolds was developed after 1934, when E. Cartan [1] published his point of view on Finsler geometry by introducing a new connection, later called the Cartan connection. In this respect we should remark the important research work carried out by Haimovici [1]–[4], Hombu [1], Wegener [1], [2], Varga [1], Davies [1], Kikuchi [1] and Rapcsàk [1]. A different approach was taken by Berwald [3], who made use of a non–metric connection, later called the Berwald connection. An important merit of Berwald's approach is that the induced second fundamental form is symmetric; thus principal curvatures, mean curvature and Gaussian curvature of a Finsler surface have been defined as in the case of Euclidean geometry. The two viewpoints were presented in Chapter V of Rund's book [3]. Also, Rund [1], [2], [4], [5], based on both the Cartan connection and his connection (called the Rund connection in our book), presented a comparison between induced and intrinsic theories of Finsler submanifolds and obtained results on curvatures of Finsler hypersurfaces. These problems have been also investigated by Brown [1], [2], Varga [2]–[5] and Comic [1]–[3].

The theory of Finsler submanifolds had an important return in 1984 when Matsumoto [4] published his excellent memoir on the basic results of the theory of Finsler hypersurfaces along with his new results. Then, new approaches to the theory of Finsler submanifolds were developed by Miron [2], Miron and Bejancu [1], [2], Bejancu [1]–[7], Dragomir [1], [2], Shen [1] and Yasuda [1]–[3].

CHAPTER 3

SPECIAL IMMERSIONS OF PSEUDO–FINSLER MANIFOLDS

The present chapter deals with the geometry of some special immersions of pseudo–Finsler manifolds. We study the geometry of an $\mathbb{F}^m$ immersed in $\widetilde{\mathbb{F}}^{m+n}$ with additional conditions on $\mathbb{F}^m$ and/or $\widetilde{\mathbb{F}}^{m+n}$. More precisely, we first present the main results on totally geodesic pseudo–Finsler submanifolds. Then we study some parallel vector bundles over pseudo–Finsler submanifolds and present the main characteristics of a Finsler immersion in a C–reducible Finsler manifold.

1. Totally Geodesic Pseudo–Finsler Submanifolds

In the first part of this section we present new characterizations of geodesics of a pseudo–Finsler manifold. Actually, these characterizations arise as a direct consequence of our general method of studying the geometry of pseudo–Finsler submanifolds via the vertical vector bundle. In the second part of the section we stress the importance of second fundamental forms, normal curvatures and normal curvature vector fields for the study of the geometry of totally geodesic pseudo–Finsler submanifolds.

Let $\widetilde{\mathbb{F}}^{m+n} = (\widetilde{M}, \widetilde{M}', \widetilde{F}^*)$ be a pseudo–Finsler manifold and $\mathbb{F}^1 = (\mathcal{C}, \mathcal{C}', F_1^*)$ be a 1–dimensional pseudo–Finsler submanifold of $\widetilde{\mathbb{F}}^{m+n}$, where $\mathcal{C}$ is locally given by the equations

$$x^i = x^i(s), \quad i \in \{1, \cdots, m+n\}, \quad s \in (a, b), \tag{1.1}$$

s being the arc length parameter on $\mathcal{C}$ (see (1.6.4)). We consider (s, v) as local coordinates on $\mathcal{C}'$ and thus (2.1.5) and (2.1.6) become

$$\frac{\partial}{\partial s} = \frac{dx^i}{ds}\frac{\partial}{\partial x^i} + v\frac{d^2x^i}{ds^2}\frac{\partial}{\partial y^i}, \tag{1.2}$$

and

$$\frac{\partial}{\partial v} = \frac{dx^i}{ds}\frac{\partial}{\partial y^i}, \tag{1.3}$$

respectively. We say that $\mathcal{C}$ is a **spacelike, timelike or lightlike (null) curve** of $\widetilde{\mathbb{F}}^{m+n}$, according as $\partial/\partial v$ from (1.3) is spacelike, timelike or lightlike (null), respectively. Now we suppose that $\mathcal{C}$ is a spacelike curve. Then by (1.1.14) and (1.6.6) we deduce that

$$\tilde{g}_{ij}(x(s), x'(s))\frac{dx^i}{ds}\frac{dx^j}{ds} = 1, \tag{1.4}$$

which is equivalent to (see (1.2.11))

$$\tilde{g}\left(\frac{\partial}{\partial v}, \frac{\partial}{\partial v}\right) = 1. \tag{1.5}$$

The induced non-linear connection $\widetilde{HC}'$ on $\mathcal{C}'$ by the canonical non-linear connection $\widetilde{GM}'$ of $\widetilde{\mathbb{F}}^{m+n}$ (see(2.1.26)) is given by

$$\frac{\tilde{\delta}}{\tilde{\delta} s} = \frac{dx^i}{ds}\frac{\delta^*}{\delta^* x^i} + \widetilde{H}^A W_A, \tag{1.6}$$

where $\{\widetilde{H}^A\}$, $A \in \{2, \cdots, m+n\}$ are the local components of the normal curvature forms of $\mathbb{F}^1$ with respect to the orthonormal basis $\{W_A\}$ of the pseudo-Finsler normal bundle $\widetilde{V}\mathcal{C}'^{\perp}$. By using (2.1.27), (2.1.2), and (1.6.11) for $\mathbb{F}^1$ in $\widetilde{\mathbb{F}}^{m+n}$, we obtain

$$\widetilde{H}^A = \widetilde{W}_i^A\left(\frac{d^2x^i}{ds^2} + 2\widetilde{G}^i(x(s), x'(s))\right)v. \tag{1.7}$$

On the other hand, (2.1.20) becomes

$$\frac{\tilde{\delta}}{\tilde{\delta} s} = \frac{\partial}{\partial s} - \tilde{f}(s, v)\frac{\partial}{\partial v}, \tag{1.8}$$

where $\tilde{f}$ is the function that locally defines $\widetilde{HC}'$. Then by using (2.1.21), (2.1.13), (1.6.30) and (1.6.11) we infer that

$$\tilde{f}(s, v) = v\tilde{g}_{kj}(x(s), x'(s))\left(\frac{d^2x^k}{ds^2} + 2\widetilde{G}^k(x(s), x'(s))\right)\frac{dx^j}{ds}. \tag{1.9}$$

Next, we consider the Cartan connection

$$\widetilde{\mathbf{FC}}^* = (\widetilde{GM}', \widetilde{\nabla}^*) = (\widetilde{G}_i^k, \widetilde{F}_i{}^{*k}{}_j, \tilde{g}_i{}^k{}_j),$$

and by using (1.3) and (1.6.22) we derive

$$\widetilde{\nabla}^{*}_{\partial/\partial v}\frac{\partial}{\partial v}=0. \tag{1.10}$$

Thus (1.8) and (1.10) yield

$$\widetilde{\nabla}^{*}_{\tilde{\delta}/\tilde{\delta}s}\frac{\partial}{\partial v}=\widetilde{\nabla}^{*}_{\partial/\partial s}\frac{\partial}{\partial v}. \tag{1.11}$$

On the other hand, by using (1.2), (1.6.18) and (1.6.11) we deduce that

$$\frac{\partial}{\partial s}=\frac{dx^{i}}{ds}\frac{\delta^{*}}{\delta^{*}x^{i}}+v\left(\frac{d^{2}x^{k}}{ds^{2}}+2\widetilde{G}^{k}(x(s),x'(s))\right)\frac{\partial}{\partial y^{k}}. \tag{1.12}$$

Then, by direct calculations using (1.12), (1.3), (1.7.29), (1.6.22) and (1.6.11) we obtain

$$\widetilde{\nabla}^{*}_{\partial/\partial s}\frac{\partial}{\partial v}=\left(\frac{d^{2}x^{k}}{ds^{2}}+2\widetilde{G}^{k}(x(s),x'(s))\right)\frac{\partial}{\partial y^{k}}. \tag{1.13}$$

As $\widetilde{\nabla}^{*}$ is a metric linear connection, it follows from (1.5) that

$$\tilde{g}\left(\widetilde{\nabla}^{*}_{\partial/\partial s}\frac{\partial}{\partial v},\frac{\partial}{\partial v}\right)=0. \tag{1.14}$$

Finally, using (1.13) and (1.3) in (1.14) and taking (1.9) into account, we obtain that $f=0$ on C'. Hence we may state the following result.

PROPOSITION 1.1 *The induced non–linear connection on $\mathbb{F}^{1}$ by the canonical non–linear connection of $\widetilde{\mathbb{F}}^{m+n}$ is locally spanned by $\partial/\partial s$ given by the equivalent formulas (1.2) and (1.12).*

Further, we note that in case of the immersion of $\mathbb{F}^{1}$ in $\widetilde{\mathbb{F}}^{m+n}$, (2.1.14) becomes

$$\frac{\partial}{\partial y^{k}}=S_{k}\frac{\partial}{\partial v}+\widetilde{W}^{A}_{k}W_{A}. \tag{1.15}$$

where $\left[S_{k}\ \widetilde{W}^{A}_{k}\right]$ is the inverse matrix of $\left[\frac{dx^{i}}{ds}\ W^{i}_{A}\right]$. Substituting from (1.15) into (1.13) and taking into account (1.14) and (1.7) we derive

$$\widetilde{\nabla}^{*}_{\partial/\partial s}\frac{\partial}{\partial v}=\frac{1}{v}\widetilde{H}^{A}W_{A}. \tag{1.16}$$

Finally, we recall that C is a non–degenerate geodesic of $\widetilde{\mathbb{F}}^{m+n}$ (see (1.6.10)) if and only if

$$\frac{d^{2}x^{i}}{ds^{2}}+2\widetilde{G}^{i}(x,x'(s))=0. \tag{1.17}$$

Remark 1.1. Let (x_0^i, y_0^i) be a fixed point of $\widetilde{M}'$. Then there exists a unique non–degenerate geodesic C passing through (x_0^i) such that (y_0^i) is tangent to C at (x_0^i). This is a direct consequence of the existence and uniqueness theorem for the solutions of the system of differential equations (1.17). ■

New characterizations for spacelike geodesics of $\widetilde{\mathbb{F}}^{m+n}$ are given in the following theorem.

THEOREM 1.1 *Let C be a spacelike curve in $\widetilde{\mathbb{F}}^{m+n}$. Then the following assertions are equivalent:*

(i) C is a spacelike geodesic of $\widetilde{\mathbb{F}}^{m+n}$;

(ii) All normal curvature forms of $\mathbb{F}^1$ in $\widetilde{\mathbb{F}}^{m+n}$ vanish;

(iii) The vector field $\partial/\partial v$ is h–parallel with respect to $\widetilde{\nabla}^*$, i.e., we have

$$\widetilde{\nabla}^*_{\partial/\partial s}\frac{\partial}{\partial v} = 0. \tag{1.18}$$

PROOF. (i)⇒ (ii). By using (1.17) in (1.7) we obtain $\widetilde{H}^A = 0$ for any $A \in \{2, \cdots, m+n\}$.

(ii) ⇒ (iii). As $\widetilde{H}^A = 0$, from (1.16) we obtain (1.18).

(iii) ⇒ (i). By using (1.18) in (1.13) we derive (1.17). ■

Now, let $\mathbb{F}^m = (M, M', F^*)$ be a pseudo–Finsler submanifold of $\widetilde{\mathbb{F}}^{m+n}$ and C be a spacelike curve in M. Suppose C is locally given by the equations

$$(a) \quad u^\alpha = u^\alpha(s) \quad \text{or} \quad (b) \quad x^i = x^i(u^\alpha(s)) = x^i(s), \; s \in (a,b). \tag{1.19}$$

Then by the general formula (2.1.2) we deduce that

$$(a) \quad v^\alpha = \frac{du^\alpha}{ds}v \quad \text{and} \quad (b) \quad y^i = \frac{dx^i}{ds}v. \tag{1.20}$$

Consider the induced non–linear connection $HM' = (N_\alpha^\gamma)$ on M' by $G\widetilde{M}' = (\widetilde{G}_i^k)$. Then HM' will define an induced non–linear connection HC' on C' locally represented by

$$\frac{\delta}{\delta s} = \frac{\partial}{\partial s} - f(s,v)\frac{\partial}{\partial v}.$$

As the induced Finsler connection $I\overset{c}{F}C = (HM', \overset{c}{\nabla})$ is a metric Finsler connection (see (iii) of Theorem 2.3.2), by repeating the calculations done in order to prove that $\tilde{f} = 0$, we obtain $f = 0$ on C'. Thus we may state the following proposition.

PROPOSITION 1.2 *Let C be a spacelike curve in $\mathbb{F}^m$. Then the induced non-linear connections HC' and $\widetilde{H}C'$ coincide.*

Next, we consider the pseudo–Finsler normal bundles $VC'^{\perp}$ and $\widetilde{V}C'^{\perp}$ for the immersion of C in M and $\widetilde{M}$ respectively. Then we take $\{B_a\}$, $a \in \{m+1, \cdots, m+n\}$ and $\{V_r\}$, $r \in \{2, \cdots, m\}$ as orthonormal basis in $\Gamma(VM'^{\perp})$ and $\Gamma(VC'^{\perp})$ respectively. Thus we obtain the orthonormal basis $\{V_r, B_a\}$ in $\Gamma(\widetilde{V}C'^{\perp})$. As HC' and $\widetilde{H}C'$ are both locally represented by $\partial/\partial s$, by using (2.1.26) we have

$$\frac{\partial}{\partial s} = \frac{du^\alpha}{ds}\frac{\delta}{\delta u^\alpha} + H^r V_r, \tag{1.21}$$

and

$$\frac{\partial}{\partial s} = \frac{dx^i}{ds}\frac{\delta^*}{\delta^* x^i} + \widetilde{H}^r V_r + H^{*a} B_a, \tag{1.22}$$

where H^r and $\{\widetilde{H}^r, H^{*a}\}$ are the normal curvature forms of the immersion of C' in M and $\widetilde{M}$ respectively. By substituting $\delta/\delta u^\alpha$ from (2.1.26) into (1.21) we infer that

$$\frac{\partial}{\partial s} = \frac{dx^i}{ds}\frac{\delta^*}{\delta^* x^i} + H^r V_r + \frac{du^\alpha}{ds} H^a_\alpha B_a, \tag{1.23}$$

where (H^a_α) are the local components of the normal curvature forms (see (2.3.24b)). Then by comparing (1.22) and (1.23) we deduce that

$$(a)\quad H^r = \widetilde{H}^r \quad \text{and} \quad (b)\quad H^{*a} = \frac{du^\alpha}{ds} H^a_\alpha. \tag{1.24}$$

Remark 1.2. All the above results on spacelike curves are also true for timelike curves. However, for lightlike curves we need a theory of lightlike Finsler submanifolds, which is beyond the scope of the present book. ■

Next, we suppose that C is an arbitrary smooth curve in M given by parametric equations $u^\alpha = u^\alpha(t)$. Then $x^i = x^i(u(t))$ and

$$(a)\quad \frac{dx^i}{dt} = B^i_\alpha \frac{du^\alpha}{dt}; \quad \text{and} \quad (b)\quad \frac{d^2x^i}{dt^2} = B^i_{\alpha\beta}\frac{du^\alpha}{dt}\frac{du^\beta}{dt} + B^i_\alpha \frac{d^2u^\alpha}{dt^2}. \tag{1.25}$$

Now, we set

$$H_0^a(u,v) = H_\alpha^a(u,v)v^\alpha, \tag{1.26}$$

and prove the following lemma.

LEMMA 1.1 *Let $\mathbb{F}^m = (M, M', F^*)$ be a pseudo–Finsler submanifold of $\widetilde{\mathbb{F}}^{m+n}$ and C be a smooth curve in M. Then we have*

$$\begin{aligned}\frac{d^2x^i}{dt^2} + 2\widetilde{G}^i(x(t), x'(t)) &= \left(\frac{d^2u^\alpha}{dt^2} + 2G^\alpha(u(t), u'(t))\right) B_\alpha^i \\ &\quad + H_0^a(u(t), u'(t)) B_a^i.\end{aligned} \tag{1.27}$$

PROOF. By direct calculations using (2.6.10), (1.7.29), and (1.25) we deduce that

$$\frac{d^2x^i}{dt^2} + 2\widetilde{G}^i(x(t), x'(t)) = \frac{d^2u^\alpha}{dt^2} B_\alpha^i + (B_{\alpha\beta}^i + \widetilde{F}_j{}^{*i}{}_k B_{\alpha\beta}^{jk}) \frac{du^\alpha}{dt}\frac{du^\beta}{dt}. \tag{1.28}$$

Contracting (2.5.2a) and (2.5.3a) by B_γ^i and B_a^i, respectively, and then adding the results we obtain

$$\overset{r}{F}{}_\alpha{}^\gamma{}_\beta B_\gamma^i + \overset{r}{H}{}_{\alpha\beta}^a B_a^i = B_{\alpha\beta}^i + \widetilde{F}_j{}^{*i}{}_k B_{\alpha\beta}^{jk}. \tag{1.29}$$

Finally, by using (1.29) in (1.28) and taking into account (2.5.4), (2.6.14) and (1.26) we derive (1.27). ■

In Section 1.6 we defined non–degenerate geodesics as extremals of the length integral and found for them the equations (1.6.10). To cover the degenerate case, we give here a general definition using an approach similar to that in pseudo–Riemannian geometry.

First, we say that a curve C in $\widetilde{M}$ is an h–**autoparallel curve** of

$$\widetilde{\mathbb{F}}^{m+n} = (\widetilde{M}, \widetilde{M}', \widetilde{F}^*)$$

with respect to the Cartan connection $\widetilde{\mathbf{FC}}^* = (G\widetilde{M}', \widetilde{\nabla}^*)$ of $\widetilde{\mathbb{F}}^{m+n}$ if we have

$$\widetilde{\nabla}^*_{\partial/\partial s} \frac{\partial}{\partial v} = 0.$$

By using (1.3), (1.12) and the local coefficients of $\widetilde{\nabla}^*$ we deduce that C is an h–autoparallel if and only if

$$\frac{d^2x^i}{ds^2} + \widetilde{F}_j{}^{*i}{}_k \frac{dx^j}{ds}\frac{dx^k}{ds} = 0.$$

By using (2.3.23) and the homogeneity of $\widetilde{G}^i$ it is easy to check that the above equations for h–autoparallel curves coincide with the equations (1.6.10) whose non–degenerate solutions are non–degenerate geodesics of $\widetilde{\mathbb{F}}^{m+n}$. This inspires the definition that, in general, a curve C in $\widetilde{M}$ is a **geodesic** of $\widetilde{\mathbb{F}}^{m+n}$ if it is an h–autoparallel curve with respect to $\widetilde{\mathbf{FC}}^*$. It is remarkable that h–autoparallels with respect to the other two classical Finsler connections, that is, the Berwald and Rund connections, coincide with the h–autoparallels with respect to $\widetilde{\mathbf{FC}}^*$. This is owed to the equalities (see (1.6.30) and (2.3.23))

$$\widetilde{G}_i{}^k{}_j y^i = \widetilde{F}_i^{*k}{}_j y^i = \widetilde{G}^k_j.$$

Thus from now on any solution of the system of differential equations (1.6.10) is a geodesic of $\widetilde{\mathbb{F}}^{m+n}$.

Next, we say that the pseudo–Finsler submanifold $\mathbb{F}^m = (M, M', F^*)$ is **geodesic** at $u \in M$ if every geodesic C of $\widetilde{\mathbb{F}}^{m+n}$ given by $x^i = x^i(s)$ with $x(0) = u$ and $x'(0) \in T_uM$ lies in M on some interval $(-\varepsilon, \varepsilon)$. Then $\mathbb{F}^m$ is said to be **totally geodesic** if it is geodesic at all points of M. It is easy to see that $\mathbb{F}^m$ is totally geodesic if and only if any geodesic of $\mathbb{F}^m$ is a geodesic of $\widetilde{\mathbb{F}}^{m+n}$.

In order to obtain a characterization of the class of totally geodesic pseudo–Finsler submanifolds, we set:

$$\mathbf{n}^a(u, v) = H^a_0(u, v)/F^*(u, v) \tag{1.30}$$

and call $\mathbf{n}^a$, $a \in \{m+1, \cdots, m+n\}$ the **normal curvatures** of $\mathbb{F}^m$. By the classical theory of curves on surfaces (see Spivak [1], Vol. 3, p. 272), it is natural to consider H^a_0 from (1.26) as normal curvatures of C. Clearly that H^a_0 and $\mathbf{n}^a$ measure the deviation of a geodesic of $\mathbb{F}^m$ from being a geodesic of $\widetilde{\mathbb{F}}^{m+n}$. This justifies giving the name 'normal curvatures' to $\mathbf{n}^a$.

THEOREM 1.2 *(Bejancu [7]). $\mathbb{F}^m$ is totally geodesic immersed in $\widetilde{\mathbb{F}}^{m+n}$, if and only if, all normal curvatures of $\mathbb{F}^m$ vanish on M'.*

PROOF. Take $(u_0, v_0) \in M'$ and suppose $\mathbb{F}^m$ is totally geodesic. Consider the geodesic C of $\mathbb{F}^m$ with initial conditions (u_0, v_0) (see Remark 1.1). Then C is a geodesic of $\widetilde{\mathbb{F}}^{m+n}$ too. Thus by using (1.17) for both $\mathbb{F}^m$ and $\widetilde{\mathbb{F}}^{m+n}$ in (1.27), we obtain $\mathbf{n}^a = 0$ for any $a \in \{m+1, \cdots, m+n\}$. The converse follows also from (1.27) via (1.17). ■

As we know from (2.3.24), both N^γ_α and H^a_α are induced by the canonical non–linear connection $G\widetilde{M}' = (\widetilde{G}^k_i)$ of $\widetilde{\mathbb{F}}^{m+n}$. Moreover, contracting (2.3.24a) and (2.3.24b) by B^k_γ and B^k_a respectively, and then, adding the results, we obtain

$$B^k_\gamma N^\gamma_\alpha + B^k_a H^a_\alpha = B^k_{\alpha 0} + B^i_\alpha \widetilde{G}^k_i. \tag{1.31}$$

We now define

$$\mathbf{n}_\alpha = H^a_\alpha B_a, \quad \alpha \in \{1, \cdots, m\}, \tag{1.32}$$

and call them the **normal curvature vector fields** of $\mathbb{F}^m$. From (1.30) and (1.32) we see that the vanishing of $\mathbf{n}_\alpha$ implies the vanishing of $\mathbf{n}^a$. The converse of this assertion is proved in the next proposition.

PROPOSITION 1.3 *Let $\mathbb{F}^m$ be a pseudo–Finsler submanifold of $\widetilde{\mathbb{F}}^{m+n}$ such that all its normal curvatures vanish on M'. Then all normal curvature vector fields of $\mathbb{F}^m$ vanish on M'.*

PROOF. First, differentiating (1.31) with respect to v^β and using (2.4.1), we obtain

$$B^k_\gamma \frac{\partial N^\gamma_\alpha}{\partial v^\beta} + \frac{\partial B^k_a}{\partial v^\beta} H^a_\alpha + B^k_a \frac{\partial H^a_\alpha}{\partial v^\beta} = B^k_{\alpha\beta} + \widetilde{G}_i{}^k{}_j B^{ij}_{\alpha\beta}. \tag{1.33}$$

Then, contracting (1.33) by $\widetilde{B}^b_k$ and taking into account (2.1.11c) and (2.1.11d), we infer that

$$\frac{\partial H^b_\alpha}{\partial v^\beta} = \widetilde{B}^b_k (B^k_{\alpha\beta} + \widetilde{G}_i{}^k{}_j B^{ij}_{\alpha\beta}) - \widetilde{B}^b_k \frac{\partial B^k_a}{\partial v^\beta} H^a_\alpha. \tag{1.34}$$

Further, contracting (1.34) with v^α and using (2.1.2a), (1.6.30), (2.3.24b) and (1.26), we deduce that

$$\frac{\partial H^b_\alpha}{\partial v^\beta} v^\alpha = H^b_\beta - \widetilde{B}^b_k \frac{\partial B^k_a}{\partial v^\beta} H^a_0. \tag{1.35}$$

On the other hand, from (1.26) we obtain

$$\frac{\partial H^b_0}{\partial v^\beta} = \frac{\partial H^b_\alpha}{\partial v^\beta} v^\alpha + H^b_\beta. \tag{1.36}$$

Thus (1.35) and (1.36) yield

$$\frac{\partial H^b_0}{\partial v^\beta} = 2H^b_\beta - \widetilde{B}^b_k \frac{\partial B^k_a}{\partial v^\beta} H^a_0. \tag{1.37}$$

Finally, if $\mathbf{n}^b = 0$ then $H^b_0 = 0$ and from (1.37) we obtain $H^b_\beta = 0$. Thus by (1.32) we have $\mathbf{n}_\beta = 0$, for any $\beta \in \{1, \cdots, m\}$. ■

Next, by using the normal curvature vector fields, we define a global section of $VM'^{\perp}$ by

$$\mathbf{n} = v^{\alpha}\mathbf{n}_{\alpha}. \tag{1.38}$$

Indeed, taking into account that (H^a_α) are the components of a mixed Finsler tensor field of type $\begin{pmatrix} 0 & 1 \\ 1 & 0 \end{pmatrix}$ on $\mathbb{F}^m$ with respect to $VM'^{\perp}$, and using (1.1.1) for $\mathbb{F}^m$, we obtain $\bar{v}^{\alpha}\bar{n}_{\alpha} = v^{\alpha}n_{\alpha}$ with respect to both, the coordinate transformation on M' and the orthogonal transformation of bases in $\Gamma(VM'^{\perp})$. By using (1.32), (1.26), (1.30) and (1.38) we infer that

$$\mathbf{n} = H^b_0 B^i_b \frac{\partial}{\partial y^i} = F^{*}\mathbf{n}^b B^i_b \frac{\partial}{\partial y^i}. \tag{1.39}$$

As the local components of $\mathbf{n}$ will determine the h–second fundamental form $\overset{b}{H} = (\overset{b}{H}{}^a{}_{\alpha\beta})$, we call it the **Berwald normal curvature vector field** of $\mathbb{F}^m$.

PROPOSITION 1.4 *Let $\mathbb{F}^m$ be a pseudo–Finsler submanifold of $\widetilde{\mathbb{F}}^{m+n}$. Then the local components of $\overset{b}{H}$ are given by*

$$\overset{b}{H}{}^a{}_{\alpha\beta} = \frac{1}{2}\widetilde{B}^a_k \frac{\partial^2(H^b_0 B^k_b)}{\partial v^{\alpha}\partial v^{\beta}}. \tag{1.40}$$

PROOF. Contracting (1.31) by v^{α} and taking into account (2.6.14) we deduce that

$$H^b_0 B^k_b = B^k_{00} + \widetilde{G}^k_i y^i - G^{\gamma}_{\mu} v^{\mu} B^k_{\gamma}. \tag{1.41}$$

Then, differentiating (1.41) and taking into account (1.6.27) and (1.6.30) we obtain

$$\frac{1}{2}\frac{\partial^2(H^b_0 B^k_b)}{\partial v^{\alpha}\partial v^{\beta}} = B^k_{\alpha\beta} + \widetilde{G}_i{}^k{}_j B^{ij}_{\alpha\beta} - G_{\alpha}{}^{\gamma}{}_{\beta} B^k_{\gamma}. \tag{1.42}$$

On the other hand, by using (2.4.5a), (2.1.12), (2.4.4a), and (2.6.51) we infer that

$$\overset{b}{H}{}^c{}_{\alpha\beta}\, B^k_c = B^k_{\alpha\beta} + \widetilde{G}_i{}^k{}_j B^{ij}_{\alpha\beta} - (G_{\alpha}{}^{\gamma}{}_{\beta} - \overset{b}{P}{}^{\gamma}{}_{\alpha\beta} - D^{\gamma}_{\alpha\|_{b}\beta}) B^k_{\gamma}. \tag{1.43}$$

Comparing (1.42) with (1.43) we derive

$$\overset{b}{H}{}^c{}_{\alpha\beta}\, B^k_c = \frac{1}{2}\frac{\partial^2(H^b_0 B^k_b)}{\partial v^{\alpha}\partial v^{\beta}} + (\overset{b}{P}{}^{\gamma}{}_{\alpha\beta} + D^{\gamma}_{\alpha\|_{b}\beta}) B^k_{\gamma}. \tag{1.44}$$

Finally, contracting (1.44) by $\widetilde{B}^a_k$ and using (2.1.11c) and (2.1.11d) we obtain (1.40). ■

THEOREM 1.3 *Let $\mathbb{F}^m$ be a pseudo–Finsler submanifold of $\widetilde{\mathbb{F}}^{m+n}$. Then the following assertions are equivalent:*

(i) $\mathbb{F}^m$ is totally geodesic;

(ii) All normal curvature vector fields vanish on M';

(iii) The Berwald normal curvature vector field of $\mathbb{F}^m$ vanishes on M';

(iv) The h–second fundamental forms of $\mathbb{F}^m$ satisfy any one (and hence all) of the conditions:

$$(a)\quad \overset{\mathrm{b}}{H}{}^{a}{}_{\alpha\beta}=0;\quad (b)\quad \overset{\mathrm{c}}{H}{}^{a}{}_{\alpha\beta}=-L^{a}{}_{\alpha\beta};\quad (c)\quad \overset{\mathrm{r}}{H}{}^{a}{}_{\alpha\beta}=-L^{a}{}_{\alpha\beta}. \tag{1.45}$$

PROOF. By Theorem 1.2 and Proposition 1.3 we have the equivalence of (i) and (ii). Next, from (1.39) it follows that $\mathbf{n}=0$ if and only if $\mathbf{n}^{a}=0$ for any $a\in\{m+1,\cdots,m+n\}$. Thus, via Theorem 1.2, (i) and (iii) are equivalent. Further, we suppose that (1.45a) is satisfied. Then from (2.4.6b) and Theorem 1.2 it follows that $\mathbb{F}^m$ is totally geodesic. Conversely, if $\mathbb{F}^m$ is totally geodesic then $H^{a}_{\alpha}=0$, and from (1.40) we deduce (1.45a). Thus we proved the equivalence of (i) and (iv). Now, suppose (iv) is satisfied. Then from (2.4.11) we derive (1.45b) via (2.4.6b). Conversely, if (1.45b) is satisfied then (2.4.11) yields

$$\overset{\mathrm{b}}{H}{}^{a}{}_{\alpha\beta}+\overset{a}{g_{b}}{}_{\alpha}H^{b}_{\beta}=0. \tag{1.46}$$

Contracting (1.46) by v^{α} and using (2.4.6b) and (2.3.32) we obtain $H^{a}_{\beta}=0$. Hence $\overset{\mathrm{b}}{H}{}^{a}{}_{\alpha\beta}=0$, and thus (1.45a) and (1.45b) are equivalent. In a similar way, using (2.5.8), the equivalence of (1.45b) and (1.45c) follows. ■

Taking into account Theorems 1.3, 2.6.4, 2.6.8, and 2.6.10 we may state the following corollary:

COROLLARY 1.1 *Let $\mathbb{F}^m$ be a totally geodesic pseudo–Finsler submanifold of $\widetilde{\mathbb{F}}^{m+n}$. Then we have:*

$$(a)\quad I\overset{\mathrm{c}}{F}C=\mathbf{FC}^{*};\quad (b)\quad I\overset{\mathrm{b}}{F}C=\mathbf{BFC}\ \textit{and}\ (c)\quad I\overset{\mathrm{r}}{F}C=\mathbf{RFC}. \tag{1.47}$$

In the Riemannian case totally geodesic submanifolds inherit most of the properties of the ambient manifold. Thus it is natural to study this phenomenon in the case of totally geodesic pseudo–Finsler submanifolds. More precisely, we prove the following theorem:

THEOREM 1.4 *Let $\mathbb{F}^m$ be a totally geodesic pseudo–Finsler submanifold of $\widetilde{\mathbb{F}}^{m+n}$. If $\widetilde{\mathbb{F}}^{m+n}$ is one of the following:*

(i) A pseudo–Finsler manifold of constant curvature $\widetilde{K}$;

(ii) A Landsberg (generalized Landsberg) manifold;

(iii) A Berwald manifold;

(iv) A locally pseudo-Minkowski manifold;

then, so is $\mathbb{F}^m$.

PROOF. By Theorem 1.2 we have $H^a_\alpha = 0$ for any $a \in \{m+1, \cdots, m+n\}$ and $\alpha \in \{1, \cdots, m\}$. Then from (2.6.6), (2.4.17b), (2.6.61), and (2.6.73) we deduce that

$$(a)\ D^\gamma_\alpha = 0; \quad (b)\ \overset{b}{D}{}_\alpha{}^\gamma{}_\beta = 0;$$

$$(c)\ \overset{r}{D}{}_\alpha{}^\gamma{}_\beta = 0, \ \forall \alpha, \beta, \gamma \in \{1, \cdots, m\}. \tag{1.48}$$

Now, suppose that $\widetilde{\mathbb{F}}^{m+n}$ is a pseudo–Finsler manifold of constant curvature $\widetilde{K}$. Then by assertion (v) of Theorem 1.8.2 we have

$$\widetilde{R}^*_{ijk} = \widetilde{K}\widetilde{F}\,(\tilde{g}_{ik}\tilde{\eta}_j - \tilde{g}_{ij}\tilde{\eta}_k)\,. \tag{1.49}$$

On the other hand, using (1.48a) in (2.6.27) and taking into account (2.3.39) and (1.49) we obtain

$$R^*_{\alpha\beta\gamma} = \widetilde{K}F\,(g_{\alpha\gamma}\eta_\beta - g_{\alpha\beta}\eta_\gamma)\,. \tag{1.50}$$

Hence by asertion (v) of Theorem 1.8.2 we deduce that $\mathbb{F}^m$ is of constant curvature $\widetilde{K}$ too. This completes the proof of (i). Next, suppose $\widetilde{\mathbb{F}}^{m+n}$ is a Landsberg manifold, that is, $\widetilde{F}_i{}^{*k}{}_j = \widetilde{G}_i{}^k{}_j$. Taking into account that $I\overset{b}{F}C=$ **BFC** and $I\overset{r}{F}C=$ **RFC** (see Corollary 1.1), we infer that $\overset{b}{F}{}_\alpha{}^\gamma{}_\beta = G_\alpha{}^\gamma{}_\beta$ and $\overset{r}{F}{}_\alpha{}^\gamma{}_\beta = F^{*\gamma}_{\alpha\ \beta}$. On the other hand, (2.4.4a) and (2.5.2a) yield $\overset{b}{F}{}_\alpha{}^\gamma{}_\beta = \overset{r}{F}{}_\alpha{}^\gamma{}_\beta$. Hence $F^{*\gamma}_{\alpha\ \beta} = G_\alpha{}^\gamma{}_\beta$, that is, $\mathbb{F}^m$ is also a Landsberg manifold. In case $\widetilde{\mathbb{F}}^{m+n}$ is a generalized Landsberg manifold, by using (1.48), (1.45a), (1.45c), (2.2.16), (1.8.41) and (2.5.6a) into (2.6.64a), and (2.6.76a), we infer that

$$H_{\gamma\varepsilon\alpha\beta} = K_{\gamma\varepsilon\alpha\beta} + A_{(\alpha\beta)}\{L^a{}_{\gamma\beta}L_{a\varepsilon\alpha}\}.$$

As a consequence of (1.8.42a), (2.3.33b), and (1.7.73) the last term in the above equation vanishes. So, $\mathbb{F}^m$ is a generalized Landsberg manifold too.

Further, we suppose that $\widetilde{\mathbb{F}}^{m+n}$ is a Berwald manifold. Then, since $\widetilde{G}_{ijkh} = 0$ and $H^a_\alpha = 0$, from (2.2.17) we obtain $\bar{P}_{ij\alpha\beta} = 0$. Thus, using (1.48b) and (1.45a) into (2.6.64b) we obtain $G_{\gamma\varepsilon\alpha\beta} = 0$. Hence $\mathbb{F}^m$ is a Berwald manifold. Finally, we suppose that $\widetilde{\mathbb{F}}^{m+n}$ is a locally pseudo–Minkowski manifold. Then by assertions (iii) and (i) we infer that $\mathbb{F}^m$ is a Berwald manifold of constant curvature $\widetilde{K} = 0$. Thus from (1.8.10) for $\mathbb{F}^m$ we deduce that $R^*_{\alpha\beta\gamma\varepsilon} = 0$, which completes the proof of assertion (iv). ∎

Example 1.1. Let $\widetilde{\mathbb{F}}^{m+n} = (\mathbb{R}^{m+n}, \widetilde{M}', \widetilde{F}^*)$ be a pseudo–Minkowski space and M be an m–dimensional non–degenerate plane given by equations

$$x^i = a^i_\alpha u^\alpha + b^i, \quad \text{rank}\,[a^i_\alpha] = m, \quad a^i_\alpha, b^i \in \mathbb{R}.$$

Then we have $B^i_\alpha = a^i_\alpha$ and $B^i_{\alpha\beta} = 0$. On the other hand, since $\widetilde{F}^*$ depends only on (y^i), from (1.6.9) and (1.6.17) we deduce that $\widetilde{G}^k_i = 0$. Hence (2.3.24b) yields $H^a_\alpha = 0$. Thus by Theorem 1.2 we may state that *any m–dimensional plane of a Minkowski space is a totally geodesic Finsler submanifold.* ∎

The converse of the above assertion is also true. More precisely, we prove the following proposition.

PROPOSITION 1.5 *Let $\widetilde{\mathbb{F}}^{m+n} = (\mathbb{R}^{m+n}, \widetilde{M}', \widetilde{F}^*)$ be a pseudo–Minkowski space and $\mathbb{F}^m = (M, M', F)$ be a totally geodesic pseudo–Finsler submanifold of $\widetilde{\mathbb{F}}^{m+n}$. Then M is $\mathbb{R}^m$ or an m-dimensional open submanifold of $\mathbb{R}^m$.*

PROOF. By assertion (iv) of Theorem 1.4 we deduce that $\mathbb{F}^m$ is a locally pseudo–Minkowski manifold. Thus $G_{\alpha\ \beta}^{\ \gamma} = 0$. As $\overset{b}{P}{}^\gamma{}_{\alpha\beta} = 0$ and $D^\gamma_\alpha = 0$, from (2.6.51) we obtain $\overset{b}{F}{}_\alpha{}^\gamma{}_\beta = 0$. Hence (2.4.4a) yields

$$\widetilde{B}^\gamma_k B^k_{\alpha\beta} = 0, \tag{1.51}$$

since $\widetilde{G}_i{}^k{}_j = 0$. Further, taking into account (1.45a) and (2.4.5a) we infer that

$$\widetilde{B}^a_k B^k_{\alpha\beta} = 0. \tag{1.52}$$

Finally, contracting (1.50) and (1.51) by B^h_γ and B^h_a respectively, and adding the results we obtain $B^k_{\alpha\beta} = 0$ via (2.1.12). Thus the local equations (2.1.1) are given by affine functions. Hence M is $\mathbb{R}^m$ or an open submanifold of $\mathbb{R}^m$. ∎

2. Parallel Vector Bundles over Pseudo–Finsler Submanifolds

In the present section we study pseudo–Finsler submanifolds endowed with h–parallel and v–parallel vertical bundle and with h–parallel and v–parallel pseudo–Finsler normal bundle. By means of h–paths and v–paths we define h–autoparallel and v–autoparallel pseudo–Finsler submanifolds. We obtain characterizations of all these classes of pseudo–Finsler submanifolds in terms of their second fundamental forms and shape operators. In particular, we consider each of the cases when the ambient pseudo–Finsler manifold is equipped with the Cartan, Berwald, or Rund connections.

Let $\mathbb{F}^{m+n} = (\widetilde{M}, \widetilde{M}', \widetilde{F}^*)$ be a pseudo–Finsler manifold endowed with an arbitrary Finsler connection

$$\widetilde{FC} = (H\widetilde{M}', \widetilde{\nabla}) = (\widetilde{N}_i^k, \widetilde{F}_i{}^k{}_j, \widetilde{C}_i{}^k{}_j),$$

and $\mathbb{F}^m = (M, M', F^*)$ be a pseudo–Finsler submanifold of $\widetilde{\mathbb{F}}^{m+n}$. Consider the induced non–linear connection $HM' = (N_\alpha^\gamma)$ and denote by h and v the projection morphisms of TM' on HM' and VM' respectively. Then we say that $\mathbb{F}^m$ has **h–parallel vertical vector bundle** with respect to $\widetilde{FC}$, if we have

$$\widetilde{\nabla}_{hX} vY \in \Gamma(VM'), \quad \forall X, Y \in \Gamma(TM'). \tag{2.1}$$

Similarly, we say that $\mathbb{F}^m$ has **v–parallel vertical vector bundle** with respect to $\widetilde{FC}$, if we have

$$\widetilde{\nabla}_{vX} vY \in \Gamma(VM'), \quad \forall X, Y \in \Gamma(TM'). \tag{2.2}$$

Now, we consider the h–and v–second fundamental forms $H = (H^a{}_{\alpha\beta})$ and $V = (V^a_{\alpha\beta})$ of $\mathbb{F}^m$, respectively. Then by using (2.1.33), (2.1.34), (2.1), and (2.2), we may state the following theorem.

THEOREM 2.1 *(i) $\mathbb{F}^m$ has h–parallel vertical vector bundle with respect to $\widetilde{FC}$ if and only if its h–second fundamental form vanishes, i.e., we have*

$$H^a{}_{\alpha\beta} = 0, \ \forall a \in \{m+1, \cdots, m+n\} \ ; \ \alpha, \beta \in \{1, \cdots, m\}. \tag{2.3}$$

(ii) $\mathbb{F}^m$ has v–parallel vertical vector bundle with respect to $\widetilde{FC}$ if and only if its v–second fundamental form vanishes, i.e., we have

$$V^a_{\alpha\beta} = 0, \ \forall a \in \{m+1, \cdots, m+n\}; \alpha, \beta \in \{1, \cdots, m\}. \tag{2.4}$$

Let $\widetilde{C}$ be a smooth curve in $\widetilde{M}'$ given locally by the equations

$$x^i = x^i(t); \ y^i = y^i(t), \quad t \in (a, b), \ i \in \{1, \cdots, m+n\}. \tag{2.5}$$

Then, taking into account (1.4.1) and (1.4.3) for $\widetilde{\mathbb{F}}^{m+n}$, the tangent vector field to $\widetilde{C}$ is written as follows

$$\frac{d}{dt} = \frac{dx^k}{dt}\frac{\delta}{\delta x^k} + \left(\frac{dy^k}{dt} + \widetilde{N}_i^k \frac{dx^i}{dt}\right)\frac{\partial}{\partial y^k}. \tag{2.6}$$

We say that $\widetilde{C}$ is an h–**curve** (resp., v–**curve**) if d/dt lies in $\Gamma(H\widetilde{M}')$ (resp., $\Gamma(V\widetilde{M}')$). Thus by (2.6) we deduce that $\widetilde{C}$ is an h–curve if and only if

$$\frac{dy^k}{dt} + \widetilde{N}_i^k\left(x(t), y(t)\right)\frac{dx^i}{dt} = 0, \tag{2.7}$$

and it is a v–curve if and only if

$$x^i = x_0^i = \text{constant on } \widetilde{C}. \tag{2.8}$$

The h–curve $\widetilde{C}$ in $\widetilde{M}'$ is called an h–**path** (h–**autoparallel**) with respect to the Finsler connection $\widetilde{FC}$ if we have

$$\widetilde{\nabla}_{d/dt}\widetilde{Q}\left(\frac{d}{dt}\right) = 0, \tag{2.9}$$

where $\widetilde{Q}$ is the associate almost product structure to the non–linear connection $H\widetilde{M}'$. By using (1.4.8), (1.5.1) and (2.7) in (2.9) we infer that a curve $\widetilde{C}$ is an h–path if and only if we have (2.7) and

$$\frac{d^2x^k}{dt^2} + \widetilde{F}_i{}^k{}_j(x(t), y(t))\frac{dx^i}{dt}\frac{dx^j}{dt} = 0. \tag{2.10}$$

In a similar way we say that a v–curve $\widetilde{C}$ in $\widetilde{M}'$ is a v–**path** (v–**autoparallel**) with respect to the Finsler connection $\widetilde{FC}$ if we have

$$\widetilde{\nabla}_{d/dt}\frac{d}{dt} = 0. \tag{2.11}$$

Thus by using (2.8) and (1.5.2) in (2.11) we deduce that $\widetilde{C}$ is a v–path if and only if we have (2.8) and

$$\frac{d^2y^k}{dt^2} + \widetilde{C}_i{}^k{}_j(x(t), y(t))\frac{dy^i}{dt}\frac{dy^j}{dt} = 0. \tag{2.12}$$

Remark 2.1. Analysing the system of differential equations (2.7) and (2.10) we deduce that there exists a unique h-path determined by initial values $(x^i(0), (dx^i/dt)_{t=0}, y^i(0))$. Similarly, from (2.8) and (2.12) it follows that the initial values $(x^i(0), y^i(0), (dy^i/dt)_{t=0})$ determine a unique v–path of $\widetilde{\mathbb{F}}^{m+n}$. ■

Next, we consider a smooth curve C in M' given locally by the equations

$$u^\alpha = u^\alpha(t); \; v^\alpha = v^\alpha(t), \quad t \in (a,b), \; \alpha \in \{1, \cdots, m\}. \tag{2.13}$$

Then C can also be considered as a curve in $\widetilde{M}'$ given by the equations

$$x^i = x^i(u(t)) \; ; \; y^i = B^i_\alpha(u(t))v^\alpha(t). \tag{2.14}$$

Differentiating both equations in (2.14) along C, we obtain (1.25) and

$$\begin{aligned} \frac{dy^i}{dt} &= B^i_{\alpha\gamma}\frac{du^\alpha}{dt}v^\gamma(t) + B^i_\alpha\frac{dv^\alpha}{dt} & (a) \\ \frac{d^2y^i}{dt^2} &= B^i_{\alpha\beta\gamma}\frac{du^\alpha}{dt}\frac{du^\beta}{dt}v^\gamma(t) + 2B^i_{\alpha\gamma}\frac{du^\alpha}{dt}\frac{dv^\gamma}{dt} + B^i_\alpha\frac{d^2v^\alpha}{dt^2}. & (b) \end{aligned} \tag{2.15}$$

When C is an h–path in M' with respect to the induced Finsler connection $IFC = (HM, \nabla) = (N^\gamma_\alpha, F_\alpha{}^\gamma{}_\beta, C_\alpha{}^\gamma{}_\beta)$ then the equations corresponding to (2.7) and (2.10) should be satisfied. Similarly, when C is a v–path then the equations corresponding to (2.8) and (2.12) must be satisfied. As a consequence we have the following two lemmas.

LEMMA 2.1 *Let C be an h–path in M' with respect to IFC. Then we have*

$$\frac{dy^k}{dt} + \widetilde{N}^k_i(x(t), y(t))\frac{dx^i}{dt} = H^a_\alpha(u(t), v(t))\frac{du^\alpha}{dt}B^k_a, \tag{2.16}$$

and

$$\frac{d^2x^k}{dt^2} + \widetilde{F}_i{}^k{}_j(x(t), y(t))\frac{dx^i}{dt}\frac{dx^j}{dt} \tag{2.17}$$

$$= \left(H^a{}_{\alpha\beta}(u(t), v(t))B^k_a - \widetilde{C}_i{}^k{}_j(x(t), y(t))B^i_\alpha B^j_a H^a_\beta(u(t), v(t))\right)\frac{du^\alpha}{dt}\frac{du^\beta}{dt}.$$

PROOF. Denote by X^k and Y^k the left sides in (2.16) and (2.17) respectively. Then by using (1.25a), (2.15a) and (1.31) for the pair $(N^\gamma_\alpha, \widetilde{N}^k_i)$ we obtain

$$X^k = H^a_\alpha(u(t), v(t))\frac{du^\alpha}{dt}B^k_a + \left(\frac{dv^\gamma}{dt} + N^\gamma_\alpha(u(t), v(t))\frac{du^\alpha}{dt}\right)B^k_\gamma. \tag{2.18}$$

Since C is an h–curve of $\mathbb{F}^m$, by (2.7) we have

$$\frac{dv^\gamma}{dt} + N^\gamma_\alpha(u(t), v(t))\frac{du^\alpha}{dt} = 0. \tag{2.19}$$

Hence (2.16) follows from (2.18) via (2.19). Next, by direct calculations using (1.25) we deduce that

$$Y^k = \left(B^k_{\alpha\beta} + \widetilde{F}_i{}^k{}_j(x(t), y(t))B^{ij}_{\alpha\beta}\right)\frac{du^\alpha}{dt}\frac{du^\beta}{dt} + B^k_\gamma\frac{d^2u^\gamma}{dt^2}. \tag{2.20}$$

Contracting (2.1.54) and (2.1.56) by B^h_γ and B^h_a respectively, and adding the results, we infer that

$$F_\alpha{}^\gamma{}_\beta B^k_\gamma + H^a{}_{\alpha\beta}B^k_a = B^k_{\alpha\beta} + \widetilde{F}_i{}^k{}_j B^{ij}_{\alpha\beta} + \widetilde{C}_i{}^k{}_j B^i_\alpha B^j_a H^a_\beta. \tag{2.21}$$

On the other hand, C being an h–path, by (2.10) we have

$$\frac{d^2u^\gamma}{dt^2} + F_\alpha{}^\gamma{}_\beta(u(t), v(t))\frac{du^\alpha}{dt}\frac{du^\beta}{dt} = 0. \tag{2.22}$$

Finally, by using (2.21) and (2.22) in (2.20) we obtain (2.17). ■

LEMMA 2.2 *Let C be a v–path in M' with respect to IFC. Then C is a v–curve in $\widetilde{M}'$ and*

$$\frac{d^2y^k}{dt^2} + \widetilde{C}_i{}^k{}_j(x(t), y(t))\frac{dy^i}{dt}\frac{dy^j}{dt} = V^a{}_{\alpha\beta}(u(t), v(t))B^k_a\frac{dv^\alpha}{dt}\frac{dv^\beta}{dt}. \tag{2.23}$$

PROOF. As C is a v–curve in M', it follows from (1.25a) that C is also a v–curve in $\widetilde{M}'$. Moreover, (2.15) becomes

$$\begin{aligned} \frac{dy^i}{dt} &= B^i_\alpha\frac{dv^\alpha}{dt} && (a) \\ \frac{d^2y^i}{dt^2} &= B^i_\alpha\frac{d^2v^\alpha}{dt^2}. && (b) \end{aligned} \tag{2.24}$$

Denote the left side in (2.23) by Z^k and by using (2.24) we derive

$$Z^k = B^k_\gamma\frac{d^2v^\gamma}{dt^2} + \widetilde{C}_i{}^k{}_j(x(t), y(t))B^{ij}_{\alpha\beta}\frac{dv^\alpha}{dt}\frac{dv^\beta}{dt}. \tag{2.25}$$

Contracting (2.1.55) and (2.1.57) by B^h_γ and B^h_a respectively, and adding the results, we deduce that

$$C_\alpha{}^\gamma{}_\beta B^k_\gamma + V^a{}_{\alpha\beta}B^k_a = \widetilde{C}_i{}^k{}_j B^{ij}_{\alpha\beta}. \tag{2.26}$$

Since C is a v–path in M', by (2.12) we have

$$\frac{d^2v^\gamma}{dt^2} + C_\alpha{}^\gamma{}_\beta(u(t), v(t))\frac{dv^\alpha}{dt}\frac{dv^\beta}{dt} = 0. \tag{2.27}$$

Thus (2.23) follows from (2.25) by using (2.26) and (2.27). ■

Next, we say that $\mathbb{F}^m = (M, M', F^*)$ is a h–**autoparallel** (resp., v–**autoparallel**) **pseudo–Finsler submanifold** of $\widetilde{\mathbb{F}}^{m+n} = (\widetilde{M}, \widetilde{M}', \widetilde{F}^*)$ with respect to $\widetilde{FC}$ if any h–path (resp., v–path) in M' with respect to IFC is an h–path (resp., v–path) in $\widetilde{M}'$ with respect to $\widetilde{FC}$. As in the case of totally geodesic pseudo–Finsler submanifolds, by means of normal curvature forms and second fundamental forms, we obtain characterization theorems for both the h–autoparallel and v–autoparallel pseudo–Finsler submanifolds.

THEOREM 2.2 *(Bejancu [7]). Let $\widetilde{\mathbb{F}}^{m+n}$ be a pseudo–Finsler manifold endowed with a Finsler connection $\widetilde{FC}$. Then a pseudo–Finsler submanifold $\mathbb{F}^m$ of $\widetilde{\mathbb{F}}^{m+n}$ is h–autoparallel with respect to $\widetilde{FC}$ if and only if we have*

$$(a) \quad H^a_\alpha = 0 \quad \text{and} \quad (b) \quad H^a{}_{\alpha\beta} + H^a{}_{\beta\alpha} = 0, \tag{2.28}$$

for any $a \in \{m+1, \cdots, m+n\}$ and $\alpha, \beta \in \{1, \cdots, m\}$.

PROOF. Suppose $\mathbb{F}^m$ is h–autoparallel and take an arbitrary point $(u_0, v_0) \in M'$. Then consider an h path C in M' determined by initial conditions $(u^\alpha(0) = u^\alpha_0, (du^\alpha/dt)_{t=0} = w^\alpha_0,\ v^\alpha(0) = v^\alpha_0)$. Then C is an h–path in $\widetilde{M}'$ and by (2.7) the left side of (2.16) vanishes. Thus from (2.16) we deduce that

$$H^a_\alpha(u_0, v_0) w^\alpha_0 = 0.$$

As this is true for any w^α_0, we obtain (2.28a) at (u_0, v_0). Similarly, by (2.10) the left side in (2.17) vanishes along C. Then by using (2.28a) in (2.17) we infer that

$$H^a{}_{\alpha\beta}(u_0, v_0) w^\alpha_0 w^\beta_0 = 0.$$

Thus we obtain (2.28b) at (u_0, v_0). The converse follows by using (2.28) and Lemma 2.1. ■

By combining Theorems 1.2 and 2.2 we state the following corollary.

COROLLARY 2.1 *Let $\widetilde{\mathbb{F}}^{m+n}$ be a pseudo–Finsler manifold endowed with a Finsler connection $\widetilde{FC} = (G\widetilde{M}', \widetilde{\nabla})$. Then any h–autoparallel pseudo–Finsler submanifold $\mathbb{F}^m$ of $\widetilde{\mathbb{F}}^{m+n}$ with respect to $\widetilde{FC}$ is totally geodesic.*

Moreover, in the case of special Finsler connections we show that the condition to be h–autoparallel is equivalent to having h–parallel vertical vector bundle.

THEOREM 2.3 *Let $\widetilde{\mathbb{F}}^{m+n}$ be a pseudo–Finsler manifold endowed with the Cartan connection $\widetilde{\mathbf{FC}}^*$ and $\mathbb{F}^m$ be a pseudo–Finsler submanifold of $\widetilde{\mathbb{F}}^{m+n}$. Then the following assertions are equivalent:*

(i) $\mathbb{F}^m$ is h–autoparallel with respect to $\widetilde{\mathbf{FC}}^*$;

(ii) The h–second fundamental form of $\mathbb{F}^m$ induced by $\widetilde{\mathbf{FC}}^*$ vanishes;

(iii) $\mathbb{F}^m$ has h–parallel vertical vector bundle with respect to $\widetilde{\mathbf{FC}}^*$.

PROOF. (*i*) $\Longrightarrow$ (*ii*). By using (2.28a) in (1.40) we deduce that $\overset{\mathrm{b}}{H}{}^a{}_{\alpha\beta} = 0$. Then from (2.4.11) we derive

$$\overset{\mathrm{c}}{H}{}^a{}_{\alpha\beta} = -\,L^a{}_{\alpha\beta}.$$

As $L^a{}_{\alpha\beta}$ are symmetric mixed Finsler tensor fields with respect to $(\alpha\beta)$, by (2.28b) we obtain $\overset{\mathrm{c}}{H}{}^a{}_{\alpha\beta} = 0$.

(*ii*) $\Longrightarrow$ (*iii*). It follows from assertion (*i*) of Theorem 2.1.

(*iii*) $\Longrightarrow$ (*i*). By (2.3) we have $\overset{\mathrm{c}}{H}{}^a{}_{\alpha\beta} = 0$, which together with (2.3.30a) implies (2.28). Hence by Theorem 2.2, $\mathbb{F}^m$ is h–autoparallel with respect to $\widetilde{\mathbf{FC}}^*$. ■

It is easy to check that Theorem 2.3 holds when we replace $\widetilde{\mathbf{FC}}^*$ by $\widetilde{\mathbf{BFC}}$ or $\widetilde{\mathbf{RFC}}$. Moreover, by using the assertion (iv) of Theorem 1.3 and Theorem 2.3 for $\widetilde{\mathbf{BFC}}$ we obtain the following corollary.

COROLLARY 2.2 *A pseudo–Finsler submanifold $\mathbb{F}^m$ of $\widetilde{\mathbb{F}}^{m+n}$ is totally geodesic if and only if it is h–autoparallel with respect to the Berwald connection.*

Remark 2.2. Theorem 2.3 holds also in case we replace h–autoparallel, h–second fundamental form and h–parallel by v–autoparallel, v–second fundamental form and v–parallel respectively (see Bejancu [7]). By Corollary 2.2 we see that the concepts we indroduced here with respect to the horizontal distribution have a geometrical meaning. We can not assert something similar for the vertical distribution. ■

As is well known, in Riemannian (pseudo–Riemannian) geometry the tangent bundle of a submanifold is parallel with respect to the Levi–Civita connection of the ambient manifold if and only if its normal bundle is parallel. We study this phenomenon in the general case of pseudo–Finsler submanifolds.

Let $\widetilde{\mathbb{F}}^{m+n} = (\widetilde{M}, \widetilde{M}', \widetilde{F}^*)$ be a pseudo–Finsler manifold endowed with a Finsler connection $\widetilde{FC} = (H\widetilde{M}', \widetilde{\nabla})$ and $\mathbb{F}^m = (M, M', F^*)$ be a pseudo–Finsler submanifold of $\widetilde{\mathbb{F}}^{m+n}$. Then we say that $\mathbb{F}^m$ has ***h*–parallel** (resp., ***v*–parallel**) **pseudo–Finsler normal bundle** with respect to $\widetilde{FC}$ if we have

$$\widetilde{\nabla}_{hX} W \in \Gamma(VM'^{\perp}) \ \ (\text{resp., } \widetilde{\nabla}_{vX} W \in \Gamma(VM'^{\perp})), \tag{2.29}$$

for any $X \in \Gamma(TM')$ and $W \in \Gamma(VM'^{\perp})$. By using (2.1.39), (2.1.40), (2.1.80) and (2.1.82) we obtain the following characterization for the above classes of pseudo–Finsler submanifolds.

THEOREM 2.4 *Let $\mathbb{F}^m$ be a pseudo–Finsler submanifold of $\widetilde{\mathbb{F}}^{m+n}$. Then we have:*

(i) $\mathbb{F}^m$ has h–parallel pseudo–Finsler normal bundle if and only if the h–shape operator vanishes, or locally,

$$H_{a\alpha\beta} + \tilde{g}_{ij|\beta} B^i_\alpha B^j_a = 0. \tag{2.30}$$

(ii) $\mathbb{F}^m$ has v–parallel pseudo–Finsler normal bundle if and only if the v–shape operator vanishes, or locally,

$$V_{a\alpha\beta} + \tilde{g}_{ij\|\beta} B^i_\alpha B^j_a = 0. \tag{2.31}$$

As the Cartan connection $\widetilde{\mathbf{FC}}^* = (G\widetilde{M}', \widetilde{\nabla}^*)$ is a metric Finsler connection with respect to $\tilde{g}$, we have

$$\tilde{g}(\widetilde{\nabla}^*_X vY, W) + \tilde{g}(vY, \widetilde{\nabla}^*_X W) = 0, \tag{2.32}$$

for any $X, Y \in \Gamma(TM')$ and $W \in \Gamma(VM'^{\perp})$. Then, by Theorems 2.3 and 2.4 we may state the following corollary:

COROLLARY 2.3 *Let $\widetilde{\mathbb{F}}^{m+n}$ be a pseudo–Finsler manifold endowed with the Cartan connection $\widetilde{\mathbf{FC}}^*$ and $\mathbb{F}^m$ be a pseudo–Finsler submanifold of $\widetilde{\mathbb{F}}^{m+n}$. Then $\mathbb{F}^m$ has h–parallel (resp., v–parallel) vertical vector bundle with respect to $\widetilde{\mathbf{FC}}^*$ if and only if it has h–parallel (resp., v–parallel) pseudo–Finsler normal bundle with respect to $\widetilde{\mathbf{FC}}^*$.* ■

3. Finsler Submanifolds of C–Reducible Finsler Manifolds

The purpose of this section is to present the basic properties of Finsler submanifolds of a C–reducible Finsler manifold. We show that the induced Finsler connections coincide with the intrinsic Finsler connections of a

Finsler submanifold, provided that the structure vector field $\tilde{\mathbf{c}}$ of the ambient manifold is tangent to the submanifold (see Theorem 3.1). In case $\tilde{\mathbf{c}}$ is normal to the submanifold, we find inter–relations between the Riemannian immersion and the Finslerian immersion for totally geodesic Finsler submanifolds (see Theorem 3.3).

Let $\widetilde{\mathbb{F}}^{m+n} = (\widetilde{M}, \widetilde{M}', \widetilde{F})$ be a C–reducible Finsler manifold whose Riemannian metric and differential 1–form are $\tilde{b}_{ij}(x)$ and $\tilde{c}_i(x)$ respectively. Then by (1.8.47) and (1.8.48) we have

$$(a)\quad \widetilde{F}(x,y) = \tilde{b}(x,y) + \tilde{c}(x,y) \quad \text{or} \quad (b)\quad \widetilde{F}(x,y) = \frac{\tilde{b}^2(x,y)}{\tilde{c}(x,y)}, \tag{3.1}$$

where we set:

$$(a)\quad \tilde{b}(x,y) = (\tilde{b}_{ij}(x)y^iy^j)^{1/2} \quad \text{and} \quad (b)\quad \tilde{c}(x,y) = \tilde{c}_i(x)y^i, \tag{3.2}$$

according as $\widetilde{F}^{m+n}$ is a Randers manifold or a Kropina manifold respectively.

Now, suppose $\mathbb{F}^m = (M, M', F)$ is a Finsler submanifold of $\widetilde{\mathbb{F}}^{m+n}$. Then $(\tilde{b}_{ij}(x))$ and $(\tilde{c}_i(x))$ will induce on M a Riemannian metric $(b_{\alpha\beta}(u))$ and a differential 1–form $(c_\alpha(u))$ given by

$$(a)\quad b_{\alpha\beta}(u) = \tilde{b}_{ij}(x(u))B^{ij}_{\alpha\beta} \quad \text{and} \quad (b)\quad c_\alpha(u) = \tilde{c}_i(x(u))B^i_\alpha, \tag{3.3}$$

respectively. On the other hand, by using (2.1.2), (2.1.3) and (3.3) we deduce that the fundamental function F of $\mathbb{F}^m$ has one of the following forms:

$$\begin{aligned} F(u,v) &= (b_{\alpha\beta}(u)v^\alpha v^\beta)^{1/2} + c_\gamma(u)v^\gamma & (a) \\ F(u,v) &= \frac{b_{\alpha\beta}(u)v^\alpha v^\beta}{c_\gamma(u)v^\gamma}. & (b) \end{aligned} \tag{3.4}$$

Hence we may state the following result:

PROPOSITION 3.1 *If $\widetilde{\mathbb{F}}^{m+n}$ is a Randers manifold, then so is the submanifold F^m. If $\widetilde{\mathbb{F}}^{m+n}$ is a Kropina manifold, then so is $\mathbb{F}^m$.*

Next, we assume that the immersion of M in $\widetilde{M}$ is given locally by the equations

$$\Phi_a(x^1, \cdots, x^{m+n}) = 0; \quad \text{rank}\left[\frac{\partial \Phi_a}{\partial x^i}\right] = n, \ a \in \{m+1, \cdots, m+n\}.$$

If we take

$$\omega_{ai}(x) = \frac{\partial \Phi_a}{\partial x^i},$$

then we define

$$(a)\quad N_a^i(x,y) = \tilde{g}^{ij}(x,y)\omega_{aj}(x) \quad \text{and} \quad (b)\quad n_a^i(x) = \tilde{b}^{ij}(x)\omega_{aj}(x). \tag{3.5}$$

It is easy to check that

$$(a)\quad N_a = N_a^i \frac{\partial}{\partial y^i} \quad \text{and} \quad (b)\quad n_a = n_a^i \frac{\bar{\partial}}{\bar{\partial} x^i}, \tag{3.6}$$

form a basis in $\Gamma(VM'^{\perp})$ and $\Gamma(TM^{\perp})$ respectively, where $\{\bar{\partial}/\bar{\partial}x^i\}$ is the natural frame field on $\widetilde{M}$, and $TM^{\perp}$ is the usual normal bundle of M as a Riemannian submanifold of $(\widetilde{M}, \tilde{b}_{ij}(x))$.

Finally, we set $\tilde{c}^i = \tilde{b}^{ij}\tilde{c}_j$ and define the vector fields

$$(a)\quad \tilde{\mathbf{c}}(x) = \tilde{c}^i(x)\frac{\bar{\partial}}{\bar{\partial} x^i}_{|x} \quad \text{and} \quad (b)\quad \tilde{\mathbf{c}}^*(x,y) = \tilde{c}^i(x)\frac{\partial}{\partial y^i}_{|(x,y)}. \tag{3.7}$$

We call $\tilde{\mathbf{c}}$ the **structure vector field** of $\widetilde{\mathbb{F}}^{m+n}$ and prove the following.

PROPOSITION 3.2 *The structure vector field $\tilde{\mathbf{c}}$ is tangent to M if and only if $\tilde{\mathbf{c}}^*$ lies in $\Gamma(VM')$.*

PROOF. By using (3.6b) we deduce that $\tilde{\mathbf{c}}$ is tangent to M if and only if

$$\tilde{b}_{ij}(x)\tilde{c}^i(x)n_a^j(x) = \tilde{b}_{ij}(x)\tilde{c}^i(x)\tilde{b}^{jk}(x)\omega_{ak}(x) = \tilde{c}^k(x)\omega_{ak}(x) = 0.$$

Similarly, by using (3.6a) we infer that $\tilde{\mathbf{c}}^*$ lies in $\Gamma(VM')$ if and only if

$$\tilde{g}_{ij}(x,y)\tilde{c}^i(x)N_a^j(x,y) = \tilde{g}_{ij}(x,y)\tilde{c}^i(x)\tilde{g}^{jk}(x,y)\omega_{ak}(x) = \tilde{c}^k(x)\omega_{ak}(x) = 0.$$

This completes the proof of our assertion. ■

PROPOSITION 3.3 *Let $\mathbb{F}^m$ be a Finsler submanifold of a C–reducible Finsler manifold $\widetilde{\mathbb{F}}^{m+n}$ such that the structure vector field $\tilde{\mathbf{c}}$ of $\mathbb{F}^m$ is tangent to M. Then we have*

$$(a)\quad \tilde{c}_i B_a^i = 0 \quad \text{and} \quad (b)\quad \widetilde{Y}_i B_a^i = 0, \quad \forall a \in \{m+1, \cdots, m+n\}, \tag{3.8}$$

where $\widetilde{Y}_i = \tilde{b}_{ij}y^j$.

PROOF. Owing to (1.8.51) and (1.8.52) the Finsler metric of $\widetilde{\mathbb{F}}^{m+1}$ is given by

$$\tilde{g}_{ij} = \frac{\widetilde{F}}{\tilde{b}}\tilde{b}_{ij} + \tilde{c}_i\tilde{c}_j + \frac{1}{\tilde{b}}\left(\tilde{c}_i\widetilde{Y}_j + \tilde{c}_j\widetilde{Y}_i\right) - \frac{\tilde{c}}{\tilde{b}^3}\widetilde{Y}_i\widetilde{Y}_j, \tag{3.9}$$

or

$$\tilde{g}_{ij} = \tau\left(2\tilde{b}_{ij} - \eta_i\tilde{c}_j - \eta_j\tilde{c}_i\right) + \eta_i\eta_j, \tag{3.10}$$

according as $\widetilde{\mathbb{F}}^{m+n}$ is a Randers manifold or a Kropina manifold, respectively. Contracting (3.9) by $\tilde{c}^i B^j_a$ and by $y^i B^j_a$ and taking into account that $\tilde{\mathbf{c}}^*$ and the Liouville vector field of $\widetilde{\mathbb{F}}^{m+n}$ lie in $\Gamma(VM')$, we deduce that

$$\begin{cases} \left(\dfrac{\widetilde{F}}{\tilde{b}} + \|\tilde{\mathbf{c}}\|^2 + \dfrac{\tilde{c}}{\tilde{b}}\right) X_a + \left(\dfrac{\|\tilde{\mathbf{c}}\|^2}{\tilde{b}} - \dfrac{\tilde{c}^2}{\tilde{b}^3}\right) Y_a = 0 \\ \tilde{b} X_a + Y_a = 0, \end{cases}$$

where we set:

$$X_a = \tilde{c}_i B^i_a; \quad Y_a = \widetilde{Y}_i B^i_a \quad \text{and} \quad \|\tilde{\mathbf{c}}\|^2 = \tilde{b}_{ij}(x)\tilde{c}^i\tilde{c}^j.$$

By direct calculations using (3.1a) and (3.2) we infer that the determinant of the coefficient matrix of the above system is $\widetilde{F}^2/\tilde{b}^2$. Hence $X_a = Y_a = 0$, for any $a \in \{m+1, \cdots, m+n\}$. Similarly, from (3.10) we obtain the system

$$\begin{cases} (2 - \eta_i \tilde{c}^i) X_a = 0, \\ \widetilde{F} X_a - 2Y_a = 0. \end{cases}$$

On the other hand, contracting (1.10) with $\tilde{c}^i y^j$ and taking into account that $\eta_j y^j = \widetilde{F}$ we obtain

$$\tilde{c}(2 - \eta_i \tilde{c}^i) = \widetilde{F}\|\tilde{\mathbf{c}}\|^2 \neq 0.$$

Hence we have again $X_a = Y_a = 0$, for any $a \in \{m+1, \cdots, m+n\}$. ■

THEOREM 3.1 *(Bejancu–Farran [6]) Let $\mathbb{F}^m$ be a Finsler submanifold of a C–reducible Finsler manifold $\widetilde{\mathbb{F}}^{m+n}$ such that the structure vector field $\tilde{\mathbf{c}}$ of $\widetilde{\mathbb{F}}^{m+n}$ is tangent to M. Then we have:*

$$(a)\quad I\overset{\mathrm{c}}{F}C = \mathbf{FC}^*; \quad (b)\quad I\overset{\mathrm{b}}{F}C = \mathbf{BFC} \text{ and } \quad (c)\quad I\overset{\mathrm{r}}{F}C = \mathbf{RFC}. \tag{3.11}$$

PROOF. By using (2.3.3a) and (1.8.45) we obtain

$$g_{a\alpha\beta} = \frac{1}{m+n+1}\left(\tilde{a}_{ij}\tilde{g}_k + \tilde{a}_{jk}\tilde{g}_i + \tilde{a}_{ki}\tilde{g}_j\right) B^i_a B^{jk}_{\alpha\beta}. \tag{3.12}$$

On the other hand, taking into account (1.7.26), (1.7.23), (2.1.2) and (2.1.9) we deduce that

$$\tilde{a}_{ij} B^i_a B^j_\alpha = 0.$$

Hence (3.12) becomes

$$g_{a\alpha\beta} = \frac{1}{m+n+1}\tilde{a}_{jk} B^{jk}_{\alpha\beta}\tilde{g}_i B^i_a. \tag{3.13}$$

Now, (1.8.49) of $\widetilde{\mathbb{F}}^{m+n}$ implies

$$(a)\quad \tilde{g}_i B^i_a = -\frac{m+n+1}{2\widetilde{F}}\tau' \tilde{c}_i B^i_a \quad \text{or} \quad \tilde{g}_i B^i_a = -\frac{m+n+1}{2\widetilde{F}}\tau \tilde{c}_i B^i_a, \quad (3.14)$$

since $\tilde{\eta}_i B^i_a = 0$. As $\tilde{\mathbf{c}}$ is assumed to be tangent to M, by Proposition 3.3 we have

$$\tilde{c}_i B^i_a = 0, \quad \forall a \in \{m+1, \cdots, m+n\}.$$

Thus from (3.14) it follows that $\tilde{g}_i B^i_a = 0$ for any $a \in \{m+1, \cdots, m+n\}$, which together with (3.13) imply that $g_{a\alpha\beta} = 0$ for any $a \in \{m+1, \cdots, m+n\}$ and $\alpha, \beta \in \{1, \cdots, m\}$. Finally, by Theorems 2.6.4, 2.6.8 and 2.6.10 we obtain the assertion of our theorem. ■

Remark 3.1. By Theorem 3.1, on any Finsler submanifold $\mathbb{F}^m$ which is tangent to $\tilde{\mathbf{c}}$ of a C–reducible Finsler manifold $\widetilde{\mathbb{F}}^{m+n}$ the induced Finsler connections of the three special Finsler connections coincide with the corresponding intrinsic Finsler connections of $\mathbb{F}^m$. The examples below show that submanifolds $\mathbb{F}^m$ tangent to $\tilde{\mathbf{c}}$ are abundant. ■

Remark 3.2. For a Finsler hypersurface $\mathbb{F}^m$ of a C–reducible Finsler manifold $\widetilde{\mathbb{F}}^{m+1}$, Shibata–Singh–Singh [1] proved that the condition $\tilde{c}_i B^i = 0$ implies that the induced and intrinsic Finsler connections coincide on $\mathbb{F}^m$. ■

Example 3.1. Let $\widetilde{M}^* = \{(y^1, y^2, y^3) \in \mathbb{R}^3;\ y^1 > 0\}$ and take $\widetilde{M} = \mathbb{R}^3$ and $\widetilde{M}' = \mathbb{R}^3 \times \widetilde{M}^*$. Then define $\widetilde{F}$ by

$$\widetilde{F}(y^1, y^2, y^3) = y^1 + \frac{(y^2)^2 + (y^3)^2}{y^1}.$$

By direct calculations we deduce that condition $(\mathbf{F_2})$ of Section 1.1 is satisfied. Hence $\widetilde{\mathbb{F}}^3 = (\widetilde{M}, \widetilde{M}', \widetilde{F})$ is a Kropina manifold. In this case we have $\tilde{b}_{ij} = \delta_{ij}$ and $\tilde{c} = (1, 0, 0)$. Now, consider a surface M of $\mathbb{R}^3$ given by the equation $\phi(x^2, x^3) = 0$. Clearly $\tilde{\mathbf{c}} = (1, 0, 0)$ is tangent to M, since $(0, \partial\phi/\partial x^2,\ \partial\phi/\partial x^3)$ is a normal vector to M with respect to the Euclidean metric on $\mathbb{R}^3$. Thus any such Finsler surface $\mathbb{F}^2 = (M, M', F)$ of $\widetilde{\mathbb{F}}^3$ satisfies the condition in Theorem 3.1. ■

Example 3.2. We consider on $\mathbb{R}^{m+1}\backslash\{0\}$ the positive smooth function $\widetilde{F}$ given by

$$\widetilde{F}(y^1 \cdots, y^{m+1}) = \left(\sum_{i=1}^{m+1} (y^i)^2\right)^{1/2} - \frac{1}{2}y^1.$$

Then by Theorem 1.1.2, $\widetilde{\mathbb{F}}^{m+1} = (\mathbb{R}^{m+1}, \mathbb{R}^{m+1} \times \mathbb{R}^{m+1}\backslash\{0\},\ \widetilde{F})$ is a Randers manifold. As in the previous example we may check that any $\mathbb{F}^m =$

(M, M', F) of $\widetilde{\mathbb{F}}^{m+1}$ with M given by an equation $\Phi(x^2, \cdots, x^{m+1}) = 0$ is tangent to $\tilde{\mathbf{c}}$. ∎

Next, we assume that $\tilde{\mathbf{c}}$ is normal to M with respect to the Riemannian metric $(\tilde{b}_{ij})$. In this case we have

$$\tilde{c}_i(x)y^i = \tilde{b}_{ij}(x)\tilde{c}^j(x)B^i_\alpha v^\alpha = 0, \tag{3.15}$$

along M. Hence by (3.1b) and (3.2b), in the case in which $\widetilde{\mathbb{F}}^{m+n}$ is a Kropina manifold $\tilde{\mathbf{c}}$ is not normal to any submanifold M of $\widetilde{M}$. When $\widetilde{\mathbb{F}}^{m+n}$ is a Randers manifold, by using (3.4a), (3.3b) and (3.15) we deduce that any $\mathbb{F}^m = (M, TM^0, F)$ with M normal to $\tilde{\mathbf{c}}$ is a Riemannian manifold with the Riemannian metric $(b_{\alpha\beta}(u))$ given by (3.3a). Summing up this discussion, we may state the following.

THEOREM 3.2 *Let $\mathbb{F}^m$ be a Finsler submanifold of a C-reducible Finsler manifold $\widetilde{\mathbb{F}}^{m+n}$ such that the structure vector field of $\widetilde{\mathbb{F}}^{m+n}$ is normal to M. Then $\widetilde{\mathbb{F}}^{m+n}$ and $\mathbb{F}^m$ must be a Randers manifold and a Riemannian manifold respectively.*

As a result of the above theorem, from now on we proceed with the study of a Riemannian manifold $\mathbb{F}^m = (M, TM^0, F)$ immersed as a Finsler submanifold of the Randers manifold $\widetilde{\mathbb{F}}^{m+n} = (\widetilde{M}, \widetilde{M}', \widetilde{F})$. First, denote by $\Gamma_{\alpha\ \beta}^{\ \gamma}$ and $\tilde{\Gamma}_{i\ j}^{\ k}$ the Christoffel symbols of M and $\widetilde{M}$, that is, we have

$$\nabla_{\partial/\partial u^\beta}\frac{\partial}{\partial u^\alpha} = \Gamma_{\alpha\ \beta}^{\ \gamma}\frac{\partial}{\partial u^\gamma} \quad \text{and} \quad (b) \ \ \tilde{\nabla}_{\partial/\partial x^j}\frac{\partial}{\partial x^i} = \tilde{\Gamma}_{i\ j}^{\ k}\frac{\partial}{\partial x^k}, \tag{3.16}$$

where ∇ and $\tilde{\nabla}$ are the Levi–Civita connections with respect to the Riemannian metrics $(b_{\alpha\beta}(u))$ and $(\tilde{b}_{ij}(x))$, respectively, and $\{\partial/\partial u^\alpha\}$ is the natural frame field on M. Suppose that $\{b_a = b^i_a(\partial/\partial x^i)\}$ is a local orthonormal basis in $\Gamma(TM^\perp)$ and denote by $\{h^a_{\alpha\beta}(u)\}$ the local components of the second fundamental form of the Riemannian immersion of M in $\widetilde{M}$. Then the classical Gauss formulas are locally expressed as follows

$$B^k_{\alpha\beta} + \tilde{\Gamma}_{i\ j}^{\ k}B^{ij}_{\alpha\beta} = \Gamma_{\alpha\ \beta}^{\ \gamma}B^k_\gamma + h^a_{\alpha\beta}(u)b^k_a. \tag{3.17}$$

Next, by (1.31) and (2.6.5) we deduce that

$$(G^\gamma_\alpha - D^\gamma_\alpha)B^k_\gamma + H^a_\alpha B^k_a = B^k_{\alpha 0} + \tilde{G}^k_i B^i_\alpha. \tag{3.18}$$

Contracting (3.18) by v^α and taking into account (2.6.13), (1.6.11), (1.6.17) and (1.26), we obtain

$$2G^\gamma B^k_\gamma + H^a_0 B^k_a = B^k_{00} + 2\tilde{G}^k. \tag{3.19}$$

As $\mathbb{F}^m$ is a Riemannian manifold, by using (2.6.2b) we derive

$$2G^{\gamma} = \Gamma_{\alpha}{}^{\gamma}{}_{\beta} v^{\alpha} v^{\beta}. \tag{3.20}$$

At this point we assume that $\tilde{\mathbf{c}}$ is a gradient vector field on $\widetilde{M}$ with the scalar function ϕ, i.e., we have

$$(a) \quad \tilde{c}^i = \tilde{b}^{ij}(x)\frac{\bar{\partial}\phi}{\bar{\partial}x^j} \quad \text{or, equivalently,} \quad (b) \quad \tilde{c}_i = \frac{\bar{\partial}\phi}{\bar{\partial}x^i}. \tag{3.21}$$

Then, by (1.8.55d), (1.8.55e), and (3.21) we infer that

$$\widetilde{F}_i^k = 0 \quad \text{and} \quad \widetilde{F}_i = 0. \tag{3.22}$$

On the other hand, by using (3.21b) we obtain $\tilde{c}_{i;j} = \tilde{c}_{j;i}$, where';' means covariant derivative on $\widetilde{M}$ with respect to the Levi–Civita connection. Then (1.8.55a) becomes for $\widetilde{\mathbb{F}}^{m+n}$

$$\widetilde{E}_{ij} = \tilde{c}_{i;j}. \tag{3.23}$$

Thus, by using (3.22) and (3.23) in the expression of $\widetilde{G}^k$ (see (1.8.54)) for $\widetilde{\mathbb{F}}^{m+n}$ we obtain

$$2\widetilde{G}^k = \left(\widetilde{\Gamma}_i{}^k{}_j + \frac{y^k}{\widetilde{F}}\tilde{c}_{i;j}\right) y^i y^j. \tag{3.24}$$

Finally, by using (3.24), (3.20) and (3.17) in (3.19), we deduce that

$$H_0^a B_a^k = \left(h^a_{\alpha\beta} b^k_a + \frac{y^k}{\widetilde{F}}\tilde{c}_{i;j} B^{ij}_{\alpha\beta}\right) v^{\alpha} v^{\beta}, \tag{3.25}$$

which enables us to prove the following important result.

THEOREM 3.3 *(Bejancu–Farran [6]). Let $\widetilde{\mathbb{F}}^{m+n} = (\widetilde{M}, \widetilde{M}', \widetilde{F})$ be a Randers manifold such that $\tilde{\mathbf{c}}$ is a gradient vector field on $\widetilde{M}$. Then any $\mathbb{F}^m = (M, TM^0, F)$ that is normal to $\tilde{\mathbf{c}}$, is totally geodesic immersed in $\widetilde{\mathbb{F}}^{m+n}$ as a Finsler submanifold, if and only if, M is totally geodesic immersed in $\widetilde{M}$ as a Riemannian submanifold.*

PROOF. Suppose that $\mathbb{F}^m$ is totally geodesic immersed in $\widetilde{\mathbb{F}}^{m+n}$ as a Finsler submanifold. Then by Theorem 1.2 we have $H_0^a = 0$ for any $a \in \{m+1, \cdots, m+n\}$. Thus by using (2.1.2a) we see that (3.25) implies

$$(h^a_{\alpha\beta}(u) v^{\alpha} v^{\beta}) b_a + \frac{1}{\widetilde{F}}(\tilde{c}_{i;j} B^{ij}_{\alpha\beta} v^{\alpha} v^{\beta}) v^{\gamma} \frac{\bar{\partial}}{\bar{\partial}u^{\gamma}} = 0. \tag{3.26}$$

As (v^α) can be considered as coordinates of any vector tangent to M at $x(u)$, from (3.26) we obtain $h^a_{\alpha\beta}(u) = 0$, since $\{b_a\}$ and $v^\alpha(\bar{\partial}/\bar{\partial}u^\alpha)$ are normal and tangent to M respectively. Hence M is totally geodesic immersed in $\widetilde{M}$ as a Riemannian submanifold. Conversely, if $h^a_{\alpha\beta}(u) = 0$ for any $a \in \{m+1, \cdots, m+n\}$ and $\alpha, \beta \in \{1, \cdots, m\}$, from (3.25) we deduce that

$$H^a_0 B_a = \frac{1}{\widetilde{F}}(\tilde{c}_{i;j} y^i y^j)\widetilde{L}, \tag{3.27}$$

where $\widetilde{L}$ is the Liouville vector field of $\widetilde{\mathbb{F}}^{m+n}$. As $\widetilde{L} \in \Gamma(VM')$ and $B_a \in \Gamma(VM'^{\perp})$ we conclude that $H^a_0 = 0$. ∎

We now show the existence of Finsler submanifolds satisfying the condition in Theorem 3.3. Let $\widetilde{\mathbb{F}}^{m+1}$ be a Randers manifold such that $\tilde{\mathbf{c}}$ is a gradient vector field on $\widetilde{M}$ with the scalar function ϕ. Then for any constant k we have a hypersurface $M(k)$ of $\widetilde{M}$ which is given by the equation $\phi(x^1, \cdots, x^{m+1}) = k$. Clearly, any $M(k)$ is normal to $\tilde{\mathbf{c}}$ and hence Theorem 3.3 holds for $\mathbb{F}^m = (M(k), TM(k)^0, F)$. Moreover, in this case we prove the following:

THEOREM 3.4 *(Matsumoto [4]).* $\mathbb{F}^m = (M(k), TM(k)^0, F)$ *is totally geodesic immersed in* $\widetilde{\mathbb{F}}^{m+1}$ *as a Finsler submanifold if and only if along* $M(k)$ *we have*

$$2\tilde{c}_{i;j} = \tilde{c}_i \tilde{d}_j + \tilde{c}_j \tilde{d}_i, \tag{3.28}$$

where $(\tilde{d}_i)$ *are the local components of a certain differential form on* $\widetilde{M}$.

PROOF. Suppose $\mathbb{F}^m$ is totally geodesic immersed in $\widetilde{\mathbb{F}}^{m+1}$. Then by Theorems 1.2 and 3.3 we have $H_0 = 0$ and $h_{\alpha\beta} = 0$. Thus from (3.25) we deduce that $\tilde{c}_{i;j} y^i y^j = 0$, which by (3.15) can be written equivalently as $\tilde{c}_{i;j} y^i y^j = (\tilde{c}_i y^i)(\tilde{d}_j y^j)$, where $(\tilde{d}_j)$ are the components of a differential form on $\widetilde{M}$. Thus we obtain (3.28). Conversely, from (3.28) via (3.15) we derive $\tilde{c}_{i;j} y^i y^j = 0$. Then by using (1.8.55b) and (2.1.2a) we infer that

$$\frac{\bar{\partial}\tilde{c}_k}{\bar{\partial}u^\beta} y^k v^\beta - \tilde{c}_k \widetilde{\Gamma}_i{}^k{}_j B^{ij}_{\alpha\beta} v^\alpha v^\beta = 0. \tag{3.29}$$

Differentiating (3.15) with respect to u^β we obtain

$$\frac{\bar{\partial}\tilde{c}_k}{\bar{\partial}u^\beta} y^k = -\tilde{c}_k B^k_{\alpha\beta} v^\alpha.$$

Hence (3.29) becomes

$$\tilde{c}_k (B^k_{\alpha\beta} + \widetilde{\Gamma}_i{}^k{}_j B^{ij}_{\alpha\beta}) v^\alpha v^\beta = 0. \tag{3.30}$$

Finally, by using (3.17) for the hypersurface $\mathbb{F}^m$, that is, b_a^k and $h_{\alpha\beta}^a$ are replaced by $\tilde{c}^k$ and $h_{\alpha\beta}$, respectively, and taking into account (3.15), from (3.30) we deduce that $h_{\alpha\beta}(u) = 0$. Hence by Theorem 3.3 $\mathbb{F}^m$ is totally geodesic immersed in $\widetilde{\mathbb{F}}^{m+n}$. ■

CHAPTER 4

GEOMETRY OF CURVES IN FINSLER MANIFOLDS

We apply here the general theory of pseudo–Finsler submanifolds developed in the previous chapters to study the geometry of curves in Finsler manifolds. We construct a Frenet frame and derive all the Frenet equations for a curve in a Finsler manifold. This enables us to state a fundamental theorem for curves in Finsler manifolds and to give theorems on the reduction of the codimension of Finsler immersions. Finally, we pay special attention to the geometry of curves in $\mathbb{F}^2$ and $\mathbb{F}^3$.

1. The Frenet Frame for a Curve in a Finsler Manifold

Let $\mathbb{F}^{m+1} = (M, M', F)$ be a Finsler manifold and $\mathbb{F}^1 = (\mathcal{C}, \mathcal{C}', F_1)$ be a 1–dimensional Finsler submanifold of $\mathbb{F}^{m+1}$, where $\mathcal{C}$ is a smooth curve in M given locally by equations (3.1.1). Denote by (s, v) the coordinates on $\mathcal{C}'$, where s is the arc length parameter on $\mathcal{C}$ (see (1.6.4)). Then (2.1.2a) becomes

$$y^i(s, v) = v\frac{dx^i}{ds}, \quad i \in \{0, \cdots, m\}. \tag{1.1}$$

Moreover, $\{\partial/\partial s, \partial/\partial v\}$ is a natural field of frames on $\mathcal{C}'$, where $\partial/\partial v$ is a unit Finsler vector field (see 3.1.5).

Remark 1.1. As $\Gamma(\mathcal{C}')$ does not intersect the zero section we are dealing with geometric objects along $\mathcal{C}$ at points $(x(s), y(s, v))$ with $v \neq 0$. Hence without loss of generality we may suppose $v > 0$ for any point of the coordinate neighborhood in the study. ■

On M', we consider the canonical field of frames $(\delta^*/\delta^* x^i, \partial/\partial y^i)$ induced by the canonical non–linear connection $GM' = (G_i^k)$ (see (2.3.20)). As a consequence of homogeneity of F it follows that g_{ij}, g^{ij} and $F_{i\ j}^{*k}$ (see

(2.3.17)) are positive homogeneous of degree zero, while G^i, G^i_j and $g_i{}^k{}_j$ are positive homogeneous of degrees 2,1, and -1 respectively. Then by using (2.3.23) and (1.6.22b) we deduce that

$$\begin{aligned} g_{ij}(x(s), vx'(s)) &= g_{ij}(x(s), x'(s)) \qquad (a) \\ g^{ij}(x(s), vx'(s)) &= g^{ij}(x(s), x'(s)), \qquad (b) \end{aligned} \tag{1.2}$$

$$x'^i(s)F_i{}^{*k}{}_j(x(s), vx'(s)) = x'^i(s)F_i{}^{*k}{}_j(x(s), x'(s)) = G^k_j(x(s), x'(s)), \tag{1.3}$$

$$x'^i(s)G^k_i(x(s), vx'(s)) = vx'^i(s)G^k_i(x(s), x'(s)) = 2vG^k(x(s), x'(s)), \tag{1.4}$$

$$\begin{aligned} x'^i(s)g_i{}^k{}_j(x(s), x'(s)) &= 0 \qquad (a) \\ g_i{}^k{}_j(x(s), vx'(s)) &= \frac{1}{v}g_i{}^k{}_j(x(s), x'(s)). \qquad (b) \end{aligned} \tag{1.5}$$

Remark 1.2. Here and in the sequel we use the vector notations $x(s)$ and $x'(s)$ to represent the vectors $(x^0(s), \cdots, x^m(s))$ and $(x'^0(s), \cdots, x'^m(s))$, respectively, where the primes denote the derivatives with respect to s. Also, if $T^{ij\cdots}_{kh\cdots}$ are the components of a geometric object T at the point $(x(s), x'(s))$ we denote them by $T^{ij\cdots}_{kh\cdots}(s)$. ■

Next, we say that a Finsler vector field X on $\mathbb{F}^{m+1}$ along C' is **projectable** on C if locally at any point $(x(s), vx'(s)) \in C'$ it is expressed as follows:

$$X(x(s), vx'(s)) = X^i(s)\frac{\partial}{\partial y^i}(x(s), vx'(s)), \tag{1.6}$$

or, equivalently, the local components of X at any point of C' depend only on the arc length parameter s of C. The above name is justified because X given by (1.6) on C' defines a vector field X^* on C by the formula

$$X^*(x(s)) = X^i(s)\frac{\partial}{\partial x^i}(x(s)).$$

Thus $X^*(x(s))$ can be considered as the projection of the Finsler vector $X(x(s), vx'(s))$ on the tangent space TM of M at $x(s) \in C$. As an example, from (3.1.3) we see that $\partial/\partial v$ is a projectable Finsler vector field. Also we shall see later in this section that a Frenet frame for a curve in a Finsler manifold contains only projectable Finsler vector fields.

Further, we consider the Cartan, Berwald, and Rund connections $\mathbf{FC}^* = (GM', \nabla^*) = (G^k_i, F_i{}^{*k}{}_j, g_i{}^k{}_j)$, $\mathbf{BFC} = (GM', \nabla') = (G^k_i, G_i{}^k{}_j, 0)$ and $\mathbf{RFC} = (GM', \nabla'') = (G^k_i, F_i{}^{*k}{}_j, 0)$ respectively. Now, we prove the following:

PROPOSITION 1.1 *The covariant derivatives of any projectable Finsler vector field X in the direction of $\partial/\partial v$ with respect to* **FC***, **BFC** *and* **RFC** *vanish identically on $\mathcal{C}'$, that is we have*

$$(\nabla_{\partial/\partial v}X)(x(s), vx'(s)) = 0, \forall s \in (-\varepsilon, \varepsilon), \tag{1.7}$$

where ∇ is ∇^*, ∇' or ∇''.

PROOF. Indeed, by using (3.1.3), (1.6), (1.5.2) and (1.5) we obtain

$$(\nabla_{\partial/\partial v}X)(x(s), vx'(s)) = X^i(s)x'^j(s)C_i{}^k{}_j(x(s), vx'(s))\frac{\partial}{\partial y^k} = 0,$$

since for the above Finsler connections we have either $C_i{}^k{}_j = g_i{}^k{}_j$ or $C_i{}^k{}_j = 0$. ■

Hence, in particular, we have

$$\nabla_{\partial/\partial v}\frac{\partial}{\partial v} = 0, \tag{1.8}$$

which enables us to state that *the vertical covariant derivatives along $\mathcal{C}$ with respect to the Cartan, Berwald, and Rund connections do not provide any Frenet frame for $\mathcal{C}$*. Hence we have to proceed with the horizontal covariant derivatives along $\mathcal{C}$. To this end we recall (see Proposition 3.1.1) that the induced non–linear connection HC' on $\mathbb{F}^1$ by the canonical non–linear connection GM' of $\mathbb{F}^{m+1}$ is locally spanned by $\partial/\partial s$ given by

$$\frac{\partial}{\partial s} = \frac{dx^i}{ds}\frac{\delta^*}{\delta^* x^i} + v\left(\frac{d^2x^k}{ds^2} + 2G^k(s)\right)\frac{\partial}{\partial y^k}. \tag{1.9}$$

In what follows we shall see that the Cartan connection is the best choice for studying the geometry of curves in a Finsler manifold. First, by direct calculations using (1.9), (3.1.3), (1.7.29), and (1.4), we deduce that

$$\nabla^*_{\partial/\partial s}\frac{\partial}{\partial v} = \left(\frac{d^2x^k}{ds^2} + 2G^k(s)\right)\frac{\partial}{\partial y^k}. \tag{1.10}$$

On the other hand, using (3.1.5) and taking into account that ∇^* is a metric linear connection we obtain

$$g\left(\nabla^*_{\partial/\partial s}\frac{\partial}{\partial v}, \frac{\partial}{\partial v}\right) = 0. \tag{1.11}$$

As a consequence of (1.11) we may set

$$\nabla^*_{\partial/\partial s}\frac{\partial}{\partial v} = k_1 N_1, \tag{1.12}$$

where N_1 is a unit Finsler vector field which lies in $\Gamma(VC'^{\perp})$ and

$$k_1 = \|\nabla^*_{\partial/\partial s}\frac{\partial}{\partial v}\|, \tag{1.13}$$

$\|,\|$, being the Finsler norm on $\mathbb{F}^{m+1}$. By (1.10) and (1.14) we infer that

$$k_1(s) = \left\{g_{ij}(s)(x''^i(s) + 2G^i(s))(x''^j(s) + 2G^j(s))\right\}^{1/2}, \tag{1.14}$$

and call it the **geodesic curvature (first curvature) function** of C. As (1.6.10) gives all the geodesics of $\mathbb{F}^{m+1}$, by (1.9) and (1.14) we may state the following theorem:

THEOREM 1.1 *Let C be a curve in M. Then the following assertions are equivalent:*

(i) C is a geodesic of $\mathbb{F}^{m+1}$;

(ii) The induced connection HC' is a vector subbundle of the vector bundle $GM'_{|C'}$;

(iii) The geodesic curvature function vanishes indentically on C'.

If $k_1(s) \neq 0$, for all $s \in (-\varepsilon, \varepsilon)$ we call

$$N_1 = \frac{1}{k_1(s)}(x''^i(s) + 2G^i(s))\frac{\partial}{\partial y^i} = N_1^i(s)\frac{\partial}{\partial y^i}, \tag{1.15}$$

the **principal (first) normal** of C. Clearly N_1 is a projectable Finsler vector field along C. Actually, this is a consequence of the following general result.

PROPOSITION 1.2 *The covariant derivative of a projectable Finsler vector field X along C with respect to the Cartan connection in the direction of $\partial/\partial s$ is a projectable Finsler vector field too, given by*

$$\nabla^*_{\partial/\partial s}X = \left(\frac{dX^i}{ds} + X^j(s)S^i_j(s)\right)\frac{\partial}{\partial y^i}, \tag{1.16}$$

where we have set

$$S^i_j(s) = G^i_j(s) + (x''^k(s) + 2G^k(s))g_j{}^i{}_k(s). \tag{1.17}$$

PROOF. The assertion follows by direct calculations using (1.9), (1.5.1), (1.5.2), (1.3), and (1.5). ∎

Next, suppose that $m+1>2$. Since ∇^* is a metric linear connection, from $g(N_1,N_1)=1$ and $g(\partial/\partial v, N_1)=0$, by using (1.12) we deduce that

$$g(\nabla^*_{\partial/\partial s}N_1, N_1)=0 \quad\text{and}\quad g\left(\nabla^*_{\partial/\partial s}N_1, \frac{\partial}{\partial v}\right)=-k_1,$$

respectively. Hence we may set

$$\nabla^*_{\partial/\partial s}N_1 = -k_1\frac{\partial}{\partial v}+N, \tag{1.18}$$

where $N\in\Gamma(VC'^{\perp})$ and it is orthogonal to N_1 at any point of C'. Thus we may define the **second curvature function** k_2 by

$$k_2=\left\|k_1(s)\frac{\partial}{\partial v}+\nabla^*_{\partial/\partial s}N_1\right\|.$$

Then by straightforward calculations using (1.15) and (1.16) we infer that

$$k_2(s)=\left\{g_{ij}(s)(k_1x'^i+N_1'^i+N_1^kS_k^i)(k_1x'^j+N_1'^j+N_1^kS_k^j)\right\}^{1/2}(s). \tag{1.19}$$

If $k_2(s)\neq 0$ for any $s\in(-\varepsilon,\varepsilon)$ we define

$$N_2(s)=\frac{1}{k_2(s)}\left(k_1(s)\frac{\partial}{\partial v}+\nabla^*_{\partial/\partial s}N_1\right). \tag{1.20}$$

Hence (1.18) becomes

$$\nabla^*_{\partial/\partial s}N_1=-k_1\frac{\partial}{\partial v}+k_2(s)N_2(s). \tag{1.21}$$

Finally, we suppose inductively that there exist orthonormal projectable Finsler vector fields $\{N_0=\partial/\partial v, N_1,\cdots,N_i\}$ and nowhere zero curvature functions $\{k_1,\cdots,k_i\}$, $1\leq i\leq m$, such that the following equations hold:

$$\begin{aligned}
&(F_1)\quad \nabla^*_{\partial/\partial s}N_0=k_1N_1,\\
&(F_2)\quad \nabla^*_{\partial/\partial s}N_1=-k_1N_0+k_2N_2,\\
&\cdots\quad\cdots\quad\cdots\quad\cdots\quad\cdots\quad\cdots\\
&(F_i)\quad \nabla^*_{\partial/\partial s}N_{i-1}=-k_{i-1}N_{i-2}+k_iN_i.
\end{aligned} \tag{1.22}$$

Then by using Proposition 1.2 and following a proof similar to that of the Riemannian case (cf., Spivak [1], vol IV, p. 32), for any $i<m$ we obtain

$$(F_{i+1})\quad \nabla^*_{\partial/\partial s}N_i=-k_iN_{i-1}+k_{i+1}N_{i+1},$$

where

$$k_{i+1} = \left\{ g_{jk}(s)(k_i N^j_{i-1} + N'^j_i + N^h_i S^j_h)(k_i N^k_{i-1} + N'^k_i + N^h_i S^k_h) \right\}^{1/2}. \tag{1.23}$$

Moreover, $\{N_0, \cdots, N_i, N_{i+1}\}$ is an orthonormal set of projectable Finsler vector fields along C.

If $i = m$, then $\{N_0, \cdots, N_m\}$ is an orthonormal basis of $\Gamma(VM'_{|C'})$ whose elements are projectable Finsler vector fields. As $\nabla^*_{\partial/\partial s} N_m$ is orthogonal to N_m, the equation (F_{i+1}) becomes

$$(F_{m+1}) \quad \nabla^*_{\partial/\partial s} N_m = -k_m N_{m-1}.$$

From equations $(F_1) - (F_{m+1})$ we see that each N_i belongs to the subspace $V_i = \text{Span}\{\xi_0, \cdots, \xi_i\}$ where we have set

$$\xi_0 = \frac{\partial}{\partial v}, \quad \xi_1 = \nabla^*_{\partial/\partial s}\xi_0, \cdots, \xi_i = \nabla^{*(i)}_{\partial/\partial s}\xi_0.$$

Thus $\{N_0, \cdots, N_m\}$ is just an orthonormal basis in $\Gamma(V\tilde{M}'_{|C'})$ obtained from $\{\xi_0, \cdots, \xi_m\}$ provided all the curvature functions are nowhere zero on $(-\varepsilon, \varepsilon)$. We call $\{N_0, \cdots, N_m\}$ and the equations $\{(F_1), \cdots, (F_{m+1})\}$ the **Frenet frame** and the **Frenet equations** along C respectively. If $k_i = 0$ on $(-\varepsilon, \varepsilon)$ for some $i < m$, then we can not define N_i. Thus the equation (F_i) becomes

$$(F_i)' \quad \nabla^*_{\partial/\partial s} N_{i-1} = -k_{i-1} N_{i-2}.$$

Therefore summing up the above results we may say that in the case in which there exist nowhere zero curvature functions $\{k_1, \cdots, k_{i-1}\}$ and k_i is everywhere zero on $(-\varepsilon, \varepsilon)$, we have constructed the Frenet frame $\{N_0, \cdots, N_{i-1}\}$ satisfying the Frenet equations $(F_1), \cdots, (F_{i-1}), (F'_i)$.

In case M is an oriented manifold and all curvature functions are nowhere zero on $(-\varepsilon, \varepsilon)$, as in the case of Riemannian manifolds, we may choose N_m as the cross product $N_0 \times \cdots \times N_{m-1}$ with respect to the metric g and taking into account the orientation of M. In this case, k_m could be positive, negative or zero.

2. A Fundamental Theorem and two Theorems on the Reduction of Codimension of a Curve in a Finsler Manifold

In the present section we show that the Frenet frame we defined in the first section is a good tool for studying the geometry of curves in a Finsler manifold. First we prove the following fundamental theorem for curves in

Finsler manifolds.

THEOREM 2.1 *(Bejancu-Deshmukh [2]). Let $\mathbb{F}^{m+1} = (M, M', F)$ be a Finsler manifold, $(x_0, y_0) = (x_0^i, y_0^i)$ be a fixed point of M', $\{V_0, \cdots, V_m\}$ be an orthonormal basis of $VM'_{(x_0,y_0)}$ and $k_1, \cdots, k_m : (-\varepsilon, \varepsilon) \to \mathbb{R}$ be everywhere positive smooth functions. Then there exists a unique curve C on M given by the equations $x^i = x^i(s)$, $s \in (-\varepsilon, \varepsilon)$, where s is the arc length parameter of C, such that $x^i(0) = x_0^i$ and $k_1, \cdots, k_m$ are the curvature functions of C with respect to the Frenet frame $\{N_0, \cdots, N_m\}$ which satisfies $N_h(0) = V_h, h \in \{0, \cdots, m\}$.*

PROOF. As a result of (1.16) we obtain some local expressions of Frenet equations which enable us to consider the following system of differential equations:

$$(F_0^*) \quad x'^i(s) = N_0^i(s),$$

$$(F_1^*) \quad N_0'^i(s) = -2G^i(s) + k_1(s)N_1^i(s),$$

$$\cdots \quad \cdots \quad \cdots \quad \cdots \quad \cdots$$

$$(F_h^*) \quad N_{h-1}'^i(s) = -N_{h-1}^j(s)S_j^i(s) - k_{h-1}(s)N_{h-2}^i(s) + k_h(s)N_h^i(s),$$

$$\cdots \quad \cdots \quad \cdots \quad \cdots \quad \cdots$$

$$(F_{m+1}^*) \quad N_m'^i(s) = -N_m^j(s)S_j^i(s) - k_m N_{m-1}^i, \tag{2.1}$$

where $G^i(s)$ and $S_j^i(s)$ are the functions from (1.6.9) and (1.17) calculated at points $(x(s), x'(s))$, and $(x^i(s), N_0^i(s), \cdots, N_m^i(s))$ are the unknown functions. Then using an argument similar to that in the case of Riemannian manifolds (see Spivak [1], vol 4, p. 35) we conclude that there exists a unique solution $(x^i(s), N_0(s), \cdots, N_m(s))$ on $(-\varepsilon, \varepsilon)$, where we set

$$N_h(s) = N_h^i(s)\frac{\partial}{\partial y^i},$$

satisfying the initial conditions $x^i(0) = x_0^i$ and $N_h(0) = V_h$ for any $h \in \{0, \cdots, m\}$. As N_h are projectable Finsler vector fields along $C(x^i = x^i(s))$, by Proposition 1.2 we see that equations in (F_h^*) are just local expressions of the Frenet equations (F_h) for $h \in \{1, \cdots, m+1\}$. It remains only to prove that $\{N_0(s), \cdots, N_m(s)\}$ is an orthonormal basis of $VM'_{(x(s),x'(s))}$. To this end we set

$$g_{hr}^*(s) = g(x(s), x'(s))(N_h(s), N_r(s)), h, r \in \{0, \cdots, m\},$$

and, taking into account that ∇^* is a metric linear connection on VM', we infer that

$$\frac{dg^*_{hr}}{ds} = g(\nabla^*_{\partial/\partial s}N_h, N_r) + g(N_h, \nabla^*_{\partial/\partial s}N_r). \tag{2.2}$$

Next, we note that the Frenet equations (F_{h+1}) can be written as follows:

$$(F_{h+1}) \quad \nabla^*_{\partial/\partial s}N_h = \sum_{k=0}^{m} a_{hk}N_k, \; h \in \{0, \cdots, m\},$$

where $[a_{hk}]$ is a skew–symmetric matrix. Thus (2.2) becomes

$$\frac{dg^*_{hr}}{ds} = \sum_{k=0}^{m} \{a_{hk}g^*_{kr} + a_{rk}g^*_{kh}\}. \tag{2.3}$$

It is easy to check that (δ_{hr}) is a solution of the system (2.3). Since $g^*_{hr}(0) = g(x(0), x'(0))(V_h, V_r) = \delta_{hr}$, we deduce that $g^*_{hr} = \delta_{hr}$. This concludes the proof of the theorem. ■

In particular we consider a curve C in a Minkowski space $\mathbb{F}^{m+1} = (M = \mathbb{R}^{m+1}, M', F)$ given locally by (3.1.1). Then we consider a vector sub–bundle D of rank p in $VM'_{|C'}$ and denote by D^* the vector space that is complementary orthogonal to $D(0)$ in $VM'_{(x(0),x'(0))}$. Then we say that D is a parallel vector bundle along C if for any $s \in (-\varepsilon, \varepsilon)$ we have

$$g(x(s), x'(s))(D(s), D^*) = 0.$$

We are now able to state two theorems on the reduction of the codimension of C in $\mathbb{F}^{m+1}$.

THEOREM 2.2 *Let C be a curve in $\mathbb{F}^{m+1} = (\mathbb{R}^{m+1}, M', F)$ for which there exists a parallel vector bundle D along C of rank $p < m+1$ such that $\partial/\partial v \in \Gamma(D)$. Then C lies in some p–dimensional plane of $\mathbb{R}^{m+1}$.*

PROOF. We choose a coordinate system (x^i, y^i) in M' such that

$$D(0) = \mathrm{Span}\{\partial/\partial y^0, \cdots, \partial/\partial y^{p-1}\}_{|(x(0),x'(0))}.$$

As $D(0)$ and $\mathrm{Span}\{\partial/\partial y^0, \cdots, \partial/\partial y^{p-1}\}_{|(x(s),x'(s))}$ have the same complementary orthogonal vector space D^* in $VM'_{|(x(0),x'(0))}$ and $VM'_{(x(s),x'(s))}$ respectively and D is parallel along C, we conclude that

$$D(s) = \mathrm{Span}\ \{\partial/\partial y^0, \cdots, \partial/\partial y^{p-1}\}_{|(x(s),x'(s))}.$$

Assume now that there exists a point $x(s) \in C$ such that $x'^k(s) \neq 0$ for certain $k > p$. Then by (3.1.3) it follows that $\partial/\partial v$ does not lie in $D(s)$, which is a contradiction. Hence at any point of C we have $x'^p = \cdots, = x'^m = 0$, which means that C lies in a p–dimensional plane of $\mathbb{R}^{m+1}$. ■

THEOREM 2.3 *Let C be a curve in $\mathbb{F}^{m+1} = (\mathbb{R}^{m+1}, M', F)$ such that the curvature functions $k_1, \cdots, k_{p-1}$ are nowhere zero and k_p is everywhere zero on $(-\varepsilon, \varepsilon)$. Then C lies in some p-dimensional plane of $\mathbb{R}^{m+1}$.*

PROOF. Denote by D the vector bundle spanned by the Frenet frame $\{N_0 = \partial/\partial v, N_1, \cdots, N_{p-1}\}$ along C. From the Frenet equations (F_1) – (F_{p-1}) and $(F_p)'$ we deduce that

$$\nabla^*_{\partial/\partial s} N_j(s) = \sum_{h=0}^{p-1} A_j^h(s) N_h(s), \ j \in \{0, \cdots, p-1\}. \tag{2.4}$$

Suppose D^* is complementary orthogonal to $D(0)$ in $VM'_{|(x(0),x'(0))}$ and take an arbitrary $W \in D^*$. Denote by the same letter the constant vector field $W(s) = W$, $s \in (-\varepsilon, \varepsilon)$. Taking into account that ∇^* is a metric linear connection and using (2.4) we infer that

$$\frac{d}{ds}(g(N_j, W)(s)) = g(\nabla^*_{\partial/\partial s} N_j, W)(s) = \sum_{h=0}^{p-1} A_j^h(s) g(N_h, W)(s). \tag{2.5}$$

Since $g(N_j, W)(0) = 0$, the uniqueness of solutions of this system implies that $g(N_j, W)(s) = 0$, for any $s \in (-\varepsilon, \varepsilon)$. Hence D is parallel along C. As $\partial/\partial v \in \Gamma(D)$, the assertion follows from Theorem 2.2. ■

3. Curves in $\mathbb{F}^2$ and $\mathbb{F}^3$

Let $\mathbb{F}^2 = (M, M', F)$ be a 2–dimensional Finsler manifold and C be a smooth curve in M. In this particular case we define the curvature of C in a slightly different way from the definition of the geodesic curvature in the general case. First we define the Finsler vector field N_1 along C by

$$N_1 = \frac{1}{\sqrt{g^*}}\left\{-(g_{12}x'^1 + g_{22}x'^2)\frac{\partial}{\partial y^1} + (g_{11}x'^1 + g_{21}x'^2)\frac{\partial}{\partial y^2}\right\}, \tag{3.1}$$

where $g^* = \det[g_{ij}]$. It is easy to check that N_1 is a unit projectable Finsler vector field along C which is orthogonal to $N_0 = \partial/\partial v$. Moreover, the bases $\{N_0, N_1\}$ and $\{\partial/\partial y^1, \partial/\partial y^2\}$ have the same orientation since by (3.1.4) we have

$$\begin{vmatrix} x'^1 & -g_{12}x'^1 - g_{22}x'^2 \\ x'^2 & g_{11}x'^1 + g_{21}x'^2 \end{vmatrix} = 1.$$

As $\nabla^*_{\partial/\partial s}N_0$ is orthogonal to N_0 we set

$$\nabla^*_{\partial/\partial s}N_0 = kN_1, \tag{3.2}$$

where k is a smooth function on $(-\varepsilon, \varepsilon)$ that we name as the **curvature** of C. Thus (3.2) is the (F_1)–Frenet equation and k might be positive, negative or zero at points of C. By some lengthy calculations using (3.1), (3.2) and (1.10) we infer that

$$k = g(\nabla^*_{\partial/\partial s}N_0, N_1) = \sqrt{g^*}\begin{vmatrix} x'^1 & x''^1 + 2G^1 \\ x'^2 & x''^2 + 2G^2 \end{vmatrix}. \tag{3.3}$$

Finally, the (F_2)–Frenet equation becomes

$$\nabla^*_{\partial/\partial s}N_1 = -kN_0. \tag{3.4}$$

We are now looking for an explicit formula for k in the case in which C is given by equations $x^i = x^i(t)$, where t is an arbitrary parameter on C. First, from the definition of the arc length we obtain

$$(a)\quad x'^i = \frac{\dot{x}^i}{F(t)} \quad \text{and} \quad (b)\quad x''^i = \frac{1}{F^2(t)}\left(\ddot{x}^i - \dot{x}^i\frac{d(\ell n\, F)}{dt}\right), \tag{3.5}$$

where $\dot{x}^i$ and $\ddot{x}^i$ denote the first and second derivatives of x^i with respect to t, and $F(t) = F(x(t), \dot{x}(t))$.

Taking into account that G^i are positive homogeneous functions of degree 2 with respect to (y^i) and using (3.5a) and (1.6.9) we deduce that

$$\begin{aligned} 2G^i(x(s), x'(s)) &= \frac{2}{F^2}G^i(x(t), \dot{x}(t)) \\ &= \frac{1}{F^2}g^{ij}\frac{\partial F}{\partial \dot{x}^j}\frac{\partial F}{\partial x^h}\dot{x}^h + \frac{g^{ij}}{F}\left(\frac{\partial^2 F}{\partial \dot{x}^j \partial x^h}\dot{x}^h - \frac{\partial F}{\partial x^j}\right). \end{aligned}$$

Then, from (1.7.23) we obtain

$$2G^i(x(s), x'(s)) = \frac{\dot{x}^i}{F^3}\frac{\partial F}{\partial x^h}\dot{x}^h + \frac{g^{ij}}{F}\left(\frac{\partial^2 F}{\partial \dot{x}^j \partial x^h}\dot{x}^h - \frac{\partial F}{\partial x^j}\right). \tag{3.6}$$

Thus using (3.5) and (3.6) in (3.3) we obtain

$$k = \frac{\sqrt{g^*}}{F^3}\left\{\ddot{x}^2\dot{x}^1 - \ddot{x}^1\dot{x}^2 + F(\dot{x}^1 g^{2j} - \dot{x}^2 g^{1j})\left(\frac{\partial^2 F}{\partial \dot{x}^j \partial x^h}\dot{x}^h - \frac{\partial F}{\partial x^j}\right)\right\}. \tag{3.7}$$

As $\partial F/\partial x^i$ are positive homogeneous of degree 1 with respect to $(\dot{x}^i)$ we have

$$\frac{\partial^2 F}{\partial x^i \partial \dot{x}^j}\dot{x}^j = \frac{\partial F}{\partial x^i}. \tag{3.8}$$

Substitute $\frac{\partial^2 F}{\partial x^1 \partial \dot{x}^1}\dot{x}^1$ and $\frac{\partial^2 F}{\partial x^2 \partial \dot{x}^2}\dot{x}^2$ from (3.8) in (3.7) and using (3.5a) and (3.1.4) we obtain

$$k = \frac{\sqrt{g^*}}{F^3}(\ddot{x}^2\dot{x}^1 - \ddot{x}^1\dot{x}^2) + \frac{1}{\sqrt{g^*}}\left(\frac{\partial^2 F}{\partial x^1 \partial \dot{x}^2} - \frac{\partial^2 F}{\partial x^2 \partial \dot{x}^1}\right). \tag{3.9}$$

As is well known (cf., Berwald [2]) we have

$$g^* = F^3 F_1, \tag{3.10}$$

where we have set

$$F_1 = \frac{1}{(\dot{x}^2)^2}\frac{\partial^2 F}{(\partial \dot{x}^1)^2} = \frac{1}{(\dot{x}^1)^2}\frac{\partial^2 F}{(\partial \dot{x}^2)^2}. \tag{3.11}$$

Thus by using (3.10) in (3.9) we obtain the final formula for k as it is stated in the next proposition.

PROPOSITION 3.1 *The curvature of a curve C in $\mathbb{F}^2$ is given by*

$$k = \frac{1}{\sqrt{F^3 F_1}}\left(F_1(\ddot{x}^2\dot{x}^1 - \ddot{x}^1\dot{x}^2) + \frac{\partial^2 F}{\partial x^1 \partial \dot{x}^2} - \frac{\partial^2 F}{\partial x^2 \partial \dot{x}^1}\right). \tag{3.12}$$

If in particular, $\mathbb{F}^2$ is a Minkowski plane, that is, $M = \mathbb{R}^2$ and F depends only on (y^1, y^2), then from (3.12) we deduce that

$$k = \frac{1}{F}\sqrt{\frac{F_1}{F}}\left(\ddot{x}^2\dot{x}^1 - \ddot{x}^1\dot{x}^2\right). \tag{3.13}$$

If, moreover, $F(y^1, y^2) = \{(y^1)^2 + (y^2)^2\}^{1/2}$ then $F_1 = 1/F^3$, and we obtain the well known expression for the curvature of a curve in a Euclidean plane (cf., Spivak [1], vol. 2, p. 11).

Next, we consider a 3–dimensional Finsler manifold $\mathbb{F}^3 = (M, M', F)$ and a smooth curve C in M given locally by the parametric equations

$$x^i = x^i(s);\ (x'^1(s), x'^2(s), x'^3(s)) \neq (0,0,0), \tag{3.14}$$

where s is the arc length parameter on C. According to (1.12) and (1.14) we have

$$\nabla^*_{\partial/\partial s}\frac{\partial}{\partial v} = k\mathbf{n}, \tag{3.15}$$

where

$$k(s) = \left\{g_{ij}(s)(x''^i(s) + 2G^i(s))(x''^j(s) + 2G^j(s))\right\}^{1/2}, \tag{3.16}$$

and $\mathbf{n}$ is a unit Finsler vector field that lies in $\Gamma(V\mathcal{C}'^{\perp})$. As in the Riemannian case we call k the **curvature** of $\mathcal{C}$. If $k(s) \neq 0$ for all $s \in (-\varepsilon, \varepsilon)$ then $\mathbf{n}$ is given by

$$\mathbf{n} = \frac{1}{k(s)}(x''^i(s) + 2G^i(s))\frac{\partial}{\partial y^i} = n^i(s)\frac{\partial}{\partial y^i}. \tag{3.17}$$

We call $\mathbf{n}$ the **principal normal Finsler vector field** of $\mathcal{C}$. In order to obtain the third Finsler vector field from the Frenet frame we set

$$x'_i = g_{ij}x'^j \quad \text{and} \quad n_i = g_{ij}n^j.$$

Then we define

$$c^1 = \begin{vmatrix} x'_2 & x'_3 \\ n_2 & n_3 \end{vmatrix}, \quad c^2 = \begin{vmatrix} x'_3 & x'_1 \\ n_3 & n_1 \end{vmatrix}, \quad c^3 = \begin{vmatrix} x'_1 & x'_2 \\ n_1 & n_2 \end{vmatrix}, \quad c = g_{ij}c^ic^j.$$

At this point we assume that M is an orientable manifold. Then we consider two overlapping coordinate systems $(\mathcal{U}; x^i)$ an $(\widetilde{\mathcal{U}}; \tilde{x}^i)$ on M such that the Jacobian determinant function

$$J(x, \tilde{x}) = \det\left[\frac{\partial \tilde{x}^i}{\partial x^j}\right],$$

is positive. By straightforward calculations we deduce that

$$c^i = \frac{1}{J}\tilde{c}^j\frac{\partial x^i}{\partial \tilde{x}^j} \quad \text{and} \quad c = \frac{1}{J^2}\tilde{c} \quad \text{on} \quad \mathcal{U} \cap \widetilde{\mathcal{U}}.$$

This enables us to define the unit Finsler vector field

$$\mathbf{b} = b^i\frac{\partial}{\partial y^i} \quad \text{where} \quad b^i = \frac{1}{\sqrt{c}}c^i, \quad i \in \{1, 2, 3\}.$$

It is easy to check that $\mathbf{b}$ is orthogonal to both $\partial/\partial v$ and $\mathbf{n}$. We call $\mathbf{b}$ the **binormal Finsler vector field** along $\mathcal{C}$. Moreover, the bases $\{\partial/\partial v, \mathbf{n}, \mathbf{b}\}$ and $\{\partial/\partial y^1, \partial/\partial y^2, \partial/\partial y^3\}$ have the same orientation. Thus we are entitled to call $\{\partial/\partial v, \mathbf{n}, \mathbf{b}\}$ the Frenet frame for the curve $\mathcal{C}$ in $\mathbb{F}^3$.

Taking into account that ∇^* is a metric linear connection and using (3.15) we obtain

$$\nabla^*_{\partial/\partial s}\mathbf{n} = -k\frac{\partial}{\partial v} + \tau\mathbf{b}, \tag{3.18}$$

and

$$\nabla^*_{\partial/\partial s}\mathbf{b} = \tau\mathbf{n}, \tag{3.19}$$

where τ is a smooth function given by

$$\tau(s) = g(\nabla^*_{\partial/\partial s}\mathbf{n}, \mathbf{b})(s) = g_{ij}(s)b^i(s)\left(\frac{dn^j}{ds} + n^k(s)S^j_k(s)\right). \tag{3.20}$$

Thus (3.15), (3.18), and (3.19) are the Frenet formulas for the curve C in the Finsler manifold $\mathbb{F}^3$. We call τ the **torsion** of C and from Theorem 2.3 we obtain the following corollary.

COROLLARY 3.1 *If C is a curve in a Minkowski space $\mathbb{F}^3$ such that its torsion is everywhere zero on $(-\varepsilon, \varepsilon)$, then C lies in a plane of $\mathbb{R}^3$.*

CHAPTER 5

PSEUDO–FINSLER HYPERSURFACES

In the first three sections of the present chapter we apply the general theory of pseudo–Finsler submanifolds developed in Chapters 2 and 3 to the particular case of pseudo–Finsler hypersurfaces. In this setting we present concepts and results owed to Berwald [3], Brown [1], [2], Matsumoto [4], Rund [1]–[5] and Varga [1], [2]. Also, we introduce the induced horizontal flag curvature of a pseudo–Finsler hypersurface $\mathbb{F}^m$ and compare it with both the intrinsic horizontal flag curvature of $\mathbb{F}^m$ and the horizontal flag curvature of the ambient manifold $\widetilde{\mathbb{F}}^{m+1}$. In particular, when $\widetilde{\mathbb{F}}^{m+1}$ is a pseudo–Minkowski space we stress the role of the structure equations induced by the Rund connection of $\widetilde{\mathbb{F}}^{m+1}$. In the last section of the chapter we present a Minkowskian approach to the theory of pseudo–Finsler hypersurfaces. Being based on a Minkowskian unit normal vector field depending on position only, the theory we develop here is completely different from what we presented till now throughout the book.

1. Induced Geometric Objects on a Pseudo–Finsler Hypersurface

Let $\mathbb{F}^m = (M, M', F^*)$ be a pseudo–Finsler hypersurface of a pseudo–Finsler manifold $\widetilde{\mathbb{F}}^{m+1} = (\widetilde{M}, \widetilde{M}', \widetilde{F}^*)$. The purpose of this section is to determine the induced geometric objects on $\mathbb{F}^m$ using the general theory we developed in Chapter 2. We will compare our findings with what is known in the literature.

1.1. THE FINSLER TENSOR FIELDS $M_{\alpha\beta}$ AND M_α

First, we assume that locally the immersion $f : M \to \widetilde{M}$ is given by the equations

$$x^i = x^i(u^\alpha), \quad 1 \leq i \leq m+1, \quad 1 \leq \alpha \leq m. \tag{1.1}$$

In this chapter we use the following range for indices: $\alpha, \beta, \gamma, \dots \in \{1, \dots, m\}$ and $i, j, k, \dots \in \{1, \dots, m+1\}$.

The Finsler normal bundle $VM'^{\perp}$ being of rank 1, we shall consider a spacelike local normal Finsler vector field $B = B^i \partial/\partial y^i$ and set $\widetilde{B}_i = \tilde{g}_{ij} B^j$. Then (2.1.9)–(2.1.12) imply

$$\begin{aligned} (a)\ \widetilde{B}_i B^i_\alpha &= 0; \quad (b)\ \widetilde{B}_i B^i = 1; \quad (c)\ \widetilde{B}^\alpha_i B^i_\beta = \delta^\alpha_\beta; \\ (d)\ \widetilde{B}^\alpha_i B^i &= 0; \quad (e)\ B^i_\alpha \widetilde{B}^\alpha_j + B^i \widetilde{B}_j = \delta^i_j. \end{aligned} \tag{1.2}$$

According to (2.3.24b) the local components of the normal curvature form of $\mathbb{F}^m$ with respect to B are given by

$$H_\alpha = \widetilde{B}_i (B^i_{\alpha 0} + B^j_\alpha \widetilde{G}^i_j). \tag{1.3}$$

Since $a = 1$ in (2.3.3), taking into account the notations used by Brown [1] we put

$$(a)\ M_{\alpha\beta} = \tilde{g}_{ijk} B^i B^{jk}_{\alpha\beta}; \quad (b)\ M_\alpha = \tilde{g}_{ijk} B^i B^j B^k_\alpha. \tag{1.4}$$

Clearly $M_{\alpha\beta}$ are symmetric Finsler tensor fields on $\mathbb{F}^m$ and we have

$$(a)\ M_{\alpha\beta} v^\alpha = 0 \quad \text{and} \quad (b)\ M_\alpha v^\alpha = 0. \tag{1.5}$$

Then (2.3.5) becomes

$$\frac{\partial \widetilde{B}^\alpha_i}{\partial v^\beta} = 2 M^\alpha_\beta \widetilde{B}_i. \tag{1.6}$$

Moreover, differentiating (1.2a) and (1.2b) with respect to v^β and using (1.6) and (1.4b) we obtain

$$(a)\ 2 M^\alpha_\beta + \widetilde{B}^\alpha_i \frac{\partial B^i}{\partial v^\beta} = 0; \quad (b)\ M_\beta + \widetilde{B}_i \frac{\partial B^i}{\partial v^\beta} = 0. \tag{1.7}$$

Contracting (1.7a) by B^j_α and multiplying (1.7b) by B^j and adding, by (1.2e) we deduce that

$$\frac{\partial B^i}{\partial v^\beta} = -2 M^\alpha_\beta B^i_\alpha - M_\beta B^i. \tag{1.8}$$

Similarly, differentiating (1.2a) and (1.2b) with respect to v^β and using (1.8) we derive

$$\frac{\partial \widetilde{B}_i}{\partial v^\beta} B^i_\alpha = 0; \quad \frac{\partial \widetilde{B}_i}{\partial v^\beta} B^i = M_\beta.$$

Thus we infer that

$$\frac{\partial \widetilde{B}_i}{\partial v^\beta} = M_\beta \widetilde{B}_i. \tag{1.9}$$

Taking into account (1.8) and (1.9) we may state the following theorem (see Theorems 2.2 and 2.3 of Brown [1]).

THEOREM 1.1 *Let $\mathbb{F}^m$ be a pseudo–Finsler hypersurface of $\widetilde{\mathbb{F}}^{m+1}$. Then the following assertions are equivalent:*

(i) The local components $(\widetilde{B}_i)$ do not depend on (v^α);

(ii) The Finsler vector fields with local components $(\partial B^i/\partial v^\alpha)$ belong to $\Gamma(VM')$;

(iii) M_α vanish identically for any $\alpha \in \{1, \cdots, m\}$.

Similarly, from (1.6) and (1.8) we deduce the following theorem (see Theorems 4.2 and 4.3 in Brown [1]):

THEOREM 1.2 *Let $\mathbb{F}^m$ be a pseudo–Finsler hypersurface of $\widetilde{\mathbb{F}}^{m+1}$. Then the following assertions are equivalent:*

(i) The local components $\widetilde{B}_i^\alpha$ do not depend on (v^α);

(ii) The Finsler vector fields with local components $(\partial B^i/\partial v^\alpha)$ lie in $\Gamma(VM'^\perp)$;

(iii) $M_{\alpha\beta}$ vanish identically for any $\alpha, \beta \in \{1, \cdots, m\}$.

1.2. THE SECOND FUNDAMENTAL FORMS AND THE SHAPE OPERATORS

We have seen that the h–second fundamental form and the v–second fundamental form for a pseudo–Finsler submanifold depend upon the choice of the Finsler connection on the ambient manifold. Thus we have obtained the h–second fundamental forms $\overset{c}{H} = (\overset{c}{H}{}^a{}_{\alpha\beta})$, $\overset{b}{H} = (\overset{b}{H}{}^a{}_{\alpha\beta})$ and $\overset{r}{H} = (\overset{r}{H}{}^a{}_{\alpha\beta})$ when the ambient manifold was equiped with the Cartan, Berwald, and Rund connections respectively. Similarly, we have obtained the v–second fundamental forms $\overset{c}{V} = (\overset{c}{V}{}^a{}_{\alpha\beta})$, $\overset{b}{V} = (\overset{b}{V}{}^a{}_{\alpha\beta})$ and $\overset{r}{V} = (\overset{r}{V}{}^a{}_{\alpha\beta})$. Since for hypersurfaces we have $a = 1$, we will write

$$(a)\quad \overset{c}{H} = (\overset{c}{H}_{\alpha\beta}); \quad (b)\quad \overset{b}{H} = (\overset{b}{H}_{\alpha\beta}); \quad (c)\quad \overset{r}{H} = (\overset{r}{H}_{\alpha\beta}),$$

and

$$(a)\quad \overset{c}{V} = (\overset{c}{V}_{\alpha\beta}); \quad (b)\quad \overset{b}{V} = (\overset{b}{V}_{\alpha\beta}); \quad (c)\quad \overset{r}{V} = (\overset{r}{V}_{\alpha\beta}).$$

Then by (2.3.30), (2.4.6b) and (2.5.4b) we obtain

$$(a)\quad \overset{c}{H}_{\alpha\beta}\, v^\alpha = H_\beta; \quad (b)\quad \overset{c}{H}_{\alpha\beta}\, v^\beta = H_\alpha + M_\alpha H_0;$$

$$(c)\ \overset{\mathrm{b}}{H}_{\alpha\beta}\, v^{\alpha} = \overset{\mathrm{r}}{H}_{\alpha\beta}\, v^{\alpha} = H_{\beta};$$

$$(d)\ \overset{\mathrm{b}}{H}_{\alpha\beta}\, v^{\beta} = \overset{\mathrm{r}}{H}_{\alpha\beta}\, v^{\beta} = H_{\alpha}, \tag{1.10}$$

and

$$\overset{\mathrm{c}}{H}_{\alpha\beta}\, v^{\alpha}v^{\beta} = \overset{\mathrm{b}}{H}_{\alpha\beta}\, v^{\alpha}v^{\beta} = \overset{\mathrm{r}}{H}_{\alpha\beta}\, v^{\alpha}v^{\beta} = H_0. \tag{1.11}$$

Then from (2.3.25), (2.4.5a), and (2.5.3a) we have

$$\begin{cases} \overset{\mathrm{c}}{H}_{\alpha\beta} = \widetilde{B}_k(B^k_{\alpha\beta} + \widetilde{F}^{*k}_{i\ j} B^{ij}_{\alpha\beta}) + M_\alpha H_\beta, & (a) \\ \overset{\mathrm{b}}{H}_{\alpha\beta} = \widetilde{B}_k(B^k_{\alpha\beta} + \widetilde{G}_{i\ j}^{\ k} B^{ij}_{\alpha\beta}), & (b) \\ \overset{\mathrm{r}}{H}_{\alpha\beta} = \widetilde{B}_k(B^k_{\alpha\beta} + \widetilde{F}^{*k}_{i\ j} B^{ij}_{\alpha\beta}), & (c) \end{cases} \tag{1.12}$$

respectively. Similarly, from (2.3.26), (2.4.5b) and (2.5.3b) we obtain

$$(a)\ \overset{\mathrm{c}}{V}_{\alpha\beta} = M_{\alpha\beta};\quad (b)\ \overset{\mathrm{b}}{V}_{\alpha\beta} = 0 \quad\text{and}\quad (c)\ \overset{\mathrm{r}}{V}_{\alpha\beta} = 0. \tag{1.13}$$

By (2.3.27), (2.4.7) and (2.5.6) the h–shape operators $\overset{\mathrm{c}}{A^h} = (\overset{\mathrm{c}}{H'^{\alpha}_{\beta}})$, $\overset{\mathrm{b}}{A^h} = (\overset{\mathrm{b}}{H'^{\alpha}_{\beta}})$ and $\overset{\mathrm{r}}{A^h} = (\overset{\mathrm{r}}{H'^{\alpha}_{\beta}})$ and the v–shape operators $\overset{\mathrm{c}}{A^v} = (\overset{\mathrm{c}}{V'^{\alpha}_{\beta}})$, $\overset{\mathrm{b}}{A^v} = (\overset{\mathrm{b}}{V'^{\alpha}_{\beta}})$ and $\overset{\mathrm{r}}{A^v} = (\overset{\mathrm{r}}{V'^{\alpha}_{\beta}})$ are given by

$$(a)\ \overset{\mathrm{c}}{H'^{\alpha}_{\beta}} = \overset{\mathrm{c}}{H^{\alpha}_{\beta}};\quad (b)\ \overset{\mathrm{b}}{H'^{\alpha}_{\beta}} = \overset{\mathrm{b}}{H^{\alpha}_{\beta}} + 2(M^\alpha H_\beta - L^\alpha_\beta);$$

$$(c)\ \overset{\mathrm{r}}{H'^{\alpha}_{\beta}} = \overset{\mathrm{r}}{H^{\alpha}_{\beta}} + 2M^\alpha H_\beta, \tag{1.14}$$

and

$$(a)\ \overset{\mathrm{c}}{V'^{\alpha}_{\beta}} = M^\alpha_\beta;\quad (b)\ \overset{\mathrm{b}}{V'^{\alpha}_{\beta}} = 2M^\alpha_\beta;\quad (c)\ \overset{\mathrm{r}}{V'^{\alpha}_{\beta}} = 2M^\alpha_\beta, \tag{1.15}$$

where according to (2.3.33b), L^α_β are given by

$$L^\alpha_\beta = \widetilde{P}^{*k}_{\ \ ij} B^i_\beta B^j \widetilde{B}^\alpha_k. \tag{1.16}$$

As usual, lowering and raising indices of tensor fields on M is done by using $g_{\alpha\beta}$ and $g^{\alpha\beta}$. Comparing with Riemannian geometry we may think of $\overset{\mathrm{c}}{H^{\alpha}_{\beta}} = g^{\alpha\gamma}\, \overset{\mathrm{c}}{H}_{\gamma\beta}$, $\overset{\mathrm{b}}{H^{\alpha}_{\beta}} = g^{\alpha\gamma}\, \overset{\mathrm{b}}{H}_{\gamma\beta}$ and $\overset{\mathrm{r}}{H^{\alpha}_{\beta}} = g^{\alpha\gamma}\, \overset{\mathrm{r}}{H}_{\gamma\beta}$ as some shape operators introduced by the Gauss formula. Apart from the Cartan connection, these Finsler tensor fields do not coincide with the shape operators introduced by the Weingarten formula. For this reason we call $\overset{b}{H}{}^{\alpha}_{\beta}$ and $\overset{r}{H}{}^{\alpha}_{\beta}$ the **Berwald–Gauss** and **Rund–Gauss shape operators**, while $\overset{b}{H'}{}^{\alpha}_{\beta}$ and $\overset{r}{H'}{}^{\alpha}_{\beta}$ will be called the **Berwald–Weingarten** and **Rund–Weingarten shape operators.**

The **normal curvature** of $\mathbb{F}^m$ in $\widetilde{\mathbb{F}}^{m+1}$ is denoted by $\mathbf{n}(u,v)$ and it is given by (see (3.1.30))

$$\mathbf{n}(u,v) = \frac{H_0(u,v)}{F^*(u,v)}, \tag{1.17}$$

where

$$H_0(u,v) = H_\alpha(u,v)v^\alpha. \tag{1.18}$$

Then by using (1.3), (1.6.17) and (1.6.11) we deduce that

$$\mathbf{n} = \frac{1}{F^*}\widetilde{B}_i(B^i_{00} + 2\widetilde{G}^i). \tag{1.19}$$

By (1.9) and (1.5b) it follows that $\widetilde{B}_i$ are positively homogeneous of degree zero with respect to (v^α). Since $\widetilde{G}^i$ and F^* are positively homogeneous of degree two, from (1.19) we deduce that the normal curvature is positively homogeneous of degree zero. Moreover, we should remark that this normal curvature is the same as that introduced by Berwald [3] (see formula (2.2a)) and considered later in several papers (see Brown [2], Cartan [1], Matsumoto [4], Rund [3], p. 184). We also note that Berwald defined a second fundamental form for $\mathbb{F}^m$ with local components (see formula (2.8) in Berwald [3])

$$\Omega_{\alpha\beta} = \frac{1}{2}\frac{\partial^2(F^*\mathbf{n})}{\partial v^\alpha \partial v^\beta} = \frac{1}{2}\frac{\partial^2 H_0}{\partial v^\alpha \partial v^\beta}. \tag{1.20}$$

It is the purpose of the remaining part of this subsection to find the relationship between our second fundamental forms and the one introduced by Berwald. First we note that by (3.1.40) we obtain for $\overset{b}{H}_{\alpha\beta}$ the following formula

$$\overset{b}{H}_{\alpha\beta} = \frac{1}{2}\widetilde{B}_k \frac{\partial^2(H_0 B^k)}{\partial v^\alpha \partial v^\beta}. \tag{1.21}$$

Next, by using (3.1.37) and (1.8) we deduce that

$$\frac{\partial H_0}{\partial v^\beta} = 2H_\beta + H_0 M_\beta. \tag{1.22}$$

Then, by direct calculations using (1.22), (1.8), (1.2a) and (1.2b) we obtain

$$\overset{b}{H}_{\alpha\beta} = \Omega_{\alpha\beta} - (H_\alpha M_\beta + H_\beta M_\alpha) - H_0 M_\alpha M_\beta + \frac{1}{2}H_0 \widetilde{B}_k \frac{\partial^2 B^k}{\partial v^\alpha \partial v^\beta}. \tag{1.23}$$

On the other hand, differentiating (1.7) with respect to v^α and using (1.8) and (1.9) we infer that

$$\widetilde{B}_k \frac{\partial^2 B^k}{\partial v^\alpha \partial v^\beta} = M_\alpha M_\beta - \frac{\partial M_\beta}{\partial v^\alpha}.$$

Thus (1.23) becomes

$$\overset{b}{H}_{\alpha\beta} = \Omega_{\alpha\beta} - (H_\alpha M_\beta + H_\beta M_\alpha) - \frac{1}{2} H_0 \left(M_\alpha M_\beta + \frac{\partial M_\beta}{\partial v^\alpha} \right). \tag{1.24}$$

The above formula is equivalent to (3.8) of Brown [1] and (6.21) of Matsumoto [4]. We remark that $\overset{b}{H}_{\alpha\beta} = \Omega_{\alpha\beta}$ for totally geodesic pseudo–Finsler hypersurfaces since by Theorem 3.1.2 and Proposition 3.1.3 we have $H_\alpha = 0$. For any other pseudo–Finsler hypersurface we prove the following.

THEOREM 1.3 *Let $\mathbb{F}^m$ be a pseudo–Finsler hypersurface of $\widetilde{\mathbb{F}}^{m+1}$. If $\mathbb{F}^m$ is not totally geodesic then $\overset{b}{H}_{\alpha\beta} = \Omega_{\alpha\beta}$ for any $\alpha, \beta \in \{1, \cdots, m\}$ if and only if $M_\alpha = 0$ for any $\alpha \in \{1, \cdots, m\}$.*

PROOF. Clearly, if $M_\alpha = 0$ then from (1.24) we obtain $\overset{b}{H}_{\alpha\beta} = \Omega_{\alpha\beta}$. Next, by contracting (1.24) by v^α, using (1.5b) and taking into account that M_β are positively homogeneous of degrees -1, we obtain

$$\overset{b}{H}_{\alpha\beta}\, v^\alpha = \Omega_{\alpha\beta} v^\alpha - \frac{1}{2} H_0 M_\beta. \tag{1.25}$$

Thus if $\overset{b}{H}_{\alpha\beta} = \Omega_{\alpha\beta}$, from (1.25) we deduce $M_\beta = 0$, since $H_0 \neq 0$ on M. ■

1.3. THE INDUCED AND THE NORMAL FINSLER CONNECTIONS

We consider on $\widetilde{\mathbb{F}}^{m+1}$ the three classical Finsler connections:

$$\widetilde{\mathbf{FC}^*} = (G\widetilde{M}', \widetilde{\nabla}^*) = (\widetilde{G}^k_i, \widetilde{F}^{*k}_{i\ j}, \tilde{g}_{i\ j}^{\ k}),$$

$$\widetilde{\mathbf{BFC}} = (G\widetilde{M}', \widetilde{\nabla}') = (\widetilde{G}^k_i, \widetilde{G}_{i\ j}^{\ k}, 0)$$

and

$$\widetilde{\mathbf{RFC}} = (G\widetilde{M}', \widetilde{\nabla}'') = (\widetilde{G}^k_i, \widetilde{F}^{*k}_{i\ j}, 0).$$

Then on $\mathbb{F}^m$ we have three induced Finsler connections:

$$I\overset{c}{F}C = (HM', \overset{c}{\nabla}) = (N^\gamma_\alpha, \overset{c}{F}{}_{\alpha\ \beta}^{\ \gamma}, \overset{c}{C}{}_{\alpha\ \beta}^{\ \gamma}),$$

$$I\overset{b}{F}C = (HM', \overset{b}{\nabla}) = (N^{\gamma}_{\alpha}, \overset{b}{F}_{\alpha}{}^{\gamma}{}_{\beta}, \overset{b}{C}_{\alpha}{}^{\gamma}{}_{\beta})$$

and

$$I\overset{r}{F}C = (HM', \overset{r}{\nabla}) = (N^{\gamma}_{\alpha}, \overset{r}{F}_{\alpha}{}^{\gamma}{}_{\beta}, \overset{r}{C}_{\alpha}{}^{\gamma}{}_{\beta}),$$

where by (2.3.24a), (2.3.35), (2.3.36), (2.4.4), and (2.5.2) we have

$$N^{\gamma}_{\alpha} = \widetilde{B}^{\gamma}_{i}(B^{i}_{\alpha 0} + B^{j}_{\alpha}\widetilde{G}^{i}_{j}), \tag{1.26}$$

$$(a)\quad \overset{c}{F}_{\alpha}{}^{\gamma}{}_{\beta} = \widetilde{B}^{\gamma}_{k}(B^{k}_{\alpha\beta} + \widetilde{F}^{*k}_{i\ j}B^{ij}_{\alpha\beta}) + M^{\gamma}_{\alpha}H_{\beta};\quad (b)\quad \overset{c}{C}_{\alpha}{}^{\gamma}{}_{\beta} = g_{\alpha}{}^{\gamma}{}_{\beta}, \tag{1.27}$$

$$(a)\quad \overset{b}{F}_{\alpha}{}^{\gamma}{}_{\beta} = \widetilde{B}^{\gamma}_{k}(B^{k}_{\alpha\beta} + \widetilde{G}_{i}{}^{k}{}_{j}B^{ij}_{\alpha\beta});\quad (b)\quad \overset{b}{C}_{\alpha}{}^{\gamma}{}_{\beta} = 0, \tag{1.28}$$

$$(a)\quad \overset{r}{F}_{\alpha}{}^{\gamma}{}_{\beta} = \widetilde{B}^{\gamma}_{k}(B^{k}_{\alpha\beta} + \widetilde{F}^{*k}_{i\ j}B^{ij}_{\alpha\beta});\quad (b)\quad \overset{r}{C}_{\alpha}{}^{\gamma}{}_{\beta} = 0. \tag{1.29}$$

We see from Corollary 3.1.1 that in case $\mathbb{F}^m$ is totally geodesic immersed in $\widetilde{\mathbb{F}}^{m+1}$ we have $I\overset{c}{F}C = \mathbf{FC}^*$, $I\overset{b}{F}C = \mathbf{BFC}$ and $I\overset{r}{F}C = \mathbf{RFC}$. It is interesting to see that when $\mathbb{F}^m$ is not totally geodesic, the Finsler tensor field with local components $M_{\alpha\beta}$ is an obstruction for the induced Finsler connections to coincide with intrinsic Finsler connections. More precisely, we prove the following theorem.

THEOREM 1.4 *Let $\mathbb{F}^m$ be a non–totally geodesic pseudo–Finsler hypersurface of $\widetilde{\mathbb{F}}^{m+1}$. Then $M_{\alpha\beta} = 0$ if and only if any one (and hence all) of the following conditions are satisfied:*

$$(a)\quad I\overset{c}{F}C = \mathbf{FC}^*;\quad (b)\quad I\overset{b}{F}C = \mathbf{BFC};\quad (c)\quad I\overset{r}{F}C = \mathbf{RFC}. \tag{1.30}$$

PROOF. By Theorem 2.6.4 we see that $I\overset{c}{F}C = \mathbf{FC}^*$ if and only if $D_{\alpha\beta} = M_{\alpha\beta}H_{\varepsilon}v^{\varepsilon} = 0$ which is equivalent to $M_{\alpha\beta} = 0$ since $\mathbb{F}^m$ is supposed to be non–totally geodesic. Similarly, from Theorems 2.6.8 and 2.6.10 follows the equivalence between $M_{\alpha\beta} = 0$ and any of the (1.30b) and (1.30c). ■

As $a = b = 1$ in (2.1.61), the induced normal Finsler connections $N\overset{c}{F}C$, $N\overset{b}{F}C$ and $N\overset{r}{F}C$ will have as local coefficients the triplets $(N^{\gamma}_{\alpha}, \overset{c}{F}_{\alpha}, \overset{c}{C}_{\alpha})$, $(N^{\gamma}_{\alpha}, \overset{b}{F}_{\alpha}, \overset{b}{C}_{\alpha})$ and $(N^{\gamma}_{\alpha}, \overset{r}{F}_{\alpha}, \overset{r}{C}_{\alpha})$, respectively. From (2.3.46) we deduce that $\overset{c}{F}_{\alpha} = \overset{c}{C}_{\alpha} = 0$, which is actually a direct consequence of the fact that $\widetilde{\nabla}^*$ is a metric linear connection on $V\widetilde{M}'$. In this case the Ricci equations (2.3.50) will provide no information for the geometry of a pseudo–Finsler hypersurface. Contrary to this situation, by (2.4.16) and (2.5.10), the local coefficients of $\overset{\mathrm{b}}{NFC}$ and $\overset{\mathrm{r}}{NFC}$ are given by

$$(a)\ \ \overset{\mathrm{b}}{F}_{\alpha} = L_{\alpha} - \mu H_{\alpha};\quad (b)\ \ \overset{\mathrm{b}}{C}_{\alpha} = -M_{\alpha}, \tag{1.31}$$

and

$$(a)\ \ \overset{\mathrm{r}}{F}_{\alpha} = -\mu H_{\alpha};\quad (b)\ \ \overset{\mathrm{r}}{C}_{\alpha} = -M_{\alpha}, \tag{1.32}$$

respectively, where we set

$$(a)\ \ L_{\alpha} = \widetilde{P}^{*k}{}_{ij}\widetilde{B}_{k}B^{i}B^{j}_{\alpha} \quad \text{and} \quad (b)\ \ \mu = \tilde{g}_{ijk}B^{i}B^{j}B^{k}. \tag{1.33}$$

Thus by using the general formulas (2.1.61) for the induced normal Finsler connection, we obtain

$$(a)\ \ \overset{\mathrm{b}}{\nabla}{}^{\perp}_{\delta/\delta u^{\alpha}} B = (L_{\alpha} - \mu H_{\alpha})\, B;\quad (b)\ \ \overset{\mathrm{b}}{\nabla}{}^{\perp}_{\partial/\partial v^{\alpha}} B = -M_{\alpha}B, \tag{1.34}$$

and

$$(a)\ \ \overset{\mathrm{r}}{\nabla}{}^{\perp}_{\delta/\delta u^{\alpha}} B = -\mu H_{\alpha}B;\quad (b)\ \ \overset{\mathrm{r}}{\nabla}{}^{\perp}_{\partial/\partial v^{\alpha}} B = -M_{\alpha}B. \tag{1.35}$$

2. The Structure Equations for a Pseudo–Finsler Hypersurface

Using the general local formulas (2.1.60a) and (2.1.69) we deduce the following *h*–**Gauss–Weingarten formulas** for a hypersurface:

$$(a)\ \ B^{i}_{\alpha|\beta} = \overset{\mathrm{c}}{H}_{\alpha\beta}B^{i};\quad (b)\ \ B^{i}{}_{|\alpha} = -\overset{\mathrm{c}}{H}{}^{\beta}{}_{\alpha}B^{i}_{\beta}, \tag{2.1}$$

$$(a)\ \ B^{i}_{\alpha|\beta} = \overset{\mathrm{b}}{H}_{\alpha\beta}B^{i};\quad (b)\ \ B^{i}{}_{|\alpha} = -\overset{\mathrm{b}}{H'}{}^{\beta}{}_{\alpha}B^{i}_{\beta}, \tag{2.2}$$

and

$$(a)\ \ B^{i}_{\alpha|\beta} = \overset{\mathrm{r}}{H}_{\alpha\beta}B^{i};\quad (b)\ \ B^{i}{}_{|\alpha} = -\overset{\mathrm{r}}{H'}{}^{\beta}{}_{\alpha}B^{i}_{\beta}, \tag{2.3}$$

where the covariant derivatives in the left hand side are considered with respect to the induced horizontal relative covariant derivatives by Cartan, Berwald, and Rund connections, respectively. Similarly, from (2.1.60b) and (2.1.71), by using (1.13) and (1.15) we deduce that the v–**Gauss–Weingarten formulas** are given by

$$(a)\qquad B^{i}_{\alpha\|\beta} = M_{\alpha\beta}B^{i};\quad (b)\ \ B^{i}{}_{\|\alpha} = -M^{\beta}_{\alpha}B^{i}_{\beta}, \tag{2.4}$$

$$(a)\qquad B^{i}_{\alpha\|\beta} = 0;\quad (b)\ \ B^{i}{}_{\|\alpha} = -2M^{\beta}_{\alpha}B^{i}_{\beta}, \tag{2.5}$$

and

$$(a) \quad B^i_{\alpha\|\beta} = 0; \qquad (b) \quad B^i{}_{\|\alpha} = -2M^\beta_\alpha B^i_\beta, \tag{2.6}$$

with respect to the induced vertical relative covariant derivatives by Cartan, Berwald, and Rund connections respectively. We should note that from any of the formulas (2.4b)–(2.6b) applying (2.1.72) we obtain a new proof for (1.8).

We say that the normal Finsler vector field B is h–**parallel** (resp., v–**parallel**) with respect to the induced relative covariant derivatives if $B^i{}_{|\alpha} = 0$ (resp., $B^i{}_{\|\alpha} = 0$). Then by using (2.4b)–(2.6b) and taking into account Theorem 1.4 we state the following theorem.

THEOREM 2.1 *Let $\widetilde{\mathbb{F}}^{m+1}$ be a pseudo–Finsler manifold endowed with the Cartan, Berwald and Rund connections, and $\mathbb{F}^m$ be a non–totally geodesic pseudo–Finsler hypersurface of $\widetilde{\mathbb{F}}^{m+1}$. Then B is v–parallel with respect to the induced relative covariant derivatives by any of these Finsler connections if and only if the induced Finsler connections coincide with the corresponding intrinsic Finsler connections.*

Since the induced Finsler connections by the above three classical Finsler connections have some torsions, we shall first present the structure equations involving these tensor fields. Among all torsion Finsler tensor fields we notice that

$$R^\gamma{}_{\alpha\beta} = \frac{\delta N^\gamma_\alpha}{\delta u^\beta} - \frac{\delta N^\gamma_\beta}{\delta u^\alpha} \tag{2.7}$$

is the same for all these Finsler connections. Then by using (2.3.39)–(2.3.42), (2.4.17) and (2.5.11) we may state the following theorem.

THEOREM 2.2 *Let $\mathbb{F}^m$ be a pseudo–Finsler hypersurface of $\widetilde{\mathbb{F}}^{m+1}$. Then the structure equations with respect to the torsion Finsler tensor fields of $I\overset{c}{F}C$, $I\overset{b}{F}C$ and $I\overset{r}{F}C$ are given by*

$$\begin{cases} (a) \quad R^\gamma{}_{\alpha\beta} = \widetilde{R}^{*k}{}_{ij} B^{ij}_{\alpha\beta} \widetilde{B}^\gamma_k + \mathcal{A}_{(\alpha\beta)}\{(L^\gamma_\alpha - \overset{c}{H}{}^\gamma_\alpha) H_\beta\}; \\ (b) \quad \overset{c}{T}_\alpha{}^\gamma{}_\beta = M^\gamma_\alpha H_\beta - M^\gamma_\beta H_\alpha; \ (c) \ \overset{c}{P}{}^\gamma{}_{\alpha\beta} = L^\gamma{}_{\alpha\beta} + H_\alpha M^\gamma_\beta; \\ (d) \quad \overset{c}{S}{}^\gamma{}_{\alpha\beta} = 0, \end{cases} \tag{2.8}$$

$$\begin{cases} (a) \quad R^\gamma{}_{\alpha\beta} = \widetilde{R}^{*k}{}_{ij} B^{ij}_{\alpha\beta} \widetilde{B}^\gamma_k + \mathcal{A}_{(\alpha\beta)}\{H_\alpha \overset{b}{H'}{}^\gamma_\beta\}; \\ (b) \quad \overset{b}{T}_\alpha{}^\gamma{}_\beta = 0; \ (c) \ \overset{b}{P}{}^\gamma{}_{\alpha\beta} = 2H_\alpha M^\gamma_\beta; \ (d) \ \overset{b}{S}{}^\gamma{}_{\alpha\beta} = 0, \end{cases} \tag{2.9}$$

and

$$\begin{cases} (a)\ R^{\gamma}{}_{\alpha\beta} = \widetilde{R}^{*k}{}_{ij} B^{ij}_{\alpha\beta} \widetilde{B}^{\gamma}_{k} + \mathcal{A}_{(\alpha\beta)}\{(L^{\gamma}_{\alpha} - \overset{\mathrm{r}}{H'}{}^{\gamma}_{\alpha}) H_{\beta}\}; \\ (b)\ \overset{\mathrm{r}}{T}_{\alpha}{}^{\gamma}{}_{\beta} = 0;\ (c)\ \overset{\mathrm{r}}{P}{}^{\gamma}{}_{\alpha\beta} = 2H_{\alpha} M^{\gamma}_{\beta} + L^{\gamma}{}_{\alpha\beta};\ (d)\ \overset{\mathrm{r}}{S}{}^{\gamma}{}_{\alpha\beta} = 0, \end{cases} \tag{2.10}$$

respectively.

We denote by $K^*(u, v; \Pi)$ the horizontal flag curvature of $\mathbb{F}^m$ with respect to the horizontal flag Π. Then by (1.8.2) we have

$$K^*(u, v; \Pi) = \frac{R^*_{\alpha\beta\gamma} X^{\alpha} v^{\beta} X^{\gamma}}{F^* a_{\alpha\gamma} X^{\alpha} X^{\gamma}}, \tag{2.11}$$

where $\{X^{\alpha}(\delta/\delta u^{\alpha}),\ v^{\alpha}(\delta/\delta u^{\alpha})\}$ is a basis of Π, $a_{\alpha\gamma}$ is the angular metric of $\mathbb{F}^m$ and

$$R^*_{\alpha\beta\gamma} = g_{\alpha\mu} R^{*\mu}{}_{\beta\gamma} = g_{\alpha\mu}\left(\frac{\delta^* G^{\mu}_{\beta}}{\delta^* u^{\gamma}} - \frac{\delta^* G^{\mu}_{\gamma}}{\delta^* u^{\beta}}\right). \tag{2.12}$$

Owing to the induced non–linear connection we have the torsion Finsler tensor field $R^{\gamma}{}_{\alpha\beta}$ which enables us to define another flag curvature of $\mathbb{F}^m$. More precisely, we call

$$K(u, v; \Pi) = \frac{R_{\alpha\beta\gamma} X^{\alpha} v^{\beta} X^{\gamma}}{F^* a_{\alpha\gamma} X^{\alpha} X^{\gamma}}, \tag{2.13}$$

the **induced horizontal flag curvature** of $\mathbb{F}^m$ with respect to the horizontal flag Π at the point $(u, v) \in M'$. Now we denote by $\widetilde{K}^*(x(u), y(u, v); \widetilde{\Pi})$ the horizontal flag curvature of $\widetilde{\mathbb{F}}^{m+1}$ with respect to the horizontal flag $\widetilde{\Pi}$ spanned by $\{\widetilde{L}, \widetilde{X}^i(\delta^*/\delta^* x^i)\}$ where $\widetilde{X}^i = B^i_{\alpha} X^{\alpha}$. Then we state the following theorems which give the relationships between K, K^* and $\widetilde{K}^*$.

THEOREM 2.3 *The induced horizontal flag curvature of* $\mathbb{F}^m$ *and the horizontal flag curvature of* $\widetilde{\mathbb{F}}^{m+1}$ *are related by any of the following formulas:*

$$\begin{aligned} K(u, v; \Pi) \ &-\ \widetilde{K}^*(x(u), y(u, v); \widetilde{\Pi}) \\ &=\ \frac{1}{\Delta}((\overset{\mathrm{c}}{H}_{\alpha\gamma} - L_{\alpha\gamma} - M_{\alpha} H_{\gamma}) H_0 - H_{\alpha} H_{\gamma}) X^{\alpha} X^{\gamma}, && (2.14) \\ &=\ \frac{1}{\Delta}((\overset{\mathrm{b}}{H}_{\alpha\gamma} - 2L_{\alpha\gamma}) H_0 - H_{\alpha} H_{\gamma}) X^{\alpha} X^{\gamma}, && (2.15) \\ &=\ \frac{1}{\Delta}((\overset{\mathrm{r}}{H}_{\alpha\gamma} - L_{\alpha\gamma}) H_0 - H_{\alpha} H_{\gamma}) X^{\alpha} X^{\gamma}, && (2.16) \end{aligned}$$

where we set

$$\Delta = F^* a_{\alpha\gamma} X^\alpha X^\gamma. \tag{2.17}$$

PROOF. By using (2.8a), (2.9a), (2.10a), (1.10)–(1.12), and (1.14), we deduce that

$$\begin{aligned} R_{\alpha\beta\gamma} X^\alpha v^\beta X^\gamma \;-\; \widetilde{R}^*_{ijk} \widetilde{X}^i y^j \widetilde{X}^k &= ((\overset{c}{H}_{\alpha\gamma} - L_{\alpha\gamma} - M_\alpha H_\gamma) H_0 - H_\alpha H_\gamma) X^\alpha X^\gamma, & (2.18)\\ &= ((\overset{b}{H}_{\alpha\gamma} - 2L_{\alpha\gamma}) H_0 - H_\alpha H_\gamma) X^\alpha X^\gamma, & (2.19)\\ &= ((\overset{r}{H}_{\alpha\gamma} - L_{\alpha\gamma}) H_0 - H_\alpha H_\gamma) X^\alpha X^\gamma. & (2.20) \end{aligned}$$

Thus the assertion of the theorem follows from (2.18)–(2.20) by using (2.13) and (1.8.2) for $\widetilde{K}^*$. ■

Similarly, by using (2.6.27) and (2.6.53) we obtain the following theorem:

THEOREM 2.4 *The induced horizontal flag curvature and the intrinsic horizontal flag curvature of $\mathbb{F}^m$ are related by any of the following formulas:*

$$K(u, v; \Pi) - K^*(u, v; \Pi) = \frac{1}{\Delta}(D_{\alpha\gamma|*\beta} v^\beta + D^\varepsilon_\alpha D_{\varepsilon\gamma}) X^\alpha X^\gamma, \tag{2.21}$$

$$= \frac{1}{\Delta}(D^\mu_{\gamma|_{\mathbf{b}}\beta} v^\beta g_{\mu\alpha} + D^\varepsilon_\alpha D_{\varepsilon\gamma}) X^\alpha X^\gamma. \tag{2.22}$$

Since $D_{\alpha\gamma|*\beta} = D_{\alpha\gamma|_{\mathbf{r}}\beta}$, we do not need a new formula of $K - K^*$ in terms of the Rund connection of $\mathbb{F}^m$.

Next, we say that the deformation Finsler tersor field $\mathbf{D} = (D^\gamma_\alpha)$ with respect to the pair (GM', HM') is parallel in the direction of the Liouville vector field L of $\mathbb{F}^m$ with respect to the Cartan connection $\mathbf{FC}^* = (GM', \nabla^*)$ of $\mathbb{F}^m$ if $\nabla^*_L \mathbf{D} = 0$. In this case we may compare K and K^* as it is stated in the next theorem.

THEOREM 2.5 *Let $\mathbb{F}^m$ be a Finsler hypersurface of the Finsler manifold $\widetilde{\mathbb{F}}^{m+1}$ such that $\mathbf{D}$ is parallel in the direction of L. Then $K \geq K^*$. The equality occurs if and only if $\mathbf{D}$ vanishes on M'.*

PROOF. For any horizontal flag Π we may take X as a unit Finsler vector field that is orthogonal to the Liouville vector field L. Hence by (1.7.26) we have

$$a_{\alpha\beta} X^\alpha X^\beta = g_{\alpha\beta} X^\alpha X^\beta - (\eta_\alpha X^\alpha)^2 = g_{\alpha\beta} X^\alpha X^\beta > 0$$

since $g_{\alpha\beta}$ is positive definite and $\eta_\alpha X^\alpha = 0$. Hence $\Delta > 0$. Then by the hypothesis, from (2.21) we obtain

$$K - K^* = \frac{1}{\Delta} g^{\mu\varepsilon}(D_{\mu\alpha}X^\alpha)(D_{\varepsilon\gamma}X^\gamma) \geq 0.$$

If $\mathbf{D}$ vanishes, clearly $K = K^*$. Conversely, if $K = K^*$ then $D_{\mu\alpha}X^\alpha = 0$ which means that $\mathbf{D}(X) = 0$ for any X from the Liouville distribution. Taking into account that $\mathbf{D}(L) = 0$ (see (2.6.13)), we deduce that $\mathbf{D} = 0$ on M'. ∎

We will now write down the equations of Gauss, Codazzi, and Ricci for $\mathbb{F}^m$ in $\widetilde{\mathbb{F}}^{m+1}$ as a particular case of the general theory we presented in Sections 2.3, 2.4 and 2.5. First, since the local coefficients of the normal Finsler connection induced by the Cartan connection vanish on M', there exist no Ricci equations for $\mathbb{F}^m$. Hence Theorem 2.3.4 can be stated as follows.

THEOREM 2.6 *Let $\widetilde{\mathbb{F}}^{m+1}$ be a pseudo-Finsler manifold endowed with the Cartan connection and $\mathbb{F}^m$ be a pseudo-Finsler hypersurface of $\widetilde{\mathbb{F}}^{m+1}$. Then we have the assertions:*

(i) The Gauss equations are given by

$$\begin{cases} \bar{R}^*_{ij\alpha\beta}B^{ij}_{\gamma\varepsilon} = \overset{c}{R}_{\gamma\varepsilon\alpha\beta} + \overset{c}{H}_{\gamma\beta}\overset{c}{H}_{\varepsilon\alpha} - \overset{c}{H}_{\gamma\alpha}\overset{c}{H}_{\varepsilon\beta}, & (a) \\ \bar{P}^*_{ij\alpha\beta}B^{ij}_{\gamma\varepsilon} = \overset{c}{P}_{\gamma\varepsilon\alpha\beta} + M_{\gamma\beta}\overset{c}{H}_{\varepsilon\alpha} - \overset{c}{H}_{\gamma\alpha}M_{\varepsilon\beta}, & (b) \\ \bar{S}^*_{ij\alpha\beta}B^{ij}_{\gamma\varepsilon} = \overset{c}{S}_{\gamma\varepsilon\alpha\beta} + M_{\gamma\beta}M_{\varepsilon\alpha} - M_{\gamma\alpha}M_{\varepsilon\beta}. & (c) \end{cases} \tag{2.23}$$

(ii) The Codazzi equations are given by

$$\begin{cases} \bar{R}^*_{ij\alpha\beta}B^i_\gamma B^j = \overset{c}{T}_\alpha{}^\mu{}_\beta\overset{c}{H}_{\gamma\mu} + R^\mu{}_{\alpha\beta}M_{\gamma\mu} + \overset{c}{H}_{\gamma\alpha|\beta} - \overset{c}{H}_{\gamma\beta|\alpha}, & (a) \\ \bar{P}^*_{ij\alpha\beta}B^i_\gamma B^j = g_\alpha{}^\mu{}_\beta\overset{c}{H}_{\gamma\mu} + P^\mu{}_{\alpha\beta}M_{\gamma\mu} + \overset{c}{H}_{\gamma\alpha\|\beta} - M_{\gamma\beta|\alpha}, & (b) \\ \bar{S}^*_{ij\alpha\beta}B^i_\gamma B^j = M_{\gamma\alpha\|\beta} - M_{\gamma\beta\|\alpha}. & (c) \end{cases} \tag{2.24}$$

In this case the mixed Finsler tensor fields $\bar{R}^*_{ij\alpha\beta}$, $\bar{P}^*_{ij\alpha\beta}$ and $\bar{S}^*_{ij\alpha\beta}$ are given by

$$\bar{R}^*_{ij\alpha\beta} = \widetilde{R}^*_{ijkh}B^{kh}_{\alpha\beta} + \widetilde{P}^*_{ijkh}B^h(B^k_\alpha H_\beta - B^k_\beta H_\alpha), \tag{2.25}$$

$$\bar{P}^*_{ij\alpha\beta} = \widetilde{P}^*_{ijkh}B^{kh}_{\alpha\beta} + \widetilde{S}^*_{ijkh}B^k H_\alpha B^h_\beta, \tag{2.26}$$

$$\bar{S}^*_{ij\alpha\beta} = \widetilde{S}^*_{ijkh}B^{kh}_{\alpha\beta}. \tag{2.27}$$

Next, we consider the Berwald connection $\widetilde{\mathbf{BFC}}$ on $\widetilde{\mathbb{F}}^{m+1}$ and we put

$$\bar{R}_{ij\alpha\beta} = \widetilde{H}_{ijkh} B^{kh}_{\alpha\beta} + \widetilde{G}_{ijkh} B^h (B^k_\alpha H_\beta - B^k_\beta H_\alpha), \tag{2.28}$$

and

$$\bar{P}_{ij\alpha\beta} = \widetilde{G}_{ijkh} B^{kh}_{\alpha\beta}, \tag{2.29}$$

where $\widetilde{H}_{ijkh}$ and $\widetilde{G}_{ijkh}$ are the h–curvature and hv–curvature Finsler tensor fields of $\widetilde{\mathbf{BFC}}$. In this case there are two major differences from the structure equations induced by the Cartan connection. Namely, there exist two types of Codazzi equations and we have some Ricci equations. This is due to the fact that $\widetilde{\mathbf{BFC}}$ is not a metric Finsler connection. More precisely, from Theorem 2.4.3 we deduce the following theorem:

THEOREM 2.7 *Let $\widetilde{\mathbb{F}}^{m+1}$ be a pseudo–Finsler manifold endowed with the Berwald connection, and $\mathbb{F}^m$ be a pseudo–Finsler hypersurface of $\widetilde{\mathbb{F}}^{m+1}$. Then we have the assertions:*

(i) The Gauss equations are given by

$$\begin{cases} \bar{R}_{ij\alpha\beta} B^{ij}_{\gamma\varepsilon} = \overset{b}{R}_{\gamma\varepsilon\alpha\beta} + \overset{b}{H}_{\gamma\beta} \overset{b}{H}{}'_{\varepsilon\alpha} - \overset{b}{H}_{\gamma\alpha} \overset{b}{H}{}'_{\varepsilon\beta} & (a) \\ \bar{P}_{ij\alpha\beta} B^{ij}_{\gamma\varepsilon} = \overset{b}{P}_{\gamma\varepsilon\alpha\beta} - 2 \overset{b}{H}_{\gamma\alpha} M_{\varepsilon\beta}. & (b) \end{cases} \tag{2.30}$$

(ii) The A–Codazzi equations and the B–Codazzi equations are given by

$$\begin{cases} \bar{R}_i{}^j{}_{\alpha\beta} B^i \widetilde{B}^\gamma_j = \overset{b}{H}{}'^\gamma{}_{\beta|\alpha} - \overset{b}{H}{}'^\gamma{}_{\alpha|\beta} - 2M^\gamma_\varepsilon R^\varepsilon{}_{\alpha\beta} & (a) \\ \bar{P}_i{}^j{}_{\alpha\beta} B^i \widetilde{B}^\gamma_j = 2M^\gamma_{\beta|\alpha} - \overset{b}{H}{}'^\gamma{}_{\alpha||\beta} - 2M^\gamma_\varepsilon \overset{b}{P}{}^\varepsilon{}_{\alpha\beta} & (b) \\ M^\alpha_{\beta||\gamma} = M^\alpha_{\gamma||\beta}, & (c) \end{cases} \tag{2.31}$$

and

$$\begin{cases} \bar{R}_i{}^j{}_{\alpha\beta} B^i_\gamma \widetilde{B}_j = \overset{b}{H}_{\gamma\alpha|\beta} - \overset{b}{H}_{\gamma\beta|\alpha} & (a) \\ \bar{P}_i{}^j{}_{\alpha\beta} B^i_\gamma \widetilde{B}_j = \overset{b}{H}_{\gamma\alpha||\beta}, & (b) \end{cases} \tag{2.32}$$

respectively.

(iii) The Ricci equations are given by

$$\begin{cases} \bar{R}_{ij\alpha\beta} B^i B^j = \overset{b}{R}{}^\perp{}_{\alpha\beta} + \overset{b}{H}{}'^\gamma{}_\beta \overset{b}{H}_{\gamma\alpha} - \overset{b}{H}{}'^\gamma{}_\alpha \overset{b}{H}_{\gamma\beta} & (a) \\ \bar{P}_{ij\alpha\beta} B^i B^j = \overset{b}{P}{}^\perp{}_{\alpha\beta} + 2M^\gamma_\beta \overset{b}{H}_{\gamma\alpha} & (b) \\ M_{\alpha||\beta} = M_{\beta||\alpha}. & (c) \end{cases} \tag{2.33}$$

According to (2.2.33a), (2.2.33b), (1.4.27) and (1.4.28), $\overset{b}{R}^{\perp}{}_{\alpha\beta}$ and $\overset{b}{P}^{\perp}{}_{\alpha\beta}$ are given by

$$\overset{b}{R}^{\perp}{}_{\alpha\beta} = \frac{\delta \overset{b}{F}_{\alpha}}{\delta u^{\alpha}} - \frac{\delta \overset{b}{F}_{\beta}}{\delta u^{\alpha}} + R^{\gamma}{}_{\alpha\beta}\, \overset{b}{C}_{\gamma}, \tag{2.34}$$

and

$$\overset{b}{P}^{\perp}{}_{\alpha\beta} = \frac{\partial \overset{b}{F}_{\alpha}}{\partial v^{\beta}} - \frac{\delta \overset{b}{C}_{\beta}}{\delta u^{\alpha}} + \frac{\partial N^{\gamma}_{\alpha}}{\partial v^{\beta}}\, \overset{b}{C}_{\gamma}, \tag{2.35}$$

respectively, where $\overset{b}{F}_{\alpha}$ and $\overset{b}{C}_{\alpha}$ are given by (1.31).

Finally, if we consider on $\widetilde{\mathbb{F}}^{m+1}$ the Rund connection $\widetilde{\mathbf{RFC}}$ we will get some structure equations that are similar to the Berwald equations. In this case, using Theorem 2.5.3, we may state the following theorem:

THEOREM 2.8 *Let $\widetilde{\mathbb{F}}^{m+1}$ be a pseudo–Finsler manifold endowed with the Rund connection, and $\mathbb{F}^{m}$ be a pseudo–Finsler hypersurface of $\widetilde{\mathbb{F}}^{m+1}$. Then we have the assertions:*

(i) The Gauss equations are given by

$$\begin{cases} \bar{R}_{ij\alpha\beta}B^{ij}_{\gamma\varepsilon} = \overset{r}{R}_{\gamma\varepsilon\alpha\beta} + \overset{r}{H}_{\gamma\beta}\overset{r}{H}'_{\varepsilon\alpha} - \overset{r}{H}_{\gamma\alpha}\overset{r}{H}'_{\varepsilon\beta}, & (a) \\ \bar{P}_{ij\alpha\beta}B^{ij}_{\gamma\varepsilon} = \overset{r}{P}_{\gamma\varepsilon\alpha\beta} - 2\,\overset{r}{H}_{\gamma\alpha}\, M_{\varepsilon\beta}. & (b) \end{cases} \tag{2.36}$$

(ii) The A–Codazzi equations and the B–Codazzi equations are given by

$$\begin{cases} \bar{R}_i{}^j{}_{\alpha\beta}B^i\widetilde{B}^{\gamma}_j = \overset{r}{H}'^{\gamma}_{\beta|\alpha} - \overset{r}{H}'^{\gamma}_{\alpha|\beta} - 2M^{\gamma}_{\varepsilon}R^{\varepsilon}{}_{\alpha\beta}, & (a) \\ \bar{P}_i{}^j{}_{\alpha\beta}B^iB^{\gamma}_j = 2M^{\gamma}_{\beta|\alpha} - \overset{r}{H}'^{\gamma}_{\alpha||\beta} - 2M^{\gamma}_{\varepsilon}\overset{r}{P}^{\varepsilon}{}_{\alpha\beta}, & (b) \\ M^{\alpha}_{\beta||\gamma} = M^{\alpha}_{\gamma||\beta}, & (c) \end{cases} \tag{2.37}$$

and

$$\begin{cases} \bar{R}_i{}^j{}_{\alpha\beta}B^i_{\gamma}\widetilde{B}_j = \overset{r}{H}_{\gamma\alpha|\beta} - \overset{r}{H}_{\gamma\beta|\alpha}, & (a) \\ \bar{P}_i{}^j{}_{\alpha\beta}B^i_{\gamma}\widetilde{B}_j = \overset{r}{H}_{\gamma\alpha||\beta}, & (b) \end{cases} \tag{2.38}$$

respectively.

(iii) The Ricci equations are given by

$$\begin{cases} \bar{R}_{ij\alpha\beta}B^iB^j = \overset{\mathrm{r}}{R}{}^{\perp}{}_{\alpha\beta} + \overset{\mathrm{r}}{H}{}'^{\gamma}_{\beta}\overset{\mathrm{r}}{H}_{\gamma\alpha} - \overset{\mathrm{r}}{H}{}'^{\gamma}_{\alpha}\overset{\mathrm{r}}{H}_{\gamma\beta}, & (a) \\ \bar{P}_{ij\alpha\beta}B^iB^j = \overset{\mathrm{r}}{P}{}^{\perp}{}_{\alpha\beta} + 2M^{\gamma}_{\beta}\overset{\mathrm{r}}{H}_{\gamma\alpha}, & (b) \\ M_{\alpha\|\beta} = M_{\beta\|\alpha}, & (c) \end{cases} \tag{2.39}$$

where we set

$$\bar{R}_{ij\alpha\beta} = \widetilde{K}_{ijkh}B^{kh}_{\alpha\beta} + \widetilde{F}_{ijkh}B^h(B^k_\alpha H_\beta - B^k_\beta H_\alpha), \tag{2.40}$$

$$\bar{P}_{ij\alpha\beta} = \widetilde{F}_{ijkh}B^{kh}_{\alpha\beta}, \tag{2.41}$$

$$\overset{\mathrm{r}}{R}{}^{\perp}{}_{\alpha\beta} = \frac{\delta\,\overset{\mathrm{r}}{F}_\alpha}{\delta u^\beta} - \frac{\delta\,\overset{\mathrm{r}}{F}_\beta}{\delta u^\alpha} + R^\gamma{}_{\alpha\beta}\,\overset{\mathrm{r}}{C}_\gamma, \tag{2.42}$$

and

$$\overset{\mathrm{r}}{P}{}^{\perp}{}_{\alpha\beta} = \frac{\partial\,\overset{\mathrm{r}}{F}_\alpha}{\partial v^\beta} - \frac{\delta\,\overset{\mathrm{r}}{C}_\beta}{\delta u^\alpha} + \frac{\partial N^\gamma_\alpha}{\partial v^\beta}\overset{\mathrm{r}}{C}_\gamma, \tag{2.43}$$

with $\overset{\mathrm{r}}{F}_\alpha$ and $\overset{\mathrm{r}}{C}_\alpha$ given by (1.32).

We should stress that all covariant derivatives which appear in Theorems 2.6, 2.7, and 2.8 are considered with respect to the relative covariant derivatives $(\overset{\mathrm{c}}{\nabla}, \overset{\mathrm{c}}{\nabla}{}^{\perp})$, $(\overset{\mathrm{b}}{\nabla}, \overset{\mathrm{b}}{\nabla}{}^{\perp})$, and $(\overset{\mathrm{r}}{\nabla}, \overset{\mathrm{r}}{\nabla}{}^{\perp})$ respectively. By using some of the above equations we may derive straight proofs for some relations that involve $M_{\alpha\beta}$ and M_α. To support this claim we consider first (2.37c), and by using the vertical relative covariant derivative induced by $(\overset{\mathrm{r}}{\nabla}, \overset{\mathrm{r}}{\nabla}{}^{\perp})$ and taking into account (1.32c), we deduce that

$$\frac{\partial M^\alpha_\beta}{\partial v^\gamma} - \frac{\partial M^\alpha_\gamma}{\partial v^\beta} = M_\alpha M_{\gamma\beta} - M_\beta M_{\gamma\alpha}. \tag{2.44}$$

We should note that by using other approaches, (2.44) was obtained by Brown [1] and Matsumoto [4]. Similarly from (2.39c) we infer that

$$\frac{\partial M_\alpha}{\partial v^\beta} = \frac{\partial M_\beta}{\partial v^\alpha}. \tag{2.45}$$

Hence we may say that (M_α) is a **vertical gradient**, that is, there exists a smooth function $h(u,v)$ such that

$$M_\alpha = \frac{\partial h}{\partial v^\alpha}. \tag{2.46}$$

Actually we may determine a function h satisfying (2.46). First, by using (1.8.46), (2.3.1), (2.1.13), (1.2e) and (1.4) we derive

$$\begin{aligned} g_\alpha = g_{\alpha\beta\gamma}g^{\beta\gamma} &= \tilde{g}_{ijk}B^{ijk}_{\alpha\beta\gamma}g^{\beta\gamma} = \tilde{g}_{ijk}B^i_\alpha\tilde{g}^{kh}(\delta^j_h - B^j\widetilde{B}_h) \\ &= \tilde{g}_iB^i_\alpha - M_\alpha. \end{aligned} \tag{2.47}$$

On the other hand, by direct calculation we obtain

$$g_\alpha = \frac{1}{2}\frac{\partial g_{\beta\gamma}}{\partial v^\alpha}g^{\beta\gamma} = \frac{1}{2}\frac{\partial}{\partial v^\alpha}\left(\ln|\det\,[g_{\alpha\beta}]|\right). \tag{2.48}$$

Thus by using (2.48) in (2.47) we deduce that

$$M_\alpha = \frac{1}{2}\frac{\partial}{\partial v^\alpha}\left(\ln\left|\frac{\det[\tilde{g}_{ij}]}{\det[g_{\alpha\beta}]}\right|\right). \tag{2.49}$$

Also, by using the structure equations we may get new information for Finsler immersions. In this respect, we prove the following theorem:

THEOREM 2.9 *Let $\widetilde{\mathbb{F}}^{m+1}$ be a Berwald manifold and $\mathbb{F}^m$ be a non–totally geodesic pseudo–Finsler hypersurface such that $I\overset{b}{F}C=$ **BFC**. Then $\mathbb{F}^m$ is also a Berwald manifold.*

PROOF. As $I\overset{b}{F}C=$ **BFC**, by Theorem 1.4 we have $M_{\alpha\beta} = 0$. Then the equation (2.30b) implies that $\overset{b}{P}_{\gamma\varepsilon\alpha\beta} = 0$ since $\widetilde{G}_{ijkh} = 0$ (see assertion (ii) in Theorem 1.8.5). On the other hand, from (2.6.55) we deduce that $\overset{b}{D} = 0$. Then, by (2.6.63), we obtain $G_{\gamma\varepsilon\alpha\beta} = \overset{b}{P}_{\gamma\varepsilon\alpha\beta} = 0$. Thus $\mathbb{F}^m$ is also a Berwald manifold. ■

Finally, it is worth remarking that the A–Codazzi equations and the Ricci equations of a pseudo–Finsler hypersurface with respect to $\widetilde{\mathbf{BFC}}$ and $\widetilde{\mathbf{RFC}}$ appear here (Theorems 2.7 and 2.8) for the first time (compare with Matsumoto [4]). This is a good indication of the merit of vectorial Finsler connections as powerful tools in the study of the geometry of pseudo–Finsler submanifolds.

3. Pseudo–Finsler Hypersurfaces in Pseudo–Minkowski Spaces

Here we stress the importance of structure equations induced by the Rund connection of the ambient manifold. By using these equations we prove the non–existence of proper totally umbilical pseudo–Finsler hypersurfaces in a pseudo-Minkowski space. Also, we pay attention to the particularities of the geometry of pseudo-Finsler hypersurfaces with vanishing $M_{\alpha\beta}$ for any $\alpha, \beta \in \{1, \cdots, m\}$.

3.1. THE GAUSS-CODAZZI-RICCI EQUATIONS

Let $\widetilde{\mathbb{F}}^{m+1} = (\mathbb{R}^{m+1}, \widetilde{M}', \widetilde{F}^*)$ be a pseudo-Minkowski space, that is, $\widetilde{F}^*$ is a function depending on $(y^1, \cdots, y^{m+1})$ alone. Then by (1.6.9), (1.6.17), (1.6.27) and (1.7.11) we have

$$(a)\ \widetilde{G}^i = 0; \quad (b)\ \widetilde{G}^i_j = 0; \quad (c)\ \widetilde{G}_i{}^k{}_j = 0 \text{ and } (d)\ \widetilde{F}^{*k}_{i\ j} = 0. \tag{3.1}$$

As a consequence, we obtain

$$(a)\ \widetilde{R}^{*k}{}_{ij} = 0; \quad (b)\ \widetilde{P}^{*k}{}_{ij} = 0; \quad (c)\ L^{\gamma}{}_{\alpha\beta} = 0; \quad (d)\ L^{\gamma}_{\alpha} = 0, \tag{3.2}$$

and

$$(a)\ \widetilde{R}^*_{ijkh} = \widetilde{P}^*_{ijkh} = 0; \quad (b)\ \widetilde{H}_{ijkh} = \widetilde{G}_{ijkh} = 0;$$
$$(c)\ \widetilde{K}_{ijkh} = \widetilde{F}_{ijkh} = 0. \tag{3.3}$$

Thus the structure equations we presented in the previous section are much simpler for a pseudo-Finlser hypersurface in a pseudo-Minkowski space. More precisely, taking into account (3.1), (3.2), and (3.3a) and Theorems 2.2 and 2.6 we may state the following theorems:

THEOREM 3.1 *Let $\mathbb{F}^m = (M, M', F^*)$ be a pseudo-Finsler hypersurface in $\widetilde{\mathbb{F}}^{m+1}$. Then the structure equations with respect to the torsion Finsler tensor fields of $I\overset{c}{F}C, I\overset{b}{F}C$ and $I\overset{r}{F}C$ are given by*

$$\begin{cases} (a)\ R^{\gamma}{}_{\alpha\beta} = \mathcal{A}_{(\alpha\beta)}\{H_\alpha \overset{c}{H}{}^{\gamma}_{\beta}\}; & (b)\ \overset{c}{T}_{\alpha}{}^{\gamma}{}_{\beta} = \mathcal{A}_{(\alpha\beta)}\{M^{\gamma}_{\alpha} H_\beta\}; \\ (c)\ \overset{c}{P}{}^{\gamma}{}_{\alpha\beta} = H_\alpha M^{\gamma}_{\beta}; & (d)\ \overset{c}{S}{}^{\gamma}{}_{\alpha\beta} = 0, \end{cases} \tag{3.4}$$

$$\begin{cases} (a)\ R^{\gamma}{}_{\alpha\beta} = \mathcal{A}_{(\alpha\beta)}\{H_\alpha \overset{b}{H}{}'^{\gamma}{}_{\beta}\}; & (b)\ \overset{b}{T}_{\alpha}{}^{\gamma}{}_{\beta} = 0; \\ (c)\ \overset{b}{P}{}^{\gamma}{}_{\alpha\beta} = 2H_\alpha M^{\gamma}_{\beta}; & (d)\ \overset{b}{S}{}^{\gamma}{}_{\alpha\beta} = 0, \end{cases} \tag{3.5}$$

and

$$\begin{cases} (a)\ R^{\gamma}{}_{\alpha\beta} = \mathcal{A}_{(\alpha\beta)}\{H_\alpha \overset{r}{H}{}'^{\gamma}{}_{\beta}\}; & (b)\ \overset{r}{T}_{\alpha}{}^{\gamma}{}_{\beta} = 0; \\ (c)\ \overset{r}{P}{}^{\gamma}{}_{\alpha\beta} = 2H_\alpha M^{\gamma}_{\beta}; & (d)\ \overset{r}{S}{}^{\gamma}{}_{\alpha\beta} = 0, \end{cases} \tag{3.6}$$

respectively.

THEOREM 3.2 *Let $\widetilde{\mathbb{F}}^{m+1}$ be a pseudo-Minkowski space endowed with the Cartan connection and $\mathbb{F}^m$ be a pseudo-Finsler hypersurface of $\widetilde{\mathbb{F}}^{m+1}$. Then we have the following assertions:*

(i) The Gauss equations are given by

$$\begin{cases} \overset{c}{R}_{\gamma\varepsilon\alpha\beta} = \overset{c}{H}_{\gamma\alpha}\overset{c}{H}_{\varepsilon\beta} - \overset{c}{H}_{\gamma\beta}\overset{c}{H}_{\varepsilon\alpha} & (a) \\ \widetilde{S}^*_{ijkh}B^{ijh}_{\gamma\varepsilon\beta}B^k H_\alpha = \overset{c}{P}_{\gamma\varepsilon\alpha\beta} + M_{\gamma\beta}\overset{c}{H}_{\varepsilon\alpha} - \overset{c}{H}_{\gamma\alpha}M_{\varepsilon\beta} & (b) \\ \widetilde{S}^*_{ijkh}B^{ijkh}_{\gamma\varepsilon\alpha\beta} = \overset{c}{S}_{\gamma\varepsilon\alpha\beta} + M_{\gamma\beta}M_{\varepsilon\alpha} - M_{\gamma\alpha}M_{\varepsilon\beta}. & (c) \end{cases} \tag{3.7}$$

(ii) The Codazzi equations are given by

$$\begin{cases} \overset{c}{H}_{\gamma\alpha|\beta} - \overset{c}{H}_{\gamma\beta|\alpha} + \overset{c}{T}_{\alpha\ \beta}^{\ \mu}\overset{c}{H}_{\gamma\mu} + R^\mu_{\ \alpha\beta}M_{\gamma\mu} = 0 & (a) \\ \widetilde{S}^*_{ijkh}B^{ih}_{\gamma\beta}B^jB^kH_\alpha = g_{\alpha\ \beta}^{\ \mu}\overset{c}{H}_{\gamma\mu} + H_\alpha M_{\gamma\mu}M^\mu_\beta & \\ \qquad + \overset{c}{H}_{\gamma\alpha\|\beta} - M_{\gamma\beta|\alpha} & (b) \\ \widetilde{S}^*_{ijkh}B^{ikh}_{\gamma\alpha\beta}B^j = M_{\gamma\alpha\|\beta} - M_{\gamma\beta\|\alpha}. & (c) \end{cases} \tag{3.8}$$

According to (3.1c) and (3.1d), the Berwald and Rund connections coincide on $\widetilde{\mathbb{F}}^{m+1}$. For this reason we will present only the structure equations induced by the Rund connection. First, from (2.40), (2.41), and (3.3c) we obtain

$$\bar{R}_{ij\alpha\beta} = \bar{P}_{ij\alpha\beta} = 0. \tag{3.9}$$

Then, by using (3.9), (3.6a), (3.6c), and taking into account Theorem 2.8, we may state the following theorem.

THEOREM 3.3 *Let $\widetilde{\mathbb{F}}^{m+1}$ be a pseudo-Minkowski space endowed with the Rund connection and $\mathbb{F}^m$ be a pseudo-Finsler hypersurface of $\widetilde{\mathbb{F}}^{m+1}$. Then we have the following assertions:*

(i) The Gauss equations are given by

$$\begin{cases} \overset{r}{R}_{\gamma\varepsilon\alpha\beta} = \overset{r}{H}_{\gamma\alpha}\overset{r}{H}'_{\varepsilon\beta} - \overset{r}{H}_{\gamma\beta}\overset{r}{H}'_{\varepsilon\alpha}, & (a) \\ \overset{r}{P}_{\gamma\varepsilon\alpha\beta} = 2\,\overset{r}{H}_{\gamma\alpha}M_{\varepsilon\beta}. & (b) \end{cases} \tag{3.10}$$

(ii) The A–Codazzi equations and the B–Codazzi equations are given by

$$\left\{\begin{array}{lcll} \overset{r}{H}'^{\gamma}{}_{\alpha|\beta} - \overset{r}{H}'^{\gamma}{}_{\beta|\alpha} & = & 2M^{\gamma}_{\varepsilon}(\overset{r}{H}'^{\varepsilon}_{\alpha}\, H_{\beta} - \overset{r}{H}'^{\varepsilon}_{\beta}\, H_{\alpha}), & (a) \\ 2M^{\gamma}{}_{\beta|\alpha} - \overset{r}{H}'^{\gamma}{}_{\alpha\|\beta} & = & 4H_{\alpha}M^{\gamma}_{\varepsilon}M^{\varepsilon}_{\beta}, & (b) \\ M^{\alpha}_{\beta\|\gamma} & = & M^{\alpha}_{\gamma\|\beta}, & (c) \end{array}\right. \tag{3.11}$$

and

$$\left\{\begin{array}{lcll} \overset{r}{H}_{\gamma\alpha|\beta} - \overset{r}{H}_{\gamma\beta|\alpha} & = & 0, & (a) \\ \overset{r}{H}_{\gamma\alpha\|\beta} & = & 0, & (b) \end{array}\right. \tag{3.12}$$

respectively.

(iii) *The Ricci equations are given by*

$$\left\{\begin{array}{lcll} \overset{r}{R}{}^{\perp}{}_{\alpha\beta} & = & \overset{r}{H}_{\beta\gamma}\overset{r}{H}'^{\gamma}_{\alpha} - \overset{r}{H}_{\alpha\gamma}\overset{r}{H}'^{\gamma}_{\beta}, & (a) \\ \overset{r}{P}{}^{\perp}_{\alpha\beta} & = & -2\overset{r}{H}_{\alpha\gamma}M^{\gamma}_{\beta}, & (b) \\ \dfrac{\partial M_{\alpha}}{\partial v^{\beta}} & = & \dfrac{\partial M_{\beta}}{\partial v^{\alpha}}. & (c) \end{array}\right. \tag{3.13}$$

Remark 3.1. We should remark that there are major differences between the structure equations induced by the Cartan connection and the Rund connection. For instance, the A–Codazzi and Ricci equations do not appear in the case of the Cartan connection. On the other hand, the equations induced by the Rund connection are very simple and easily handled in our study. ■

We should stress that according to (2.5.16) and (2.5.17) the index $a = 1$ is a lower index and an upper index in (3.11) and (3.12), respectively. Thus, by using (1.5.20) and (1.5.21) with respect to the pair $(I\overset{r}{F}C,\ N\overset{r}{F}C)$ and taking into account (1.32) and (1.29b) we deduce that the covariant derivatives from (3.11) and (3.12) are given by

$$\left\{\begin{array}{lcll} \overset{r}{H}'^{\gamma}{}_{\alpha|\beta} & = & \dfrac{\delta \overset{r}{H}'^{\gamma}_{\alpha}}{\delta u^{\beta}} + \mu\, \overset{r}{H}'^{\gamma}_{\alpha}\, H_{\beta} + \overset{r}{H}'^{\varepsilon}_{\alpha}\overset{r}{F}_{\varepsilon}{}^{\gamma}{}_{\beta} - \overset{r}{H}'^{\gamma}_{\varepsilon}\overset{r}{F}_{\alpha}{}^{\varepsilon}{}_{\beta}, & (a) \\ \overset{r}{H}'^{\gamma}{}_{\alpha\|\beta} & = & \dfrac{\partial\, \overset{r}{H}'^{\gamma}_{\alpha}}{\partial v^{\beta}} + \overset{r}{H}'^{\gamma}_{\alpha}\, M_{\beta}, & (b) \\ M^{\alpha}_{\beta\|\gamma} & = & \dfrac{\partial M^{\alpha}_{\beta}}{\partial v^{\gamma}} + M^{\alpha}_{\beta}M_{\gamma}, & (c) \end{array}\right. \tag{3.14}$$

and

$$\begin{cases} \overset{\mathrm{r}}{H}_{\gamma\alpha|\beta} = \dfrac{\delta \overset{\mathrm{r}}{H}_{\gamma\alpha}}{\delta u^{\beta}} - \mu \overset{\mathrm{r}}{H}_{\gamma\alpha} H_{\beta} - \overset{\mathrm{r}}{H}_{\varepsilon\alpha} \overset{\mathrm{r}}{F}_{\gamma}{}^{\varepsilon}{}_{\beta} - \overset{\mathrm{r}}{H}_{\gamma\varepsilon} \overset{\mathrm{r}}{F}_{\alpha}{}^{\varepsilon}{}_{\beta}, & (a) \\ \overset{\mathrm{r}}{H}_{\gamma\alpha||\beta} = \dfrac{\partial \overset{\mathrm{r}}{H}_{\gamma\alpha}}{\partial v^{\beta}} - \overset{\mathrm{r}}{H}_{\gamma\alpha} M_{\beta}, & (b) \end{cases} \tag{3.15}$$

respectively.

The study which follows in this subsection will show the importance of (3.12b). First, following the terminology from Riemannian (pseudo–Riemannian) geometry we say that $\mathbb{F}^m$ is r–**totally umbilical** if there exists a smooth function ρ on M' such that on each coordinate neighborhood $\mathcal{U}'$ we have

$$\overset{\mathrm{r}}{H}_{\gamma\alpha} = \rho g_{\gamma\alpha}, \ \forall \gamma, \alpha \in \{1, \cdots, m\}. \tag{3.16}$$

Next, taking into account (3.2d) and Theorem 3.1.3 we may state the following theorem.

THEOREM 3.4 *Let $\mathbb{F}^m$ be a pseudo–Finsler hypersurface of a pseudo–Minkowski space $\widetilde{\mathbb{F}}^{m+1}$. Then $\mathbb{F}^m$ is totally geodesic if and only if the second fundamental forms of $\mathbb{F}^m$ satisfy any one (and hence all) of the conditions*

$$(a) \ \overset{\mathrm{b}}{H}_{\alpha\beta} = 0; \quad (b) \ \overset{\mathrm{c}}{H}_{\alpha\beta} = 0; \quad (c) \ \overset{\mathrm{r}}{H}_{\alpha\beta} = 0. \tag{3.17}$$

Therefore, by (3.16) and (3.17c) we see that any totally geodesic pseudo–Finsler hypersurface is r–totally umbilical. We shall see here that the converse is also true, unless $\mathbb{F}^m$ is a pseudo–Riemannian manifold.

THEOREM 3.5 *Let $\mathbb{F}^m$ be an r–totally umbilical pseudo–Finsler hypersurface of a pseudo–Minkowski space $\widetilde{\mathbb{F}}^{m+1}$. Then at any point of M' we have:*

$$(a) \ \overset{\mathrm{r}}{H}_{\alpha\beta} = 0, \ \forall \alpha, \beta \in \{1, \cdots, m\} \ \text{ or}$$

$$(b) \ g_{\alpha\beta\gamma} = 0, \ \forall \alpha, \beta, \gamma \in \{1, \cdots, m\}. \tag{3.18}$$

PROOF. We take the vertical covariant derivative of (3.16) with respect to $I\overset{\mathrm{r}}{F}C$ and by using (3.12b), (3.15b), and (1.6.21) we obtain

$$\rho_{||\beta} g_{\gamma\alpha} + 2\rho g_{\gamma\alpha\beta} = 0. \tag{3.19}$$

Contracting (3.19) by v^{α} and taking into account (1.6.22a) we deduce that

$$\rho_{||\beta} g_{\gamma\alpha} v^{\alpha} = 0. \tag{3.20}$$

As M' does not intersect the zero section of TM and $g_{\gamma\alpha}$ are the local components of a pseudo–Riemannian metric on VM', from (3.20) we obtain $\rho_{||\beta} = 0$ at any point of M'. Hence (3.19) becomes

$$\rho g_{\alpha\beta\gamma} = 0, \quad \forall \alpha, \beta, \gamma \in \{1, \cdots, m\},$$

which implies (3.18) via (3.16). ∎

The Theorems 3.4 and 3.5 yield the following corollary.

COROLLARY 3.1 *Let $\mathbb{F}^m = (M, TM^0, F^*)$ be an r–totally umbilical pseudo–Finsler hypersurface of a pseudo–Minkowski space $\widetilde{\mathbb{F}}^{m+1}$. If $\mathbb{F}^m$ is not geodesic at any point of M then it must be a pseudo–Riemannian manifold.*

Taking into account (3.1c) and (3.1d), from (1.12) we deduce that

$$(a) \quad \overset{\mathrm{r}}{H}_{\alpha\beta} = \overset{\mathrm{b}}{H}_{\alpha\beta} \quad \text{and} \quad (b) \quad \overset{\mathrm{c}}{H}_{\alpha\beta} = \overset{\mathrm{r}}{H}_{\alpha\beta} + M_\alpha H_\beta. \tag{3.21}$$

Then we say that $\mathbb{F}^m$ is *c*–**totally umbilical** if we have

$$\overset{\mathrm{c}}{H}_{\alpha\beta} = \rho g_{\alpha\beta}, \tag{3.22}$$

where ρ is a smooth function on M'. Then by using (3.21) and (3.22) and taking into account that $g_{\alpha\beta}$ and $\overset{\mathrm{r}}{H}_{\alpha\beta}$ are symmetric Finsler tensor fields, we obtain

$$M_\alpha H_\beta = M_\beta H_\alpha. \tag{3.23}$$

Contracting (3.23) by v^α and using (1.5b), (1.10) and (1.11) we derive

$$M_\beta H_0 = 0, \ \forall \beta \in \{1, \cdots, m\}. \tag{3.24}$$

Thus by Theorem 3.1.2 we infer that $M_\beta = 0$ for all $\beta \in \{1, \cdots, m\}$ at any point $(u, v) \in M'$ provided $\mathbb{F}^m$ is not geodesic at $u \in M$. Hence (3.21) yields $\overset{\mathrm{c}}{H}_{\alpha\beta} = \overset{\mathrm{r}}{H}_{\alpha\beta}$, that is, $\mathbb{F}^m$ is *r*–totally umbilical.

We now study the existence of totally umbilical pseudo–Finsler submanifolds with respect to the shape operators. According to (1.14a) we see that (3.22) is equivalent to

$$\overset{\mathrm{c}}{H}{}'^{\alpha}_{\beta} = \rho\delta^\alpha_\beta.$$

On the other hand, (1.14b), (1.14c), and (3.2d) imply

$$\overset{\mathrm{b}}{H}{}'^{\alpha}_{\beta} = \overset{\mathrm{r}}{H}{}'^{\alpha}_{\beta} = \overset{\mathrm{r}}{H}{}^{\alpha}_{\beta} + 2M^\alpha H_\beta. \tag{3.25}$$

Then we say that $\mathbb{F}^m$ is **r'–totally umbilical** if there exists a smooth function ρ on M' such that locally we have

$$\overset{r}{H}{}'^{\alpha}_{\beta} = \rho\delta^{\alpha}_{\beta}, \quad \forall\alpha, \beta \in \{1, \cdots, m\}. \tag{3.26}$$

By the same arguments we used in case of c–totally umbilical pseudo–Finsler hypersurfaces, it is easy to show that an r'–totally umbilical pseudo–Finsler hypersurface is r–totally umbilical.

As a conclusion of the above study we may state the following result.

THEOREM 3.6 *Let $\widetilde{\mathbb{F}}^{m+1}$ be a pseudo–Minkowski space endowed with the Cartan, Berwald, and Rund connections. Suppose $\mathbb{F}^m$ is a pseudo–Finsler hypersurface of $\widetilde{\mathbb{F}}^{m+1}$ that satisfies one of the following conditions:*

(i) The second fundamental forms are proportional to the Finsler metric of $\mathbb{F}^m$, i.e., they satisfy (3.16) or (3.22).

(ii) The shape operators are proportional to the identity morphism, i.e., they satisfy relations like (3.26).

Then $\mathbb{F}^m$ is either totally geodesic or a pseudo–Riemannian manifold.

Therefore, according to the present theory, there exist no proper totally umbilical pseudo–Finsler hypersurfaces in a pseudo–Minkowski space. However, in the Minkowskian approach to the theory of pseudo–Finsler hypersurfaces (see Section 4), we introduce Minkowskian totally umbilical pseudo–Finsler hypersurfaces and give examples.

As we noticed by (2.46) the Finsler covector field (M_α) is a vertical gradient. It is interesting to note that by using again (3.12b) we will obtain a new formula for (M_α) as a vertical gradient. First, by (3.12b), and (3.15b) we infer that

$$\frac{\partial \overset{r}{H}_{\gamma\alpha}}{\partial v^\beta} = \overset{r}{H}_{\gamma\alpha} M_\beta. \tag{3.27}$$

Then we denote by $\overset{r}{H}{}^{\gamma\alpha}_{*}$ the cofactors of $\overset{r}{H}_{\gamma\alpha}$ in the matrix $[\overset{r}{H}_{\gamma\alpha}]$ and by using (3.27) we obtain

$$\begin{aligned}\frac{\partial}{\partial v^\beta}(\det[\overset{r}{H}_{\gamma\alpha}]) &= \frac{\partial \overset{r}{H}_{\gamma\alpha}}{\partial v^\beta}\, \overset{r}{H}{}^{\gamma\alpha}_{*} \\ &= \overset{r}{H}_{\gamma\alpha}\overset{r}{H}{}^{\gamma\alpha}_{*} M_\beta = m \det[\overset{r}{H}_{\gamma\alpha}] M_\beta.\end{aligned}$$

Thus we have the desired formula

$$M_\beta = \frac{1}{m}\frac{\partial}{\partial v^\beta}(\ln|\det[\overset{r}{H}_{\gamma\alpha}]|), \tag{3.28}$$

provided $\overset{r}{H}= (\overset{r}{H}_{\gamma\alpha})$ is nonsingular at any point of M'. Finally, due to (3.27) and Theorem 3.4 we may state the following.

COROLLARY 3.2 *Let $\mathbb{F}^m$ be a pseudo-Finsler hypersurface of $\widetilde{\mathbb{F}}^{m+1}$ which is not geodesic at any point of M. Then $M_\alpha = 0$ for any $\alpha \in \{1, \cdots, m\}$ if and only if $\overset{r}{H}_{\alpha\beta}, \alpha, \beta \in \{1, \cdots, m\}$ depend only on $(u^1, \cdots, u^m)$.*

3.2. PSEUDO-FINSLER HYPERSURFACES WITH $M_{\alpha\beta} = 0$

The material we developed till now shows that $M_\alpha, M_{\alpha\beta}$ and μ (see (1.4) and (1.33b)) have an important role in the theory of pseudo-Finlser hypersurfaces. Moreover, there exists a large class of pseudo-Finsler hypersurfaces in a pseudo-Minkowski space (see Examples 3.3.1 and 3.3.2) for which $M_{\alpha\beta} = 0$ for any $\alpha, \beta \in \{1, \cdots, m\}$.

Our purpose in this subsection is to study pseudo-Finsler hypersurfaces $\mathbb{F}^m = (M, M', F^*)$ of a pseudo-Minkowski space $\widetilde{\mathbb{F}}^{m+1} = (\mathbb{R}^{m+1}, \widetilde{M}', \widetilde{F}^*)$, such that $M_{\alpha\beta} = 0$ for any $\alpha, \beta \in \{1, \cdots, m\}$. First we show that these hypersurfaces fall in a particular class of pseudo-Finsler manifolds.

THEOREM 3.7 *Any pseudo-Finsler hypersurface $\mathbb{F}^m$ of $\widetilde{\mathbb{F}}^{m+1}$ with $M_{\alpha\beta} = 0$ for any $\alpha, \beta \in \{1, \cdots, m\}$ is a Berwald manifold.*

PROOF. First, from (3.10b) we obtain $\overset{r}{P}_{\gamma\varepsilon\alpha\beta}= 0$. Then by using (1.5.25) and taking into account that $\overset{r}{C}_{\alpha}{}^{\gamma}{}_{\beta}= 0$ we deduce that $\overset{r}{F}_{\alpha}{}^{\gamma}{}_{\beta}$ are functions of (u^α) only. As $I\overset{r}{FC}= \mathbf{RFC}$ (see Theorem 1.4) we conclude that $F^{*\gamma}_{\alpha\ \beta}$ are functions of (u^α) only. Thus the assertion (v) of Theorem 1.8.5 completes the proof of the theorem. ∎

By using Theorem 1.4 and (1.27)–(1.29) we deduce that

$$(a) \quad I\overset{c}{FC}= \mathbf{FC}^* \ ; \quad (b) \quad I\overset{b}{FC}= \mathbf{BFC}; \quad (c) \quad I\overset{r}{FC}= \mathbf{RFC}, \tag{3.29}$$

and

$$F^{*\gamma}_{\alpha\ \beta} = \overset{c}{F}_{\alpha}{}^{\gamma}{}_{\beta} = \overset{b}{F}_{\alpha}{}^{\gamma}{}_{\beta} = \overset{r}{F}_{\alpha}{}^{\gamma}{}_{\beta}\,. \tag{3.30}$$

Since any Berwald manifold is a Landsberg manifold we have **BFC** = **RFC**. Hence from (3.29b) and (3.29c) we obtain

$$\overset{\mathrm{b}}{IFC} = \overset{\mathrm{r}}{IFC} = \mathbf{BFC} = \mathbf{RFC}. \tag{3.31}$$

In this case the structure equations induced by the Cartan and Rund connections become very simple, as the following theorems show.

THEOREM 3.8 *Let $\widetilde{\mathbb{F}}^{m+1}$ be a pseudo-Minkowski space endowed with the Cartan connection and $\mathbb{F}^m$ be a pseudo-Finsler hypersurface of $\widetilde{\mathbb{F}}^{m+1}$ such that $M_{\alpha\beta} = 0$ for all $\alpha, \beta \in \{1, \cdots, m\}$. Then we have the following assertions:*

(i) The Gauss equations are given by

$$\begin{cases} R^*_{\gamma\varepsilon\alpha\beta} = \overset{\mathrm{c}}{H}_{\gamma\alpha}\overset{\mathrm{c}}{H}_{\varepsilon\beta} - \overset{\mathrm{c}}{H}_{\gamma\beta}\overset{\mathrm{c}}{H}_{\varepsilon\alpha}, & (a) \\ S^*_{\gamma\varepsilon\alpha\beta} = \widetilde{S}^*_{ijkh}B^{ijkh}_{\gamma\varepsilon\alpha\beta}. & (b) \end{cases} \tag{3.32}$$

(ii) The Codazzi equations are given by

$$\begin{cases} \overset{\mathrm{c}}{H}_{\gamma\alpha|\beta} - \overset{\mathrm{c}}{H}_{\gamma\beta|\alpha} = 0, & (a) \\ \widetilde{S}^*_{ijkh}B^{ih}_{\gamma\beta}B^jB^kH_\alpha = g_{\alpha\ \beta}^{\ \mu}\overset{\mathrm{c}}{H}_{\gamma\mu} + \overset{\mathrm{c}}{H}_{\gamma\alpha||\beta}, & (b) \\ \widetilde{S}^*_{ijkh}B^{ikh}_{\gamma\alpha\beta}B^j = 0. & (c) \end{cases} \tag{3.33}$$

PROOF. By (3.29a) we have $\overset{\mathrm{c}}{R}_{\gamma\varepsilon\alpha\beta} = R^*_{\gamma\varepsilon\alpha\beta}$ and $\overset{\mathrm{c}}{S}_{\gamma\varepsilon\alpha\beta} = S^*_{\gamma\varepsilon\alpha\beta}$. Thus (3.32) follows from (3.7a) and (3.7c). Next, from (3.4b) we deduce that $\overset{\mathrm{c}}{T}_{\alpha\ \beta}^{\ \gamma} = 0$. Hence (3.33) follows from (3.8). ∎

Similarly, from Theorem 3.3 we deduce the following theorem.

THEOREM 3.9 *Let $\widetilde{\mathbb{F}}^{m+1}$ be a pseudo-Minkowski space endowed with the Rund connection and $\mathbb{F}^m$ be a pseudo-Finsler hypersurface of $\widetilde{\mathbb{F}}^{m+1}$ such that $M_{\alpha\beta} = 0$ for all $\alpha, \beta \in \{1, \cdots, m\}$. Then we have the following assertions:*

(i) The Gauss equations are given by

$$K_{\gamma\varepsilon\alpha\beta} = \overset{\mathrm{r}}{H}_{\gamma\alpha}\overset{\mathrm{r}}{H}'_{\varepsilon\beta} - \overset{\mathrm{r}}{H}_{\gamma\beta}\overset{\mathrm{r}}{H}'_{\varepsilon\alpha} \tag{3.34}$$

(ii) The A–Codazzi equations and the B–Codazzi equations are given by

$$\begin{cases} \overset{r}{H}{}'^{\gamma}_{\beta|\alpha} - \overset{r}{H}{}'^{\gamma}_{\alpha|\beta} = 0 & (a) \\ \overset{r}{H}{}'^{\gamma}_{\alpha||\beta} = 0, & (b) \end{cases} \tag{3.35}$$

and (3.12), respectively.

(iii) The Ricci equations are given by (3.13a), (3.13c) and

$$\overset{r}{P}{}^{\perp}_{\alpha\beta} = 0. \tag{3.36}$$

Under some additional conditions, $\mathbb{F}^m$ becomes a pseudo–Riemannian manifold, as we prove in the next theorem.

THEOREM 3.10 *Let* $\mathbb{F}^m = (M, TM^0, F^*)$ *be a pseudo–Finsler hypersurface of a pseudo–Minkowski space* $\widetilde{\mathbb{F}}^{m+1}$*, satisfying the following conditions:*

(i) $M_{\alpha\beta}$ *and* M_α *vanish on* TM^0 *for any* $\alpha, \beta \in \{1, \cdots, m\}$.

(ii) The horizontal second fundamental form $\overset{r}{H} = (\overset{r}{H}_{\alpha\beta})$ *is nonsingular at each point of* M*. Then* $\mathbb{F}^m$ *is a pseudo–Riemannian manifold.*

PROOF. From (1.14c) we deduce that

$$\overset{r}{H}_{\alpha\varepsilon} = \overset{r}{H}{}'_{\alpha\varepsilon}, \quad \forall \alpha, \varepsilon \in \{1, \cdots, m\}, \tag{3.37}$$

since $M_\alpha = 0$ for any $\alpha \in \{1, \cdots, m\}$. Also, (3.11b) and $M_{\alpha\beta} = 0$ for any $\alpha, \beta \in \{1, \cdots, m\}$ imply

$$\overset{r}{H}{}^{\gamma}_{\varepsilon||\beta} = \overset{r}{H}{}'^{\gamma}_{\varepsilon||\beta} = 0, \quad \forall \beta, \gamma, \varepsilon \in \{1, \cdots, m\}. \tag{3.38}$$

Then by using (3.12b), (3.38), and (2.5.5b) we obtain

$$0 = \overset{r}{H}_{\alpha\varepsilon||\beta} = (g_{\alpha\gamma} \overset{r}{H}{}^{\gamma}_{\varepsilon})_{||\beta} = 2 g_{\alpha\beta\gamma} \overset{r}{H}{}^{\gamma}_{\varepsilon}.$$

As $[\overset{r}{H}_{\alpha\beta}]$ is nonsingular at each point of M it follows that $[\overset{r}{H}{}^{\alpha}_{\beta}]$ is nonsingular at each point $(u, v) \in TM^0$. Thus $g_{\alpha\beta\gamma} = 0$ for any $\alpha, \beta, \gamma \in \{1, \cdots, m\}$, that is $\mathbb{F}^m$ is a pseudo–Riemannian manifold. ■

4. A Minkowskian Approach to the Theory of Pseudo–Finsler Hypersurfaces

The Minkowskian approach we present in this section is based on the construction of a Minkowskian unit normal vector field depending on position only. In this way all the induced geometric objects and structure equations live on the base manifold M of a pseudo–Finsler hypersurface $\mathbb{F}^m = (M, M', F^*)$. We define the Minkowskian totally umbilical pseudo–Finsler hypersurfaces and find examples in a pseudo–Minkowski space.

4.1. THE HAMILTONIAN FUNDAMENTAL FUNCTION

Let $\widetilde{\mathbb{F}}^{m+1} = (\widetilde{M}, \widetilde{M}', \widetilde{F}^*)$ be a pseudo–Finsler manifold with spacelike Liouville vector field. At any point $x \in \widetilde{M}$ we define the mapping

$$\mathbf{L}_x : T_x\widetilde{M} \to T_x^*\widetilde{M};\ \mathbf{L}_x(y) = \theta, \quad \theta_i = \tilde{g}_{ij}(x, y)y^j, \tag{4.1}$$

which is known in the literature as the **Legendre transformation** (see Miron [3], Shen [1]). By using the homogeneity property of the Cartan tensor field we deduce that

$$\frac{\partial \theta_i}{\partial y^k} = 2\tilde{g}_{ijk}(x, y)y^j + \tilde{g}_{ik}(x, y) = \tilde{g}_{ik}(x, y). \tag{4.2}$$

As det $[\tilde{g}_{ik}(x, y)] \neq 0$, we infer that the Legendre transformation is one-to-one.

Next, we denote $\widetilde{M}_x^* = \mathbf{L}(\widetilde{M}_x')$ and consider

$$\widetilde{M}^* = \bigcup_{x \in \widetilde{M}} \widetilde{M}_x^*.$$

It is easy to see that $\widetilde{M}^*$ is an open submanifold of the cotangent bundle $T^*\widetilde{M}$. Also, $\widetilde{M}^*$ is the total space of a fibre bundle over $\widetilde{M}$ which we call the **dual bundle** of $\widetilde{M}'$. We note that $\Gamma(\widetilde{M}^*)$ does not intersect the zero section of $T^*\widetilde{M}$ since $\Gamma(\widetilde{M}')$ has this property with respect to $T\widetilde{M}$. Moreover, each fibre $\widetilde{M}_x^*$ is a positive conic set in $T_x^*\widetilde{M}$. Indeed, if $\theta \in \widetilde{M}_x^*$, then for any $k > 0$ it follows that $k\theta \in \widetilde{M}_x^*$, since by homogeneity of the pseudo–Finsler metric $\tilde{g}$ we have

$$\tilde{g}_{ij}(x, ky)ky^j = \tilde{g}_{ij}(x, y)ky^j = k\theta_i.$$

We now define the smooth function $\widetilde{H}^*$ on $\widetilde{M}^*$ by

$$\widetilde{H}^*(x, \theta) = \widetilde{F}^*(x, y), \quad \forall \theta \in \widetilde{M}_x^*, \tag{4.3}$$

where $y = \mathbf{L}_x^{-1}(\theta)$. As the Liouville vector field is spacelike we have $\widetilde{F}^* = \widetilde{F}^2$ and hence $\widetilde{H}^* = \widetilde{H}^2$, where $\widetilde{H}$ is given by

$$\widetilde{H}(x,\theta) = \widetilde{F}(x,y), \forall\theta \in \widetilde{M}_x^*. \tag{4.4}$$

We call $\widetilde{H}$ the **Hamiltonian fundamental function** of $\widetilde{\mathbb{F}}^{m+1}$.

In what follows in this subsection we shall define some new geometrical objects which we find useful in the approach we develop in the section. First, by using (1.7.23) and (4.1) we deduce that

$$\theta_i = \widetilde{F}\frac{\partial \widetilde{F}}{\partial y^i} = \widetilde{F}\eta_i. \tag{4.5}$$

Also, by differentiating (4.4) with respect to (y^k) and using (4.2), (4.4) and (1.7.23) we obtain

$$y^i = \widetilde{H}(x,\theta)\frac{\partial \widetilde{H}}{\partial \theta_i}. \tag{4.6}$$

Now we define

$$\widetilde{\omega}^{ij}(x,\theta) = \frac{1}{2}\frac{\partial^2 \widetilde{H}^*}{\partial\theta_i \partial\theta_j}. \tag{4.7}$$

Then by direct calculations using (4.2), (4.3), (1.1.7), (4.7), (1.6.21), and (4.6) we derive

$$\tilde{g}_{ij} = \widetilde{\omega}^{hk}\tilde{g}_{hi}\tilde{g}_{kj} + 2\tilde{g}_{ijk}\frac{\partial \widetilde{H}}{\partial \theta_k}\widetilde{H} = \widetilde{\omega}^{hk}\tilde{g}_{hi}\tilde{g}_{kj},$$

which is equivalent to

$$\delta_i^k = \tilde{g}_{ih}(x,y)\widetilde{\omega}^{hk}(x,\theta). \tag{4.8}$$

Thus $\widetilde{\omega}^{hk}(x,\theta)$ are the contravariant components of a pseudo–Riemannian metric $\widetilde{\omega} = (\widetilde{\omega}_{ij}(x,\theta))$ on the vertical vector bundle of $\widetilde{M}^*$. Moreover, (4.8) implies that

$$\widetilde{\omega}^{hk}(x,\theta) = \tilde{g}^{hk}(x,y). \tag{4.9}$$

Thus by using (4.1) and (4.9) we deduce that

$$y^i = \widetilde{\omega}^{ij}(x,\theta)\theta_j, \tag{4.10}$$

which is going to be the starting point of our Minkowskian approach to the geometry of pseudo–Finlser hypersurfaces. Finally, by using (4.10) and (4.1) we obtain

$$\widetilde{\omega}^{ij}(x,\theta)\theta_i\theta_j = \widetilde{F}^*(x,y) = \widetilde{H}^*(x,\theta). \tag{4.11}$$

4.2. A UNIT NORMAL VECTOR FIELD DEPENDING ON POSITION ONLY

Let M be an orientable manifold and $\mathbb{F}^m = (M, M', F^*)$ be a pseudo–Finsler hypersurface of $\widetilde{\mathbb{F}}^{m+1} = (\widetilde{M}, \widetilde{M}', \widetilde{F}^*)$. Suppose that locally on $\mathcal{U} \subset M$ the immersion is given by the equations:

$$(a)\ \ x^i = x^i(u^1, \cdots, u^m) \text{ or } \ (b)\ \ f(x^1, \cdots, x^{m+1}) = 0, \qquad (4.12)$$

where x^i and f are smooth functions such that

$$\text{rank } [B^i_\alpha] = m \text{ and rank } [\partial f/\partial x^i] = 1$$

at any point of $\mathcal{U}$. Then $\theta_i = (\partial f/\partial x^i)(x(u))$ are the local components of a 1–form θ on $\widetilde{M}$ defined at points of M. Moreover, we have $\theta(X) = 0$ for any $X \in \Gamma(M')$. Indeed, by differentiating (4.12b) with respect to (u^α) we derive

$$\theta_i B^i_\alpha = 0. \qquad (4.13)$$

Next, by (4.10) we obtain on $\mathcal{U}$ the vector field

$$n^i(x) = \widetilde{\omega}^{ij}(x, \theta)\theta_j(x). \qquad (4.14)$$

Also by (4.1) we deduce that

$$\theta_i(x) = \tilde{g}_{ij}(x, n(x))n^j(x). \qquad (4.15)$$

Finally, we define the vector field N on $\widetilde{M}$ along $\mathcal{U}$ with local components

$$N^i(x) = \frac{1}{\widetilde{F}(x, n(x))} n^i(x). \qquad (4.16)$$

Then by using (4.14), (4.16), (4.9), (4.11) and taking into account the homogeneity of $\tilde{g}_{ij}$ we obtain

$$(a)\ \ \tilde{g}_{ij}(x, N)N^i B^j_\alpha = 0 \ \text{ and } \ (b)\ \ \tilde{g}_{ij}(x, N)N^i N^j = 1. \qquad (4.17)$$

Here and in the sequel, to simplify expressions we shortly write (x, N) instead of $(x(u), N(x(u)))$. We notice that (4.17b) is equivalent to

$$\widetilde{F}(x, N) = 1. \qquad (4.18)$$

As M is supposed to be orientable, we have a global section N of the vector bundle $T\widetilde{M}_{|M}$, which we call the **Minkowskian unit normal vector field** of $\mathbb{F}^m$.

Thus, instead of a cone of normals $B(x,y) = B^i(x,y)(\partial/\partial y^i)$ at each point $x \in M$, we have now a single normal $N(x)$. We find here the relations between these normal directions. First, from (1.2a) and (4.13) we see that $\widetilde{B}_i$ and θ_i are collinear, that is, we have

$$\widetilde{B}_i(x,y) = \lambda(x,y)\theta_i(x), \tag{4.19}$$

where λ is a smooth function on a coordinate neighborhood $\mathcal{U}'$ in M'. Then contracting (4.19) by $n^i(x)$ and using (4.15) and (4.16) we deduce that

$$\lambda(x,y) = \frac{1}{\widetilde{F}(x,n)}\widetilde{B}_i(x,y)N^i(x). \tag{4.20}$$

Finally, taking into acount that $\widetilde{B}_i$ are the covariant components of B, and using (4.19), (4.20), (4.15), and (4.16) we obtain

$$B^i(x,y) = S^i_j(x,y)N^j(x), \tag{4.21}$$

where we set

$$S^i_j(x,y) = \widetilde{B}_k(x,y)N^k(x)\tilde{g}^{ih}(x,y)\tilde{g}_{hj}(x,N). \tag{4.22}$$

From (4.22) it follows that $[S^i_j(x,y)]$ is a nonsingular matrix at any point $(x,y) \in \mathcal{U}'$. Thus by (4.21) we may say that any normal B from the normal cone is obtained from the Minkowskian normal N by a 'Finsler rotation'.

4.3. GEOMETRIC OBJECTS INDUCED BY THE MINKOWSKIAN NORMAL

First, by using the Minkowskian normal $N(x)$ we define on M the tensor field g^N with local components

$$g^N_{\alpha\beta}(u) = \tilde{g}_{ij}(x,N)B^{ij}_{\alpha\beta}. \tag{4.23}$$

If $\tilde{g}$ is a Riemannian metric on $V\widetilde{M}'$(*i.e.* $\widetilde{\mathbb{F}}^{m+1}$ is a Finsler manifold), then g^N is a Riemannian metric on M. Since we are concerned only with the non-degenerate case, in general, we suppose that g^N is a pseudo-Riemannian metric on M.

We denote by $\{\bar{\partial}/\bar{\partial}u^\alpha\}$ and $\{\bar{\partial}/\bar{\partial}x^i\}$ the natural fields of frames on M and $\widetilde{M}$ respectively. We put a bar on ∂ to distinguish the vector fields $\partial/\partial u^\alpha$ and $\partial/\partial x^i$ on M' and $\widetilde{M}'$ from $\bar{\partial}/\bar{\partial}u^\alpha$ and $\bar{\partial}/\bar{\partial}x^i$ on M and $\widetilde{M}$ respectively. Then we have

$$(a) \quad \frac{\bar{\partial}}{\bar{\partial}u^\alpha} = B^i_\alpha\frac{\bar{\partial}}{\bar{\partial}x^i} \quad \text{and} \quad (b) \quad N = N^i(x)\frac{\bar{\partial}}{\bar{\partial}x^i}, \tag{4.24}$$

where N^i are given by (4.16). Clearly $\{\bar{\partial}/\bar{\partial}u^\alpha, N\}$ is also a field of frames on $\widetilde{M}$ but defined at points of M. Thus we set

$$\frac{\partial}{\partial x^i} = b_i^\alpha \frac{\bar{\partial}}{\bar{\partial} u^\alpha} + \widetilde{N}_i N, \tag{4.25}$$

where $[b_i^\alpha \widetilde{N}_i]$ is the inverse matrix of $[B_\alpha^i N^i]$, that is we have

$$\begin{aligned} &(a)\ \ \widetilde{N}_i B_\alpha^i = 0; \quad (b)\ \ \widetilde{N}_i N^i = 1; \quad (c)\ \ b_i^\alpha B_\beta^i = \delta_\beta^\alpha; \\ &(d)\ \ b_i^\alpha N^i = 0; \quad (e)\ \ B_\alpha^i b_j^\alpha + N^i \widetilde{N}_j = \delta_j^i. \end{aligned} \tag{4.26}$$

Comparing (4.26b) with (4.17b) we deduce that

$$\widetilde{N}_i = \tilde{g}_{ij}(x, N) N^j. \tag{4.27}$$

Also, contracting (4.23) by $g^{N\beta\gamma} b_k^\alpha$ and using (4.26a) and (4.17a) we deduce that

$$b_k^\gamma = g^{N\gamma\beta} B_\beta^j \tilde{g}_{jk}(x, N). \tag{4.28}$$

Finally, by using (4.28) it is easy to check that

$$g^{N\alpha\gamma} = \tilde{g}^{ij}(x, N) b_i^\alpha b_j^\gamma. \tag{4.29}$$

To relate the geometric objects on M with those of the ambient pseudo–Finsler manifold, we consider the latter defined on the line bundle $M'^{\perp}$ spanned by N over M. However, when we write the local components for the geometric objects at points of $M'^{\perp}$, we omit the coordinates (x, N). First, we would like to relate the Levi–Civita connection ∇_* on M with the Cartan connection of $\widetilde{\mathbb{F}}^{m+1}$. To obtain this we denote by $\mathbf{F}_{\alpha\ \beta}^{\ \gamma}$ the Christoffel symbols with respect to g^N, that is, we have (see (1.3.30)).

$$\mathbf{F}_{\alpha\ \beta}^{\ \gamma} = \frac{1}{2} g^{N\gamma\varepsilon} \left(\frac{\partial g_{\varepsilon\alpha}^N}{\partial u^\beta} + \frac{\partial g_{\varepsilon\beta}^N}{\partial u^\alpha} - \frac{\partial g_{\alpha\beta}^N}{\partial u^\varepsilon} \right). \tag{4.30}$$

Also, we set

$$\widetilde{N}_j^i = \frac{\partial N^i}{\partial x^j} + \widetilde{G}_j^i. \tag{4.31}$$

Then we may state the following result.

PROPOSITION 4.1 *The local coefficients of the Levi–Civita connection on M and of the Cartan connection of $\widetilde{\mathbb{F}}^{m+1}$ are related by*

$$\mathbf{F}_{\alpha\ \beta}^{\ \gamma} = \left(\widetilde{F}_{i\ j}^{*k} + \tilde{g}_{i\ h}^{\ k} \widetilde{N}_j^h + \tilde{g}_{j\ h}^{\ k} \widetilde{N}_i^h - \tilde{g}_{ijh} \widetilde{N}_s^h \tilde{g}^{sk} \right) B_{\alpha\beta}^{ij} b_k^\gamma + b_k^\gamma B_{\alpha\beta}^k. \tag{4.32}$$

PROOF. By direct calculations using derivatives of (4.23) in (4.30) we obtain

$$\mathbf{F}_{\alpha\ \beta}^{\ \gamma} = \tfrac{1}{2} g^{N\gamma\varepsilon} B^{ijk}_{\alpha\beta\varepsilon} \left(\frac{\partial \tilde{g}_{ki}}{\partial x^j} + \frac{\partial \tilde{g}_{kj}}{\partial x^i} - \frac{\partial \tilde{g}_{ij}}{\partial x^k} + \frac{\partial \tilde{g}_{ki}}{\partial y^h}\frac{\partial N^h}{\partial x^j} + \frac{\partial \tilde{g}_{kj}}{\partial y^h}\frac{\partial N^h}{\partial x^i} - \frac{\partial \tilde{g}_{ij}}{\partial y^h}\frac{\partial N^h}{\partial x^k} \right) + g^{N\gamma\varepsilon} \tilde{g}_{ij} B^i_{\varepsilon} B^j_{\alpha\beta}. \tag{4.33}$$

Then, taking into account (1.6.21), (1.7.11), (2.3.20) and (4.31), we see that (4.33) becomes

$$\mathbf{F}_{\alpha\ \beta}^{\ \gamma} = g^{N\gamma\varepsilon} B^{ijk}_{\alpha\beta\varepsilon} (\tilde{g}_{kh} \widetilde{F}^{*h}_{i\ j} + \tilde{g}_{ikh} \widetilde{N}^h_j + \tilde{g}_{jkh} \widetilde{N}^h_i - \tilde{g}_{ijh} \widetilde{N}^h_k) + g^{N\gamma\varepsilon} \tilde{g}_{ij} B^i_{\varepsilon} B^j_{\alpha\beta}. \tag{4.34}$$

Finally, (4.32) follows from (4.34) by using (4.29), (4.26c) and (4.17a). ■

Remark 4.1. We should remark that the Levi–Civita connection on M is not an intrinsic geometric object of $\mathbb{F}^m$. This is because the pseudo–Riemannian metric g^N is defined by means of the Minkowskian unit normal vector field. ■

To define and study other induced geometric objects on M, we first introduce two projection morphisms as follows. Consider the vector bundles $T^1_1(V\widetilde{M}'_{|M'^{\perp}})$ and $T^1_1(G\widetilde{M}'_{|M'^{\perp}})$ of tensors of type (1,1) on the vector bundles $V\widetilde{M}'_{|M'^{\perp}}$ and $G\widetilde{M}'_{|M'^{\perp}}$ respectively. Then define the projection morphisms:

$$\begin{cases} \widetilde{P}^V \left(T^i_j(x,N) \left(\dfrac{\partial}{\partial y^i} \otimes \delta^* y^j \right)_{|(x,N)} \right) = T^i_j(x,N) \left(\dfrac{\bar{\partial}}{\bar{\partial} x^i} \otimes dx^j \right)_{|x}, & (a) \\ \widetilde{P}^H \left(T^i_j(x,N) \left(\dfrac{\delta^*}{\delta^* x^i} \otimes dx^j \right)_{|(x,N)} \right) = T^i_j(x,N) \left(\dfrac{\bar{\partial}}{\bar{\partial} x^i} \otimes dx^j \right)_{|x}, & (b) \end{cases} \tag{4.35}$$

where

$$\delta^* y^j = dy^j + \widetilde{G}^j_k dx^k.$$

Also, we define two lifts of tensor fields on $\widetilde{M}$ along M to tensor fields on vector bundles $V\widetilde{M}'_{|M'^{\perp}}$ and $G\widetilde{M}'_{|M'^{\perp}}$ as follows. Let

$$T = T^i_j(x) \left(\frac{\bar{\partial}}{\bar{\partial} x^i} \otimes dx^j \right)_{|x(u)},$$

be a tensor field of type (1,1) on $\widetilde{M}$ along M. Then the **vertical** (resp., **horizontal**) **lift** of T is denoted by T^V (resp., T^H) and it is defined by

$$\begin{cases} T^V = T^i_j(x)\left(\dfrac{\partial}{\partial y^i}\otimes \delta^* y^j\right)_{|(x,N)}, & (a) \\ T^H = T^i_j(x)\left(\dfrac{\delta^*}{\delta^* x^i}\otimes dx^j\right)_{|(x,N)}. & (b) \end{cases} \tag{4.36}$$

It is easy to see that $\widetilde{P}^V$ and $\widetilde{P}^H$ are the inverse operators of the vertical and horizontal lifts respectively.

In particular, the Minkowskian unit normal vector field N has vertical and horizontal lifts, given by

$$(a)\ N^V = N^i(x)\frac{\partial}{\partial y^i}_{|(x,N)} \quad \text{and} \quad (b)\ N^H = N^i(x)\frac{\delta^*}{\delta^* x^i}_{|(x,N)}, \tag{4.37}$$

respectively. The covariant derivatives of N^V with respect to the classical Finsler connections are given in the next proposition.

PROPOSITION 4.2 (i) *The vertical covariant derivatives of N^V with respect to $\widetilde{\mathbf{FC}}^*$, $\widetilde{\mathbf{BFC}}$ and $\widetilde{\mathbf{RFC}}$ vanish on $M'^{\perp}$.*

(ii) *The horizontal covariant derivatives of N^V with respect to $\widetilde{\mathbf{FC}}^*$, $\widetilde{\mathbf{BFC}}$ and $\widetilde{\mathbf{RFC}}$ coincide, and they are given by*

$$N^i{}_{|j} = \widetilde{N}^i_j. \tag{4.38}$$

PROOF. The vertical covariant derivatives of N^V with respect to $\widetilde{\mathbf{BFC}}$ and $\widetilde{\mathbf{RFC}}$ vanish because N^i do not depend on (y^i) and $\widetilde{C}_i{}^k{}_j = 0$ for both Finsler connections. Regarding $\widetilde{\mathbf{FC}}^*$ we have

$$N^i{}_{\|*j} = N^k \widetilde{g}_k{}^i{}_j(x, N) = 0,$$

which completes the proof of the first assertion. Next, by using (1.7.29) we obtain

$$\begin{aligned} N^i{}_{|\mathbf{r}j} = N^i{}_{|*j} &= \frac{\delta^* N^i}{\delta^* x^j} + N^k \widetilde{F}^{*i}_k{}_j(x, N) \\ &= \frac{\partial N^i}{\partial x^j} + \widetilde{G}^i_j(x, N) = \widetilde{N}^i_j. \end{aligned}$$

Similarly, by using (1.6.30) we deduce that $N^i{}_{|_{\mathbf{b}}j}$ is also given by (4.38). ■

In what follows in this subsection we find some induced linear connections on M and compare them with the Levi–Civita connection of M. Also we define some second fundamental forms and shape operators on M.

Let $\widetilde{\mathbb{F}}^{m+1} = (\widetilde{M}, \widetilde{M}', \widetilde{F}^*)$ be a pseudo–Finsler manifold endowed with a Finsler connection $\widetilde{FC} = (G\widetilde{M}', \widetilde{\nabla}) = (\widetilde{G}^k_i, \widetilde{F}_i{}^k{}_j, \widetilde{C}_i{}^k{}_j)$ and $\mathbb{F}^m = (M, M', F^*)$ be a pseudo–Finsler hypersurface of $\widetilde{\mathbb{F}}^{m+1}$. Consider a vector field $X = X^\alpha(\bar{\partial}/\bar{\partial}u^\alpha)$ on M. Then by using (4.36) for tensor fields on $\widetilde{M}$ we define the **vertical** and **horizontal lifts** of X by

$$\left\{\begin{array}{lll} X^V & = & X^\alpha B^i_\alpha(\dfrac{\partial}{\partial x^i})^V = X^\alpha B^i_\alpha \dfrac{\partial}{\partial y^i}_{|(x,N)}, \qquad (a) \\ X^H & = & X^\alpha B^i_\alpha(\dfrac{\partial}{\partial x^i})^H = X^\alpha B^i_\alpha \dfrac{\delta^*}{\delta^* x^i}_{|(x,N)}. \qquad (b) \end{array}\right. \tag{4.39}$$

This can naturally be extended to any tensor field on M. Also, for any smooth function f on M we define its vertical and horizontal lifts on $M'^{\perp}$ as follows

$$f^V(x,N) = f^H(x,N) = f(x). \tag{4.40}$$

Now, we see that (4.25) implies the decomposition

$$\widetilde{M}'_{|M'^{\perp}} = M' \oplus M'^{\perp}. \tag{4.41}$$

Because of (4.41) we may set

$$\widetilde{P}^V(\widetilde{\nabla}_{X^H} Y^V) = \nabla_X Y + h(X,Y)N, \tag{4.42}$$

and

$$\widetilde{P}^V(\widetilde{\nabla}_{X^H} N^V) = -AX + \omega(X)N, \tag{4.43}$$

for any $X, Y \in \Gamma(M')$, where $\nabla_X Y$ and AX lie in $\Gamma(M')$, while $h(X,Y)$ and $\omega(X)$ are smooth functions on M. Taking into account that $\widetilde{\nabla}$ is a linear connection on $V\widetilde{M}'$ and using (4.39) and (4.35a) we deduce that ∇ is a linear connection on M and A, h and ω are tensor fields on M of type (1,1), (0, 2) and (0, 1) respectively. We call (4.42) and (4.43) the **Gauss** and **Weingarten formulas** for the Minkowskian approach of the immersion of $\mathbb{F}^m$ in $\widetilde{\mathbb{F}}^{m+1}$. We also call ∇ the **induced linear connection** on M.

To deduce local formulas for (4.42) and (4.43) we first note that for any smooth function f on M we have

$$\frac{\delta^*}{\delta^* x^i}(f^H) = \frac{\partial}{\partial x^j}(f^H) = \frac{\bar{\partial} f}{\bar{\partial} x^j}. \tag{4.44}$$

Then we set

$$(a) \quad \nabla_{\bar{\partial}/\bar{\partial}u^\beta} \frac{\bar{\partial}}{\bar{\partial}u^\alpha} = F_\alpha{}^\gamma{}_\beta \frac{\bar{\partial}}{\bar{\partial}u^\gamma} \quad \text{and} \quad (b) \ h\left(\frac{\bar{\partial}}{\bar{\partial}u^\beta}, \frac{\bar{\partial}}{\bar{\partial}u^\alpha}\right) = h_{\alpha\beta}. \tag{4.45}$$

Now, we replace X and Y from (4.42) by $\bar{\partial}/\bar{\partial}u^\beta$ and $\bar{\partial}/\bar{\partial}u^\alpha$, respectively, and by using (4.24a), (4.44), (2.1.41), (4.39), and (4.35a) we obtain

$$(B^k_{\alpha\beta} + B^{ij}_{\alpha\beta}\widetilde{F}_i{}^k{}_j)\frac{\bar{\partial}}{\bar{\partial}x^k} = F_\alpha{}^\gamma{}_\beta\frac{\bar{\partial}}{\bar{\partial}u^\gamma} + h_{\alpha\beta}N. \tag{4.46}$$

Finally, by using (4.25) in (4.46) and equating the components from M' and $M'^\perp$ we infer that

$$F_\alpha{}^\gamma{}_\beta = b^\gamma_k(B^k_{\alpha\beta} + B^{ij}_{\alpha\beta}\widetilde{F}_i{}^k{}_j(x,N)), \tag{4.47}$$

and

$$h_{\alpha\beta} = \widetilde{N}_k(B^k_{\alpha\beta} + B^{ij}_{\alpha\beta}\widetilde{F}_i{}^k{}_j(x,N)). \tag{4.48}$$

Similarly, we replace X by $\bar{\partial}/\bar{\partial}u^\alpha$ in (4.43) and by using (4.39b), (4.37a), (4.44) and (4.35a) we derive

$$\left(\frac{\bar{\partial}N^k}{\bar{\partial}u^\alpha} + N^i\widetilde{F}_i{}^k{}_j(x,N)B^j_\alpha\right)\frac{\bar{\partial}}{\bar{\partial}x^k} = -A^\beta_\alpha\frac{\bar{\partial}}{\bar{\partial}u^\beta} + \omega_\alpha N, \tag{4.49}$$

where we set

$$(a)\;\; A\left(\frac{\bar{\partial}}{\bar{\partial}u^\alpha}\right) = A^\beta_\alpha\frac{\bar{\partial}}{\bar{\partial}u^\beta} \quad\text{and}\quad (b)\;\; \omega\left(\frac{\bar{\partial}}{\bar{\partial}u^\alpha}\right) = \omega_\alpha. \tag{4.50}$$

Then, using (4.25) in (4.49) we deduce that

$$(a)\;\; A^\beta_\alpha = -b^\beta_k N^k_\alpha \quad\text{and}\quad (b)\;\; \omega_\alpha = \widetilde{N}_k N^k_\alpha, \tag{4.51}$$

where we put

$$N^k_\alpha = \frac{\bar{\partial}N^k}{\bar{\partial}u^\alpha} + N^i\widetilde{F}_i{}^k{}_j(x,N)B^j_\alpha. \tag{4.52}$$

The tensor fields h and A on M with local components $h_{\alpha\beta}$ and A^β_α are called the **Minkowskian second fundamental form** and the **Minkowskian shape operator** of the immersion of $\mathbb{F}^m$ in $\widetilde{\mathbb{F}}^{m+1}$ induced by the Finsler connection $\widetilde{FC}$.

Keeping the notations from (2.1.6), where we put c, r, and b on the top of the induced geometric objects by the Cartan, Rund, and Berwald connections, respectively, from (4.47) and (4.48) we obtain

$$\begin{cases} \overset{c}{F}_\alpha{}^\gamma{}_\beta = \overset{r}{F}_\alpha{}^\gamma{}_\beta = b^\gamma_k(B^k_{\alpha\beta} + B^{ij}_{\alpha\beta}\widetilde{F}^{*}_i{}^k{}_j(x,N)), & (a) \\ \overset{b}{F}_\alpha{}^\gamma{}_\beta = b^\gamma_k(B^k_{\alpha\beta} + B^{ij}_{\alpha\beta}\widetilde{G}_i{}^k{}_j(x,N)), & (b) \end{cases} \tag{4.53}$$

and

$$\begin{cases} \overset{c}{h}_{\alpha\beta} = \overset{r}{h}_{\alpha\beta} = \widetilde{N}_k(B^k_{\alpha\beta} + B^{ij}_{\alpha\beta}\widetilde{F}^{*k}_{i\ j}(x, N)), & (a) \\ \overset{b}{h}_{\alpha\beta} = \widetilde{N}_k(B^k_{\alpha\beta} + B^{ij}_{\alpha\beta}\widetilde{G}_i{}^k{}_j(x, N)), & (b) \end{cases} \tag{4.54}$$

respectively. What is important and surprising, as well, in the Minkowskian theory we present in this section, is the following result.

THEOREM 4.1 *The Minkowskian shape operators of $\mathbb{F}^m$ induced by the Cartan, Rund, and Berwald connections of $\widetilde{\mathbb{F}}^{m+1}$ coincide, and have the local components given by*

$$A^\beta_\alpha = -b^\beta_k \widetilde{N}^k_i B^i_\alpha. \tag{4.55}$$

PROOF. By (1.7.29) and (1.6.30) we see that (4.52) becomes

$$N^k_\alpha = N^k{}_{|i} B^i_\alpha. \tag{4.56}$$

Thus the assertion follows from (4.51a) by using (4.56). ■

From now on (4.55) gives us the local components of the Minkowskian shape operator we use further in the Minkowskian theory of $\mathbb{F}^m$.

THEOREM 4.2 *The second fundamental forms induced by the Cartan and Rund connections coincide and they are related with the shape operator of $\mathbb{F}^m$ as follows*

$$(a)\ \ h_{\alpha\beta} = g^N_{\alpha\gamma} A^\gamma_\beta \ \ \text{or, equivalently,} \ \ (b)\ \ A^\gamma_\beta = g^{N\gamma\alpha} h_{\alpha\beta}. \tag{4.57}$$

PROOF. By differentiating (4.17a) with respect to u^β and by using (2.3.20), (1.6.22a), and (4.27), we deduce that

$$\frac{\delta^* \tilde{g}_{ij}}{\delta^* x^k}(x, N) B^{ik}_{\alpha\beta} N^j + \widetilde{N}_i(x, N) B^i_{\alpha\beta} + \tilde{g}_{ij}(x, N) B^i_\alpha \frac{\partial N^j}{\partial u^\beta} = 0. \tag{4.58}$$

As both the Cartan and Rund connections are h–metric Finsler connections, we have

$$\frac{\delta^* \tilde{g}_{ij}}{\delta^* x^k} = \tilde{g}_{hj} \widetilde{F}^{*h}_{i\ k} + \tilde{g}_{ih} \widetilde{F}^{*h}_{j\ k}. \tag{4.59}$$

Then by using (4.59) and (4.38) in (4.58) we obtain

$$\widetilde{N}_h(B^h_{\alpha\beta} + \widetilde{F}^{*h}_{i\ k} B^{ik}_{\alpha\beta}) + \tilde{g}_{ih} N^h{}_{|k} B^{ik}_{\alpha\beta} = 0. \tag{4.60}$$

On the other hand, from (4.28) we derive

$$\tilde{g}_{hi} B^i_\alpha = g^N_{\alpha\gamma} b^\gamma_h. \tag{4.61}$$

Finally, by using (4.61) in (4.60) and taking into account (4.54a) and (4.55) we obtain (4.57). ∎

Remark 4.2. Since the Berwald connection is not an h–metric Finsler connection (see (1.7.88)), it is easy to check that $\overset{b}{h}_{\alpha\beta}$ are not related to the Minkowskian shape operator by (4.57). As (4.57) is similar to the well known relation between the second fundamental form and the shape operator of a Riemannian submanifold (see Chen [1], p. 41 and Spivak [1], Vol. 3, p. 22), from now on, $h_{\alpha\beta}$ induced by the Cartan and Rund connections give us the local components of the Minkowskian second fundamental form of $\mathbb{F}^m$. ∎

Also, the Weingarten equations (4.43) take the usual form from the theory of Riemannian hypersurfaces, provided the ambient manifold is equipped with the Cartan or Rund connections.

PROPOSITION 4.3 *Let $\widetilde{\mathbb{F}}^{m+1}$ be equipped with the Cartan and Rund connections and $\mathbb{F}^m$ be a pseudo–Finsler hypersurface of $\widetilde{\mathbb{F}}^{m+1}$. Then we have $\overset{c}{\omega} = \overset{r}{\omega} = 0$ on M.*

PROOF. By differentiating (4.17b) with respect to u^β and by using (2.3.20), (1.6.22a), and (4.27) we obtain

$$\frac{\delta^* \tilde{g}_{ij}}{\delta^* x^k}(x, N)N^i N^j B^k_\beta + 2\widetilde{N}_i \frac{\bar{\partial} N^i}{\bar{\partial} u^\beta} = 0. \tag{4.62}$$

Then taking into account (1.7.28a), (1.7.76a) and using (4.51b), (4.52) and (4.62) we deduce that $\overset{c}{\omega} = \overset{r}{\omega} = 0$ on M. ∎

As we have seen (cf., (4.53a)), the induced linear connections by the Cartan and Rund connections coincide. We denote them by $\nabla = (F_\alpha{}^\gamma{}_\beta)$ and consider a relative covariant derivative on M defined by $(F_\alpha{}^\gamma{}_\beta(u), \widetilde{F}^{*k}_{i\ j}(x, N))$ as follows. Suppose S is a mixed thensor field on M with local components $S^{i\alpha}_{j\beta}$. Then define the **relative covariant derivative** of S by

$$\begin{aligned} S^{i\alpha}_{j\beta|\gamma} &= \frac{\bar{\partial} S^{i\alpha}_{j\beta}}{\bar{\partial} u^\gamma} + S^{h\alpha}_{j\beta} \widetilde{F}^{*i}_{h\ \gamma} - S^{i\alpha}_{h\beta} \widetilde{F}^{*h}_{j\ \gamma} \\ &\quad + S^{i\varepsilon}_{j\beta} F_\varepsilon{}^\alpha{}_\gamma - S^{i\alpha}_{h\varepsilon} F_\beta{}^\varepsilon{}_\gamma, \end{aligned} \tag{4.63}$$

where we set

$$\widetilde{F}^{*i}_{h\ \gamma} = \widetilde{F}^{*i}_{h\ k} B^k_\gamma. \tag{4.64}$$

Then by using (4.63), (4.53a) and (4.54a) we obtain

$$B^i_{\alpha|\beta} = h_{\alpha\beta} N^i. \tag{4.65}$$

Similarly, using (4.63) and (4.52) we deduce that

$$N^i{}_{|\alpha} = N^i_\alpha = \frac{\partial N^i}{\partial u^\alpha} + \widetilde{G}^i_j B^j_\alpha. \tag{4.66}$$

On the other hand, contracting (4.55) by B^i_β and using (4.26e) and (4.56) we derive

$$A^\beta_\alpha B^i_\beta = -N^i_\alpha, \tag{4.67}$$

since by Proposition 4.3 we have $\widetilde{N}_i N^i{}_{|\alpha} = \omega_\alpha = 0$. Hence (4.66) and (4.67) imply

$$N^i{}_{|\alpha} = -A^\beta_\alpha B^i_\beta. \tag{4.68}$$

Clearly (4.65) and (4.68) are nothing but the local expressions of the Gauss and Weingarten formulas (4.42) and (4.43) with respect to the Rund connection of $\widetilde{\mathbb{F}}^{m+1}$.

A major difference between the geometry of a pseudo–Riemannian hypersurface and the geometry of a pseudo–Finsler hypersurface is that in the latter, the induced linear connection is not a metric connection. More precisely, we prove the following theorem.

THEOREM 4.3 *The Minkowskian induced linear connection* $\nabla = (F_\alpha{}^\gamma{}_\beta)$ *of* $\mathbb{F}^m$ *by the Cartan and Rund connections of* $\widetilde{\mathbb{F}}^{m+1}$ *is torsion-free and the covariant derivative of* g^N *with respect to* ∇ *is given by*

$$\overset{N}{g}_{\alpha\beta|\gamma} = -2\tilde{g}_{ijk}(x, N) B^{ijk}_{\alpha\beta\varepsilon} A^\varepsilon_\gamma. \tag{4.69}$$

PROOF. The first assertion of the theorem follows from (4.53a) since we have

$$\widetilde{F}^{*k}_{i\ j}(x, N) = \widetilde{F}^{*k}_{j\ i}(x, N).$$

Next, we differentiate (4.23) with respect to u^γ and by using (2.3.20), (1.6.21) and (4.38) we deduce that

$$\frac{\partial \overset{N}{g}_{\alpha\beta}}{\partial u^\gamma} = \left(\frac{\delta^* \tilde{g}_{ij}}{\delta^* x^h} + 2\tilde{g}_{ijk} N^k{}_{|h}\right) B^{ijh}_{\alpha\beta\gamma} + \tilde{g}_{ij} B^i_{\alpha\gamma} B^j_\beta + \tilde{g}_{ij} B^i_\alpha B^j_{\beta\gamma}. \tag{4.70}$$

Then by direct calculations using (4.63), (4.70) and (4.53a) we obtain

$$\begin{aligned} \overset{N}{g}_{\alpha\beta|\gamma} &= \left(\frac{\delta^* \tilde{g}_{ij}}{\delta^* x^h} + 2\tilde{g}_{ijk} N^k{}_{|h}\right) B^{ijh}_{\alpha\beta\gamma} + \tilde{g}_{ij} B^i_{\alpha\gamma} B^j_\beta + \tilde{g}_{ij} B^i_\alpha B^j_{\beta\gamma} \\ &\quad - \overset{N}{g}_{\varepsilon\beta} b^\varepsilon_k (B^k_{\alpha\gamma} + \widetilde{F}^{*k}_{i\ j} B^{ij}_{\alpha\gamma}) - \overset{N}{g}_{\alpha\varepsilon} b^\varepsilon_k (B^k_{\beta\gamma} + \widetilde{F}^{*k}_{i\ j} B^{ij}_{\beta\gamma}). \end{aligned} \tag{4.71}$$

Finally, by using (4.61), (4.56), (4.66), and (4.68), and taking into account that $\tilde{g}_{ij|\mathrm{r}h} = 0$, we obtain (4.69). ■

It follows from Theorem 4.3 that the Minkowskian induced linear connection $\nabla = (F_{\alpha}{}^{\gamma}{}_{\beta})$, in general, does not coincide with the Levi–Civita connection $\nabla_* = (\mathbf{F}_{\alpha}{}^{\gamma}{}_{\beta})$ on M with respect to the pseudo–Riemannian metric $g^N_{\alpha\beta}$. On this matter we prove the following.

THEOREM 4.4 *The linear connection* $\nabla = (F_{\alpha}{}^{\gamma}{}_{\beta})$ *induced on* M *by the Cartan and Rund connections of* $\widetilde{\mathbb{F}}^{m+1}$ *is related with the Levi–Civita connection* $\nabla_* = (\mathbf{F}_{\alpha}{}^{\gamma}{}_{\beta})$ *on* M *as follows*

$$\mathbf{F}_{\alpha}{}^{\gamma}{}_{\beta} = F_{\alpha}{}^{\gamma}{}_{\beta} + \tilde{g}_{ish} B^h_{\mu} g^{N\gamma\varepsilon}(A^{\mu}_{\varepsilon} B^{is}_{\alpha\beta} - A^{\mu}_{\alpha} B^{is}_{\beta\varepsilon} - A^{\mu}_{\beta} B^{is}_{\alpha\varepsilon}). \tag{4.72}$$

PROOF. Using (4.53a) into (4.32) we deduce that

$$\mathbf{F}_{\alpha}{}^{\gamma}{}_{\beta} = F_{\alpha}{}^{\gamma}{}_{\beta} + (\tilde{g}_i{}^k{}_h \widetilde{N}^h_j + \tilde{g}_j{}^k{}_h \widetilde{N}^h_i - \tilde{g}_{ijh} \widetilde{N}^h_s \tilde{g}^{sk}) B^{ij}_{\alpha\beta} b^{\gamma}_k. \tag{4.73}$$

On the other hand, by using (4.38), (4.56), (4.66), (4.68), and (4.28) we obtain

$$\tilde{g}_i{}^k{}_h \widetilde{N}^h_j B^{ij}_{\alpha\beta} b^{\gamma}_k = -\tilde{g}_{ish} B^{ish}_{\alpha\mu\varepsilon} A^{\varepsilon}_{\beta} g^{N\mu\gamma}. \tag{4.74}$$

Then (4.72) follows from (4.73) by using (4.74). ∎

Finally, by using (4.69) in (4.72) we infer that

$$\mathbf{F}_{\alpha}{}^{\gamma}{}_{\beta} = F_{\alpha}{}^{\gamma}{}_{\beta} + \frac{1}{2} g^{N\gamma\varepsilon}(g^N_{\varepsilon\alpha|\beta} + g^N_{\varepsilon\beta|\alpha} - g^N_{\alpha\beta|\varepsilon}). \tag{4.75}$$

This equality can also be checked directly by using (4.63) for the covariant derivative of $g^N_{\alpha\beta}$ and (1.3.30) for the Christoffel symbols of ∇_*.

4.4. THE MINKOWSKIAN GAUSS–CODAZZI EQUATIONS FOR A PSEUDO–FINSLER HYPERSURFACE

Let $\widetilde{\mathbb{F}}^{m+1} = (\widetilde{M}, \widetilde{M}', \widetilde{F}^*)$ be a pseudo–Finsler manifold endowed with the Rund connection $\widetilde{\mathbf{RFC}} = (\widetilde{G}^k_i, \widetilde{F}^{*k}_{i\ j}, 0)$. Until now the Cartan and Rund connections have been equally used to obtain information on the induced geometric objects of a pseudo–Finsler hypersurface $\mathbb{F}^m = (M, M', F^*)$ of $\widetilde{\mathbb{F}}^{m+1}$. Our choice for the Rund connection is owed to the simplicity of the structure equations by means of the geometric objects related to $\widetilde{\mathbf{RFC}}$. All the geometric objects of $\widetilde{\mathbb{F}}^{m+1}$ are here again considered at points $(x(u), N(x(u)))$ of $M'^{\perp}$.

To get the structure equations for the Minkowskian immersion of $\mathbb{F}^m$ in $\widetilde{\mathbb{F}}^{m+1}$, we first use (4.63) and obtain

$$B^i_{\alpha|\beta} = B^i_{\alpha\beta} + \widetilde{F}^{*i}_{h\ k} B^{hk}_{\alpha\beta} - B^i_{\varepsilon} F_{\alpha}{}^{\varepsilon}{}_{\beta}. \tag{4.76}$$

Then we set $B^i_{\alpha|\beta\gamma} = B^i_{\alpha|\beta|\gamma}$ and by direct calculations using (4.63), (4.76), (2.3.20) and (1.7.29) we deduce that

$$\begin{aligned} B^i_{\alpha|\beta\gamma} &= B^i_{\alpha\beta\gamma} + \left(\frac{\delta^* \widetilde{F}^{*i}_{j\ k}}{\delta^* x^h} + \widetilde{F}^{*t}_{j\ k} \widetilde{F}^{*i}_{t\ h} \right) B^{jkh}_{\alpha\beta\gamma} + \frac{\partial \widetilde{F}^{*i}_{j\ k}}{\partial y^t} B^{jk}_{\alpha\beta} N^t_{|\gamma} \qquad (4.77) \\ &+ \widetilde{F}^{*i}_{j\ k} (B^j_{\alpha\gamma} B^k_\beta + B^j_{\alpha\beta} B^k_\gamma + B^j_\alpha B^k_{\beta\gamma} - B^{jk}_{\gamma\varepsilon} F_{\alpha\ \beta}^{\ \varepsilon} - B^{jk}_{\beta\varepsilon} F_{\alpha\ \gamma}^{\ \varepsilon}) \\ &- B^i_{\gamma\varepsilon} F_{\alpha\ \beta}^{\ \varepsilon} - B^i_{\beta\varepsilon} F_{\alpha\ \gamma}^{\ \varepsilon} - B^i_\varepsilon \left(\frac{\partial F_{\alpha\ \gamma}^{\ \varepsilon}}{\partial u^\gamma} - F_{\alpha\ \gamma}^{\ \mu} F_{\mu\ \beta}^{\ \varepsilon} \right) - B^i_{\alpha|\varepsilon} F_{\beta\ \gamma}^{\ \varepsilon}. \end{aligned}$$

Thus taking into account (4.77), (1.7.77a), (1.7.77b), (1.3.28), and (4.68) we derive

$$\begin{aligned} B^i_{\alpha|\beta\gamma} - B^i_{\alpha|\gamma\beta} &= \widetilde{K}_{j\ kh}^{\ i} B^{jkh}_{\alpha\beta\gamma} + \widetilde{F}_{j\ kh}^{\ i} B^{jh}_{\alpha\varepsilon} (A^\varepsilon_\beta B^k_\gamma - A^\varepsilon_\gamma B^k_\beta) \\ &- R_{\alpha\ \beta\gamma}^{\ \varepsilon} B^i_\varepsilon. \end{aligned} \qquad (4.78)$$

On the other hand, by differentiating (4.65) covariantly and using (4.68) and (4.57b) we infer that

$$B^i_{\alpha|\beta\gamma} - B^i_{\alpha|\gamma\beta} = (h_{\alpha\beta|\gamma} - h_{\alpha\gamma|\beta}) N^i + (h_{\alpha\gamma} h_{\beta\mu} - h_{\alpha\beta} h_{\gamma\mu}) g^{\mu\varepsilon} B^i_\varepsilon. \qquad (4.79)$$

Then taking into account (4.25) and (4.61) and comparing (4.78) and (4.79) we obtain

$$R_{\alpha\varepsilon\beta\gamma} = \widetilde{K}_{ijkh} B^{ijkh}_{\alpha\varepsilon\beta\gamma} + \widetilde{F}_{ijkh} B^{ijh}_{\alpha\varepsilon\mu} (A^\mu_\beta B^k_\gamma - A^\mu_\gamma B^k_\beta) + h_{\alpha\beta} h_{\varepsilon\gamma} - h_{\alpha\gamma} h_{\varepsilon\beta}, \qquad (4.80)$$

and

$$h_{\alpha\beta|\gamma} - h_{\alpha\gamma|\beta} = \left(\widetilde{K}_{ijkh} B^{ikh}_{\alpha\beta\gamma} + \widetilde{F}_{ijkh} B^{ih}_{\alpha\mu} (A^\mu_\beta B^k_\gamma - A^\mu_\gamma B^k_\beta) \right) N^j. \qquad (4.81)$$

We call (4.80) and (4.81) the **Minkowskian Gauss equations** and the **Minkowskian Codazzi equations** of the immersion of $\mathbb{F}^m$ in $\widetilde{\mathbb{F}}^{m+1}$.

Next, we consider the tensor field $\mathbf{D} = (D_{\alpha\ \beta}^{\ \gamma})$ on M given by

$$D_{\alpha\ \beta}^{\ \gamma} = \frac{1}{2} g^{N\gamma\varepsilon} (g^N_{\varepsilon\alpha|\beta} + g^N_{\varepsilon\beta|\alpha} - g^N_{\alpha\beta|\varepsilon}), \qquad (4.82)$$

and call it the **deformation tensor field** of the pair of linear connections (∇, ∇_*) (see Vaisman [1]). Thus (4.75) becomes

$$\mathbf{F}_{\alpha\ \beta}^{\ \gamma} = F_{\alpha\ \beta}^{\ \gamma} + D_{\alpha\ \beta}^{\ \gamma}. \qquad (4.83)$$

By direct calculations using (1.3.28) and (4.83) we deduce that

$$\mathbf{R}_{\alpha\varepsilon\beta\gamma} = R_{\alpha\varepsilon\beta\gamma} + \mathcal{A}_{(\beta\gamma)}\left\{D_{\alpha\varepsilon\beta|\gamma} - D_{\alpha}{}^{\mu}{}_{\beta}g^{N}_{\mu\varepsilon|\gamma} + D_{\alpha}{}^{\mu}{}_{\beta}D_{\mu\varepsilon\gamma}\right\}, \tag{4.84}$$

where $\mathbf{R}$ is the curvature tensor of ∇_* and

$$D_{\alpha\varepsilon\beta} = D_{\alpha}{}^{\mu}{}_{\beta}\, g^{N}_{\mu\varepsilon}.$$

Thus by (4.84), the Minkowskian Gauss equations of $\mathbb{F}^m$ can be expressed in terms of the curvature tensor fields of both ∇_* and $\widetilde{\mathbf{RFC}}$ as follows

$$\begin{aligned}\mathbf{R}_{\alpha\varepsilon\beta\gamma} &= \widetilde{K}_{ijkh}B^{ijkh}_{\alpha\varepsilon\beta\gamma} \\ &+\mathcal{A}_{(\beta\gamma)}\{\widetilde{F}_{ijkh}B^{ijhk}_{\alpha\varepsilon\mu\gamma}A^{\mu}_{\beta} + h_{\alpha\beta}h_{\varepsilon\gamma} \\ &+D_{\alpha\varepsilon\beta|\gamma} - D_{\alpha}{}^{\mu}{}_{\beta}g^{N}_{\mu\varepsilon|\gamma} + D_{\alpha}{}^{\mu}{}_{\beta}D_{\mu\varepsilon\gamma}\}.\end{aligned} \tag{4.85}$$

In particular, if $\widetilde{\mathbb{F}}^{m+1}$ is a locally pseudo–Minlowski manifold, by Theorem 1.8.6 we have $\widetilde{K}_{ijkh} = \widetilde{F}_{ijkh} = 0$. Thus by (4.80), (4.81), and (4.85) we may state the following.

THEOREM 4.5 *Let $\mathbb{F}^m$ be a pseudo–Finsler hypersurface of a locally pseudo–Minkowski manifold $\widetilde{\mathbb{F}}^{m+1}$. Then we have the assertions:*

(i) The Minkowskian Gauss equations are given by

$$R_{\alpha\varepsilon\beta\gamma} = h_{\alpha\beta}h_{\varepsilon\gamma} - h_{\alpha\gamma}h_{\varepsilon\beta}, \tag{4.86}$$

or equivalently by

$$\mathbf{R}_{\alpha\varepsilon\beta\gamma} = \mathcal{A}_{(\beta\gamma)}\{h_{\alpha\beta}h_{\varepsilon\gamma} + D_{\alpha\varepsilon\beta|\gamma} - D_{\alpha}{}^{\mu}{}_{\beta}g^{N}_{\mu\varepsilon|\gamma} + D_{\alpha}{}^{\mu}{}_{\beta}D_{\mu\varepsilon\gamma}\}. \tag{4.87}$$

(ii) The Minkowskian Codazzi equations are given by

$$h_{\alpha\beta|\gamma} - h_{\alpha\gamma|\beta} = 0. \tag{4.88}$$

4.5. MINKOWSKIAN TOTALLY UMBILICAL PSEUDO–FINSLER HYPERSURFACES

In the present subsection we define totally umbilical pseudo–Finlser hypersurfaces and give examples. Actually the examples are consisting of all non–degenerate indicatrices of a pseudo–Minkowski space.

Let $\mathbb{F}^m = (M, M', F^*)$ be a pseudo–Finsler hypersurface of a pseudo–Finsler manifold $\widetilde{\mathbb{F}}^{m+1} = (\widetilde{M}, \widetilde{M}', \widetilde{F}^*)$. Then we say that $\mathbb{F}^m$ is a **Minkowskian totally umbilical pseudo–Finsler hypersurface** if there exists a smooth function ρ on M such that on each coordinate neighborhood $\mathcal{U} \subset M$ we have

$$h_{\alpha\beta} = \rho g_{\alpha\beta}, \quad \forall \alpha, \beta \in \{1, \cdots, m\}. \tag{4.89}$$

By using (4.57) and (4.89) it is easy to see that $\mathbb{F}^m$ is Minkowskian totally umbilical if and only if the shape operator A of $\mathbb{F}^m$ satisfies

$$A_\alpha^\beta = \rho \delta_\alpha^\beta, \tag{4.90}$$

on each $\mathcal{U} \subset M$. When ρ is nowhere zero on $\mathcal{U}$ we say that $\mathbb{F}^m$ is **Minkowskian proper totally umbilical**. Then by using (4.69), (4.90), and (4.75) we obtain the following theorem:

THEOREM 4.6 *Let $\mathbb{F}^m$ be a Minkowskian proper totally umbilical pseudo–Finsler hypersurface of $\widetilde{\mathbb{F}}^{m+1}$. Then the induced linear connection coincides with the Levi–Civita connection of $\mathbb{F}^m$ if and only if we have*

$$\tilde{g}_{ijk}(x, N) B^{ijk}_{\alpha\beta\gamma} = 0, \quad \forall \alpha, \beta, \gamma \in \{1, \cdots, m\}. \tag{4.91}$$

Also we prove the following:

THEOREM 4.7 *Suppose $\mathbb{F}^m = (M, TM^0, F^*)$ is Minkowskian totally umbilical in $\widetilde{\mathbb{F}}^{m+1}$ and $\nabla = \nabla_*$ on M. Then (M, g^N) is a pseudo–Riemannian manifold of constant curvature, provided $m > 2$.*

PROOF. Since $\nabla = \nabla_*$ the deformation tensor field $\mathbf{D}$ vanishes on M. Hence taking into account (4.89) we see that (4.87) becomes

$$\mathbf{R}_{\alpha\varepsilon\beta\gamma} = \rho \left(g_{\alpha\beta} g_{\varepsilon\gamma} - g_{\alpha\gamma} g_{\varepsilon\beta} \right).$$

As $\mathbf{R}$ is the curvature tensor of the Levi–Civita connection on M, by Schur's theorem we have the assertion of the theorem. ∎

We now show that a large class of pseudo–Finsler hypersurfaces in a pseudo–Minkowski space are Minkowskian proper totally umbilical. Suppose $\widetilde{\mathbb{F}}^{m+1} = (\widetilde{M} = \mathbb{R}^{m+1}, \widetilde{M}' = \mathbb{R}^{m+1} \backslash \{0\}, \widetilde{F}^*)$ is a pseudo–Minkowski space whose Liouville vector field is spacelike, that is,

$$\widetilde{F}^*(y) = \tilde{g}_{ij}(y) y^i y^j > 0. \tag{4.92}$$

Thus by (1.1.9) we have $\widetilde{F}^*(y) = \widetilde{F}^2(y)$, where $\widetilde{F}$ is the fundamental function of $\widetilde{\mathbb{F}}^{m+1}$. Then we consider a hypersurface $M(a)$ of $\widetilde{M} = \mathbb{R}^{m+1}$ given by the equation

$$(a) \;\; \widetilde{F}^2(x^1, \cdots, x^{m+1}) = a^2 \;\text{ or equivalently }\; (b) \;\; \tilde{g}_{ij}(x) x^i x^j = a^2, \tag{4.93}$$

where a is a positive constant. Suppose that the parameteric equations of $M(a)$ are given by

$$x^i = x^i(u^1, \cdots, u^m), \quad i \in \{1, \cdots, m+1\}. \tag{4.94}$$

Then differentiating (4.93b) with respect to (u^α) and using (1.6.21) and (1.6.22) we deduce that

$$\tilde{g}_{ij}(x) B^i_\alpha x^j = 0. \tag{4.95}$$

By (4.95) and (4.93b) we see that (4.17a) and (4.17b) are satisfied for the vector field N with local components

$$N^i = \frac{1}{a} x^i. \tag{4.96}$$

As in the case of the pseudo–Minkowski space $\widetilde{\mathbb{F}}^{m+1}$ we have $\widetilde{G}^i_j = 0$, from (4.66) and (4.96) we infer that

$$N^i{}_{|\alpha} = \frac{1}{a} B^i_\alpha. \tag{4.97}$$

Thus, comparing (4.97) with (4.68) we derive

$$\frac{1}{a} B^i_\alpha = -A^\beta_\alpha B^i_\beta. \tag{4.98}$$

Contracting (4.98) by b^γ_i and using (4.26c) we obtain

$$A^\beta_\alpha = -\frac{1}{a} \delta^\beta_\alpha. \tag{4.99}$$

Hence $\mathbb{F}^m$ is Minkowskian proper totally umbilical immersed in $\widetilde{\mathbb{F}}^{m+1}$ with $\rho = -1/a$. Thus (4.86) becomes

$$R_{\alpha\epsilon\beta\gamma} = \frac{1}{a^2} \left(g_{\alpha\beta} g_{\epsilon\gamma} - g_{\alpha\gamma} g_{\epsilon\beta} \right). \tag{4.100}$$

It follows that the sectional curvature of $M(a)$ with respect to the induced linear connection ∇ is a positive constant $k = 1/a^2$. However, in general, $\nabla \neq \nabla_*$ on $M(a)$ as we prove in the next theorem.

THEOREM 4.8 *Suppose that there exists $a > 0$ such that the induced linear connection coincides with the Levi–Civita connection of $M(a)$. Then the ambient manifold $\widetilde{\mathbb{F}}^{m+1}$ must be a semi–Euclidean space.*

PROOF. As $\mathbb{F}^m$ is proper totally umbilical, from Theorem 4.6 we deduce that

$$\tilde{g}_{ijk}(x)B^{ijk}_{\alpha\beta\gamma} = 0, \quad \forall\alpha,\beta,\gamma \in \{1,\cdots,m\}, \tag{4.101}$$

at any point $x(u) \in M(a)$. On the other hand, by (1.6.22a) we have

$$\tilde{g}_{ijk}(x)B^{ij}_{\alpha\beta}x^k = 0, \quad \forall\alpha,\beta \in \{1,\cdots,m\}. \tag{4.102}$$

Contracting (4.101) and (4.102) by b^γ_h and $\tilde{g}_{sh}x^s$, respectively, and then adding we obtain

$$\tilde{g}_{ijh}(x)B^{ij}_{\alpha\beta} = 0, \quad \forall\alpha,\beta \in \{1,\cdots,m\},\ h \in \{1,\cdots,m+1\}.$$

Thus by a similar method as above we infer that

$$\tilde{g}_{ijh}(x) = 0, \quad \forall i,j,h \in \{1,\cdots,m+1\}, \tag{4.103}$$

at any point $x(u) \in M(a)$. Now suppose $x_0 \in \mathbb{R}^{m+1}\backslash\{0\}$ does not lie in $M(a)$ and set $a_0 = \widetilde{F}(x^1_0,\cdots,x^{m+1}_0)$. Then it is easy to see that $(\frac{a}{a_0}x^i_0)$ is a point of $M(a)$. Hence by (4.103) and homogeneity of $\tilde{g}_{ijh}$ we have

$$0 = \tilde{g}_{ijh}\left(\frac{a}{a_0}x_0\right) = \frac{a_0}{a}\tilde{g}_{ijh}(x_0),$$

that is, $\tilde{g}_{ijh}(x_0) = 0$. Thus $\widetilde{\mathbb{F}}^{m+1}$ must be a semi–Euclidean space. ■

Finally, we make some historical remarks on the new approach we developed in the present section. It was Rund [1], [2], who first considered the normal $N(x(u))$ on $\mathbb{F}^m$. More about this approach can be found in Section 7 of Chapter V of his book [3]. Also, Matsumoto [5] has considered $N(x(u))$ and used it in developing a theory of minimal hypersurfaces in a Finsler manifold. However, it is not clear if the above approach can be used to shed more light on the theory of Finsler submanifolds in general.

CHAPTER 6

FINSLER SURFACES

The Berwald and Moor frames are used to study $M_\alpha, M_{\alpha\beta}$ and μ for a Finsler surface $\mathbb{F}^2 = (M, M', F)$ in a 3–dimensional Finsler manifold $\widetilde{\mathbb{F}}^3 = (\widetilde{M}, \widetilde{M}', \widetilde{F})$. In particular, when $\widetilde{\mathbb{F}}^3$ is a Minkowski space we state a *Theorema Egregium* for $\mathbb{F}^2$ in $\widetilde{\mathbb{F}}^3$, provided the induced and intrinsic Finsler connections coincide on $\mathbb{F}^2$. Also, we show that the mean curvature of a proper Finsler surface $\mathbb{F}^2$ in a Minkowski space $\widetilde{\mathbb{F}}^3$ does not vanish identically on M', unless $\mathbb{F}^2$ is a Minkowski space.

1. Special Features of the Geometry of Finsler Manifolds of Dimensions 2 and 3

Here we construct both the Berwald frame and the Moor frame on a Finsler manifold of dimension two and three respectively. By using these frame fields we present the special features of the curvature and torsion Finsler tensor fields of some classical Finsler connections on these Finsler manifolds.

Throughout the chapter we use the following range for indices: $\alpha, \beta, \gamma, \dots \in \{1, 2\}$ and $i, j, k, \dots \in \{1, 2, 3\}$.

Let $\mathbb{F}^2 = (M, M', F^*)$ be a 2–dimensional Finsler manifold, and $L = v^\alpha(\partial/\partial v^\alpha)$ be the Liouville vector field on M'. Then $F^* = F^2$ and

$$\ell = \frac{1}{F}\, L = \frac{v^\alpha}{F}\frac{\partial}{\partial v^\alpha} = \ell^\alpha \frac{\partial}{\partial v^\alpha}, \tag{1.1}$$

is the unit Liouville vector field of $\mathbb{F}^2$. Consider the Riemannian metric $g = (g_{\alpha\beta})$ induced by F^* on the vertical bundle VM' and set

$$\ell_\alpha = g_{\alpha\beta}\ell^\beta .$$

Then by direct calculations we determine a unique Finsler vector field m such that

$$(a)\ \ g(\ell, m) = 0 \quad \text{and} \quad (b)\ \ g(m, m) = 1. \tag{1.2}$$

Explicitly, m is given by

$$m = \frac{1}{\sqrt{g_*}}\left(-\ell_2\frac{\partial}{\partial v^1} + \ell_1\frac{\partial}{\partial v^2}\right) = m^\alpha\frac{\partial}{\partial v^\alpha}, \tag{1.3}$$

where $g_* = \det\,[g_{\alpha\beta}]$. Thus if we put

$$m_\alpha = g_{\alpha\beta}m^\beta,$$

we obtain

$$(a)\ \ \ell_\alpha\ell^\alpha = 1; \quad (b)\ \ \ell_\alpha m^\alpha = m_\alpha\ell^\alpha = 0; \quad (c)\ \ m_\alpha m^\alpha = 1. \tag{1.4}$$

Therefore $\{\ell, m\}$ is an orthonormal basis in $\Gamma(VM'_{|\mathcal{U}'})$, where $\mathcal{U}'$ is a coordinate neighborhood in M'. The frame $\{\ell, m\}$ will be called the **Berwald frame** because it was introduced by Berwald [2], [4].

Remark 1.1. The above construction of the Berwald frame also holds for the case when $\mathbb{F}^2$ is a pseudo–Finsler manifold, provided L is a spacelike or timelike Finsler vector field. On the other hand, we should stress that there exists no coordinate neighborhood in M' on which L is a lightlike Finsler vector field. To show this we assume that

$$g_{\alpha\beta}(u,v)v^\alpha v^\beta = 0 \ \text{ on } \mathcal{U}' \subset M'.$$

Then differentiate it with respect to (v^γ) and taking into account (1.6.22a) we deduce that $g_{\alpha\beta}(u,v)v^\beta = 0$ on $\mathcal{U}'$ for all $\alpha \in \{1,2\}$. This is a contradiction since $g_* \neq 0$ on $\mathcal{U}'$. ■

The Cartan tensor field C of type (1, 2) of $\mathbb{F}^2$ is a Finsler tensor field (see. (1.6.21)) defined as follows

$$\begin{cases} C : \Gamma(VM') \times \Gamma(VM') \to \Gamma(VM'), \\ C(X,Y) = g_{\alpha\ \beta}^{\ \gamma} X^\alpha Y^\beta \dfrac{\partial}{\partial v^\gamma}, \ \forall X, Y \in \Gamma(VM'). \end{cases}$$

Denote also by C the Cartan tensor field of type (0,3) defined by

$$\begin{cases} C : \Gamma(VM') \times \Gamma(VM') \times \Gamma(VM') \to \mathcal{F}(M'), \\ C(X,Y,Z) = g(C(X,Y),Z), \ \forall X, Y, Z \in \Gamma(VM'). \end{cases}$$

Then it is easy to see that

$$C\left(\frac{\partial}{\partial v^\gamma}, \frac{\partial}{\partial v^\beta}, \frac{\partial}{\partial v^\alpha}\right) = g_{\alpha\beta\gamma}, \tag{1.5}$$

and

$$\begin{cases} C(\ell,\ell,\ell) &= C(\ell,\ell,m) = C(\ell,m,m) = 0, \quad (a) \\ C(m,m,m) &= g_{\alpha\beta\gamma}m^{\alpha}m^{\beta}m^{\gamma}. \quad (b) \end{cases} \tag{1.6}$$

On the other hand, the natural frame field $\{\partial/\partial v^{\alpha}\}$ on $\Gamma(VM'_{|\mathcal{U}'})$ is related to the Berwald frame as follows:

$$\frac{\partial}{\partial v^{\alpha}} = \ell_{\alpha}\ell + m_{\alpha}m. \tag{1.7}$$

Now we put

$$\mathbf{I} = FC(m,m,m), \tag{1.8}$$

and call **I** the **main scalar field** of $\mathbb{F}^2$. Then by using (1.5)–(1.8) we deduce that

$$(a)\quad g_{\alpha\beta\gamma} = \frac{\mathbf{I}}{F}m_{\alpha}m_{\beta}m_{\gamma} \quad \text{and} \quad (b)\quad \mathbf{I} = Fg_{\alpha\beta\gamma}m^{\alpha}m^{\beta}m^{\gamma}. \tag{1.9}$$

Remark 1.2. Important research work has been done on the geometry of $\mathbb{F}^2$ with a particular main scalar field. In this respect, Berwald ([2], [4]) determined all 2–dimensional Finsler manifolds $\mathbb{F}^2 = (M, TM^0, F)$ for which **I** is a constant function on TM^0 or depends on position only. More results on the geometry of $\mathbb{F}^2$ can be found in Asanov's book [2]. ■

Next, we set

$$g^{\alpha} = g^{\alpha\gamma}g^{\beta\mu}g_{\gamma\beta\mu}. \tag{1.10}$$

Then by using (1.6.22) it is easy to check that $g^{\alpha}(\partial/\partial v^{\alpha})$ is orthogonal to ℓ and therefore it is collinear to m. More precisely, by using (1.10), (1.9a) and (1.4c) we deduce that

$$g^{\alpha} = \frac{\mathbf{I}}{F}m^{\alpha}. \tag{1.11}$$

Now we consider the Cartan connection

$$\mathbf{FC}^* = (GM', \nabla^*) = (G^{\gamma}_{\alpha}, F^{*\gamma}_{\alpha\ \beta}, g_{\alpha\ \beta}^{\ \gamma})$$

on $\mathbb{F}^2$. Then from (1.7.34) we deduce that

$$\nabla^*_X\ell = 0, \quad \forall X \in \Gamma(GM'). \tag{1.12}$$

By using (1.12) and taking into account that ∇^* is a metric linear connection on VM', we obtain

$$\nabla^*_X m = 0, \quad \forall X \in \Gamma(GM'). \tag{1.13}$$

Thus the Berwald frame is h–parallel with respect to the Cartan connection, i.e., locally we have

$$(a)\quad \ell^{\alpha}{}_{|*\beta} = 0 \quad \text{and} \quad (b)\quad m^{\alpha}{}_{|*\beta} = 0. \tag{1.14}$$

To examine the vertical covariant derivatives of ℓ and m with respect to $\mathbf{FC}^*$ we decompose any $X \in \Gamma(VM')$ as follows:

$$X = g(X,\ell)\ell + g(X,m)m. \tag{1.15}$$

Then by direct calculations using (1.7.34), (1.7.19), and (1.10) we derive

$$\nabla^*_X \ell = \frac{1}{F} g(X,m)m, \quad \forall X \in \Gamma(VM'). \tag{1.16}$$

Following a procedure similar to that used to deduce (1.13), we obtain

$$\nabla^*_X m = -\frac{1}{F} g(X,m)\ell, \quad \forall X \in \Gamma(VM'). \tag{1.17}$$

Locally (1.16) and (1.17) are expressed as follows:

$$(a)\quad \ell^{\alpha}{}_{||*\beta} = \frac{1}{F} m_{\beta} m^{\alpha} \quad \text{and} \quad (b)\quad m^{\alpha}{}_{||*\beta} = -\frac{1}{F} m_{\beta} \ell^{\alpha}, \tag{1.18}$$

respectively. Matsumoto [6], p. 183 obtained (1.14) and (1.18) using a different approach.

The curvature Finsler tensor fields of $\mathbf{FC}^*$ on $\mathbb{F}^2$ have some special forms as follows. First, by direct calculations using (1.9a) and (1.7.75) we deduce that the v–curvature Finsler tensor field of $\mathbf{FC}^*$ vanishes, i.e., we have

$$S^*_{\alpha\beta\gamma\delta} = 0, \quad \forall \alpha,\beta,\gamma,\delta \in \{1,2\}. \tag{1.19}$$

To deduce the expressions of the other curvature tensor fields of $\mathbf{FC}^*$ we first note that the local components $g_{\alpha\beta}$ of the Riemannian metric g on VM' are given by

$$g_{\alpha\beta} = g\left(\frac{\partial}{\partial v^{\alpha}}, \frac{\partial}{\partial v^{\beta}}\right) = \ell_{\alpha}\ell_{\beta} + m_{\alpha}m_{\beta}. \tag{1.20}$$

Also, the local components of any Finsler covector field (ω_{α}) can be expressed as follows

$$\omega_{\alpha} = (\omega_{\mu}\ell^{\mu})\ell_{\alpha} + (\omega_{\mu}m^{\mu})m_{\alpha}. \tag{1.21}$$

Then by using (1.5.23a), (1.7), (1.7.52a) and (1.7.53a) we obtain the h–curvature Finsler tensor field as follows

$$\begin{aligned} R^*_{\alpha\beta\gamma\delta} &= g\left(R^*\left(\frac{\partial}{\partial v^{\delta}}, \frac{\partial}{\partial v^{\gamma}}\right)\frac{\partial}{\partial v^{\beta}}, \frac{\partial}{\partial v^{\alpha}}\right) \\ &= g(R^*(\ell,m)\ell,m)(m_{\alpha}\ell_{\beta} - m_{\beta}\ell_{\alpha})(m_{\gamma}\ell_{\delta} - m_{\delta}\ell_{\gamma}). \end{aligned} \tag{1.22}$$

Finally, by direct calculations using (1.20), we see that (1.22) becomes

$$R^*_{\alpha\beta\gamma\delta} = \mathbf{R}^*(g_{\alpha\gamma}g_{\beta\delta} - g_{\alpha\delta}g_{\beta\gamma}), \tag{1.23}$$

where $\mathbf{R}^* = g(R^*(\ell, m)\ell, m)$. Next, by (1.7.74) the *hv*–curvature Finsler tensor field of $\mathbf{FC}^*$ is written as follows

$$P^*_{\alpha\beta\gamma\delta} = \mathcal{A}_{(\alpha\beta)}\{g_{\beta\gamma\delta|*\alpha} + g_{\alpha}{}^{\mu}{}_{\gamma}g_{\mu\beta\delta|*\nu}v^{\nu}\}. \tag{1.24}$$

On the other hand, taking the *h*–covariant derivative of (1.9a) with respect to $\mathbf{FC}^*$ and using (1.14b) we deduce that

$$g_{\alpha\beta\gamma|*\delta} = \frac{\mathbf{I}_{|*\delta}}{F}m_\alpha m_\beta m_\gamma, \tag{1.25}$$

since $F_{|*\delta} = 0$ (see (1.7.38b)). Then it is easy to check that

$$\mathcal{A}_{(\alpha\beta)}\{g_{\alpha}{}^{\mu}{}_{\gamma}g_{\mu\beta\delta|*\nu}v^{\nu}\} = 0,$$

via (1.25). Thus by using (1.25) in (1.24) we infer that

$$P^*_{\alpha\beta\gamma\delta} = \frac{1}{F}m_\gamma m_\delta \mathcal{A}_{(\alpha\beta)}\{\mathbf{I}_{|*\alpha}m_\beta\}. \tag{1.26}$$

Finally, using (1.21) for $I_{|*\alpha}$ and taking into account (1.9a) in (1.26) we derive

$$P^*_{\alpha\beta\gamma\delta} = \mathbf{P}^*\left(\ell_\alpha g_{\beta\gamma\delta} - \ell_\beta g_{\alpha\gamma\delta}\right), \tag{1.27}$$

where $\mathbf{P}^*$ is given by

$$\mathbf{P}^* = \frac{1}{\mathbf{I}}\mathbf{I}_{|*\alpha}\ell^\alpha. \tag{1.28}$$

Next, we deal with torsion Finsler tensor fields of $\mathbf{FC}^*$. First, by using (1.23) and (1.7.56b) for $\mathbb{F}^2$ we obtain

$$R^*_{\beta\gamma\delta} = \mathbf{R}^*F(\ell_\gamma g_{\beta\delta} - \ell_\delta g_{\beta\gamma}). \tag{1.29}$$

Contracting (1.29) by $m^\beta v^\gamma m^\delta$ and using (1.4) we deduce that $\mathbb{F}^2$ is a Finsler manifold of scalar curvature

$$\mathbf{R}^* = \frac{1}{F^*}R^*_{\beta\gamma\delta}m^\beta v^\gamma m^\delta. \tag{1.30}$$

It is interesting to note that the local components of the other two surviving torsion Finsler tensor fields of $\mathbf{FC}^*$ are proportional. Indeed, contracting (1.25) by v^δ and using (1.7.73), (1.9a) and (1.28) we infer that

$$P^*_{\alpha\beta\gamma} = \mathbf{P}^*Fg_{\alpha\beta\gamma}. \tag{1.31}$$

Finally, by using (1.31), (1.7.73) and the assertion (ii) of Theorem 1.8.4 for $\mathbb{F}^2$ we may state the following theorem:

THEOREM 1.1 *Any proper Finsler manifold $\mathbb{F}^2$ is a Landsberg manifold if and only if $\mathbf{P}^*$ vanishes identically on M'.*

We now prove a theorem that characterizes 2–dimensional generalized Landsberg manifolds in terms of $\mathbf{P}^*$.

THEOREM 1.2 *Let $\mathbb{F}^2$ be a proper Finsler manifold. Then $\mathbb{F}^2$ is a generalized Landsberg manifold if and only if the function $\mathbf{P}^*$ is a solution of the following partial differential equation*

$$\mathbf{P}^*{}_{|*\alpha}\ell^\alpha + (\mathbf{P}^*)^2 = 0. \tag{1.32}$$

PROOF. By Theorem 1.8.7 we deduce that $\mathbb{F}^2$ is a generalized Landsberg manifold if and only if

$$P^*_{\beta\gamma\mu}P^{*\mu}{}_{\alpha\delta} - P^*_{\beta\delta\mu}P^{*\mu}{}_{\alpha\gamma} = 0, \tag{1.33}$$

and

$$P^*_{\alpha\beta\gamma|*\delta} - P^*_{\alpha\beta\delta|*\gamma} = 0, \tag{1.34}$$

are satisfied. From (1.31) and (1.9a) we deduce that (1.33) is always satisfied on $\mathbb{F}^2$. Next, taking into account (1.31) we see that (1.34) is equivalent to

$$g_{\alpha\beta\gamma}\mathbf{P}^*_{|*\delta} - g_{\alpha\beta\delta}\mathbf{P}^*_{|*\gamma} = \mathbf{P}^*(g_{\alpha\beta\delta|*\gamma} - g_{\alpha\beta\gamma|*\delta}). \tag{1.35}$$

Since the last term in (1.24) vanishes, (1.35) becomes

$$g_{\alpha\beta\gamma}\mathbf{P}^*_{|*\delta} - g_{\alpha\beta\delta}\mathbf{P}^*_{|*\gamma} = \mathbf{P}^*P^*_{\gamma\delta\alpha\beta}. \tag{1.36}$$

Thus by (1.27) we write (1.36) as follows

$$g_{\alpha\beta\gamma}A_\delta - g_{\alpha\beta\delta}A_\gamma = 0, \tag{1.37}$$

where we set

$$A_\gamma = \mathbf{P}^*_{|*\gamma} + (\mathbf{P}^*)^2\ell_\gamma. \tag{1.38}$$

Contracting (1.37) by ℓ^γ and using (1.6.22a), (1.38), and (1.4a) we obtain (1.32) since $\mathbb{F}^2$ is not a Riemannian manifold. Conversely, suppose that (1.32) is satisfied. Then $A_\gamma\ell^\gamma = 0$ and by (1.21) we have

$$A_\alpha = (A_\gamma m^\gamma)m_\alpha. \tag{1.39}$$

Thus by using (1.9a) and (1.39) we deduce that (1.37) is satisfied. Hence (1.34) holds, that is, $\mathbb{F}^2$ is a generalized Landsberg manifold. ■

Next, we consider on $\mathbb{F}^2$ the Rund connection and look for special expressions of its curvature Finsler tensor fields. First, by (1.7.81) the h–curvature Finsler tensor field of **RFC** is given by

$$K_{\alpha\beta\gamma\delta} = R^*_{\alpha\beta\gamma\delta} - g_{\alpha\beta\mu}R^{*\mu}{}_{\gamma\delta}. \tag{1.40}$$

Then by using (1.23) and (1.29) into (1.40) we infer that

$$K_{\alpha\beta\gamma\delta} = \mathbf{R}^* \left\{g_{\alpha\gamma}g_{\beta\delta} - g_{\alpha\delta}g_{\beta\gamma} + F\left(g_{\alpha\beta\gamma}\ell_\delta - g_{\alpha\beta\delta}\ell_\gamma\right)\right\}. \tag{1.41}$$

According to (1.7.82) the hv–curvature Finsler tensor field of **RFC** is given by

$$F_{\alpha\beta\gamma\delta} = P^*_{\alpha\beta\gamma\delta} + g_{\alpha\beta\delta|*\gamma} - g_{\alpha\beta\mu}P^{*\mu}{}_{\gamma\delta}. \tag{1.42}$$

Thus by using (1.27), (1.31), (1.25), and (1.9a) in the right hand side of (1.42) we deduce that

$$\begin{aligned} F_{\alpha\beta\gamma\delta} \quad &= \quad \frac{\mathbf{P}^*\mathbf{I}}{F}(\ell_\alpha m_\beta - \ell_\beta m_\alpha)m_\gamma m_\delta \\ &+ \frac{\mathbf{I}_{|*\gamma}}{F} m_\alpha m_\beta m_\delta - \frac{\mathbf{I}^2}{F^*} m_\alpha m_\beta m_\gamma m_\delta. \end{aligned} \tag{1.43}$$

Now we consider a 3–dimensional Finsler manifold $\widetilde{\mathbb{F}}^3 = (\widetilde{M}, \widetilde{M}', \widetilde{F}^*)$ with Liouville vector field $\widetilde{L} = y^i(\partial/\partial y^i)$. As in the case of $\mathbb{F}^2$ we have $\widetilde{F}^* = \widetilde{F}^2$ and the unit Liouville vector field is given by

$$\tilde{\ell} = \frac{1}{\widetilde{F}}\widetilde{L} = \frac{y^i}{\widetilde{F}}\frac{\partial}{\partial y^i} = \tilde{\ell}^i\frac{\partial}{\partial y^i}. \tag{1.44}$$

It was Moor [1] who first constructed an intrinsic orthonormal frame field on $\widetilde{\mathbb{F}}^3$. First, we note that $\tilde{h} = \tilde{g}^i(\partial/\partial y^i)$ is orthogonal to $\tilde{\ell}$, where we set

$$\tilde{g}^i = \tilde{g}^{ij}\tilde{g}^{kh}\tilde{g}_{jkh}. \tag{1.45}$$

Then we consider

$$\tilde{m} = \frac{1}{||\tilde{h}||}\tilde{g}^i\frac{\partial}{\partial y^i} = \tilde{m}^i\frac{\partial}{\partial y^i}. \tag{1.46}$$

Finally, since the fibres of $V\widetilde{M}'$ are 3–dimensional we may choose the third Finsler vector field $\tilde{n} = \tilde{n}^i(\partial/\partial y^i)$ such that $\{\tilde{\ell}, \tilde{m}, \tilde{n}\}$ is a positively oriented orthonormal frame field. From now on we call $\{\tilde{\ell}, \tilde{m}, \tilde{n}\}$ the **Moor frame** of $\widetilde{\mathbb{F}}^3$. We should note that from (1.45) it follows that the Moor frame does not exist for Riemannian manifolds.

Next, we look for covariant derivatives of the Finsler vector fields from the Moor frame with respect to the Cartan connection $\widetilde{\mathbf{FC}}^* = (G\widetilde{M}', \widetilde{\nabla}^*) = (\widetilde{G}_i^k, \widetilde{F}_{i\ j}^{*k}, \widetilde{g}_{i\ j}^{\ k})$. First, from (1.7.34) we obtain

$$\widetilde{\nabla}_X^* \tilde{\ell} = 0, \quad \forall X \in \Gamma(G\widetilde{M}'). \tag{1.47}$$

To derive the horizontal covariant derivatives of $\tilde{m}$ and $\tilde{n}$ we define on $\widetilde{M}'$ the differential 1–form

$$\tilde{\omega}(X) = \tilde{g}(\widetilde{\nabla}_X^* \tilde{m}, \tilde{n}), \quad \forall X \in \Gamma(T\widetilde{M}'). \tag{1.48}$$

Also, we write any section Y of $V\widetilde{M}'$ as follows

$$Y = \tilde{g}(Y, \tilde{\ell})\tilde{\ell} + \tilde{g}(Y, \tilde{m})\tilde{m} + \tilde{g}(Y, \tilde{n})\tilde{n}. \tag{1.49}$$

Then by direct calculations using (1.47)–(1.49) and taking into account that $\widetilde{\nabla}^*$ is a metric linear connection on $V\widetilde{M}'$ we derive

$$(a) \quad \widetilde{\nabla}_X^* \tilde{m} = \tilde{\omega}(X)\tilde{n} \quad \text{and} \quad (b) \quad \widetilde{\nabla}_X^* \tilde{n} = -\ \tilde{\omega}(X)\tilde{m}, \tag{1.50}$$

for any $X \in \Gamma(G\widetilde{M}')$. Due to (1.7.19) and (1.7.34) we see that

$$\widetilde{\nabla}_Y^* \tilde{\ell} = \frac{1}{\widetilde{\widetilde{F}}}(Y - \tilde{\eta}(Y)\tilde{\ell}), \ \forall Y \in \Gamma(V\widetilde{M}'), \tag{1.51}$$

where $\tilde{\eta}(Y) = \tilde{g}(Y, \tilde{\ell})$. By similar arguments as for (1.50) we infer that

$$\widetilde{\nabla}_Y^* \tilde{m} = -\frac{1}{\widetilde{\widetilde{F}}}\tilde{g}(Y, \tilde{m})\tilde{\ell} + \tilde{\omega}(Y)\tilde{n}, \tag{1.52}$$

and

$$\widetilde{\nabla}_Y^* \tilde{n} = -\frac{1}{\widetilde{\widetilde{F}}}\tilde{g}(Y, \tilde{n})\tilde{\ell} - \tilde{\omega}(Y)\tilde{m}, \tag{1.53}$$

for any $Y \in \Gamma(V\widetilde{M}')$. Take $X = \delta^*/\delta^* x^i$ and $Y = \partial/\partial y^i$ in the sets of equations (1.47), (1.50), and (1.51)–(1.53) and obtain

$$(a) \ \tilde{\ell}^i{}_{|*j} = 0; \quad (b) \ \tilde{m}^i{}_{|*j} = \ \tilde{\omega}_j \tilde{n}^i; \quad (c) \ \tilde{n}^i{}_{|*j} = -\ \tilde{\omega}_j \tilde{m}^i, \tag{1.54}$$

and

$$(a) \ \tilde{\ell}^i{}_{||*j} = \frac{1}{\widetilde{\widetilde{F}}}(\delta_j^i - \ell_j \tilde{\ell}^i); \quad (b) \ \tilde{m}^i{}_{||*j} = -\frac{1}{\widetilde{\widetilde{F}}}\tilde{m}_j \tilde{\ell}^i + \tilde{\omega}'_j \tilde{n}^i;$$

$$(c) \ \tilde{n}^i{}_{||*j} = -\frac{1}{\widetilde{\widetilde{F}}}\tilde{n}_j \tilde{\ell}^i - \tilde{\omega}'_j \tilde{m}^i, \tag{1.55}$$

where we set $\tilde{\omega}_i = \tilde{\omega}(\delta^*/\delta^* x^i)$ and $\tilde{\omega}'_i = \tilde{\omega}(\partial/\partial y^i)$.

Finally, we will present some special formulas for local components of both the Cartan Finsler tensor field and the v–curvature Finsler tensor field of $\widetilde{\mathbf{FC}}^*$ on $\widetilde{\mathbb{F}}^3$. First, from (1.49) we obtain

$$\frac{\partial}{\partial y^i} = \tilde{\ell}_i\tilde{\ell} + \tilde{m}_i\tilde{m} + \tilde{n}_i\tilde{n}. \tag{1.56}$$

Then by direct calculations, using (1.6.22) and (1.56) we deduce that the local components of the Cartan tensor field $\widetilde{C}$ of $\widetilde{\mathbb{F}}^3$ are given by

$$\begin{aligned}\tilde{g}_{ijk} &= \widetilde{C}\left(\frac{\partial}{\partial y^k}, \frac{\partial}{\partial y^j}, \frac{\partial}{\partial y^i}\right) \\ &= \widetilde{C}(\tilde{m},\tilde{m},\tilde{m})\tilde{m}_i\tilde{m}_j\tilde{m}_k + \widetilde{C}(\tilde{n},\tilde{n},\tilde{m})\sum_{(i,j,k)}\{\tilde{n}_i\tilde{n}_j\tilde{m}_k\} \\ &\quad + \widetilde{C}(\tilde{n},\tilde{n},\tilde{n})\tilde{n}_i\tilde{n}_j\tilde{n}_k + \widetilde{C}(\tilde{m},\tilde{m},\tilde{n})\sum_{(i,j,k)}\{\tilde{m}_i\tilde{m}_j\tilde{n}_k\}.\end{aligned} \tag{1.57}$$

It is important to note that the scalar fields in (1.57) should satisfy some relations. Indeed, contracting (1.57) by $\tilde{g}^{jk}$ and using (1.45) and (1.46) we obtain

$$\begin{aligned}\tilde{m}_i\|\tilde{h}\| &= (\widetilde{C}(\tilde{m},\tilde{m},\tilde{m}) + \widetilde{C}(\tilde{n},\tilde{n},\tilde{m}))\tilde{m}_i \\ &\quad + (\widetilde{C}(\tilde{n},\tilde{n},\tilde{n}) + \widetilde{C}(\tilde{m},\tilde{m},\tilde{n}))\tilde{n}_i.\end{aligned}$$

Hence we have

$$\widetilde{C}(\tilde{m},\tilde{m},\tilde{m}) + \widetilde{C}(\tilde{n},\tilde{n},\tilde{m}) = \|\tilde{h}\|, \tag{1.58}$$

and

$$\widetilde{C}(\tilde{n},\tilde{n},\tilde{n}) + \widetilde{C}(\tilde{m},\tilde{m},\tilde{n}) = 0. \tag{1.59}$$

Thus (1.57) becomes

$$\begin{aligned}\tilde{g}_{ijk} &= \widetilde{\mathbf{H}}\tilde{m}_i\tilde{m}_j\tilde{m}_k + \widetilde{\mathbf{I}}\sum_{(i,j,k)}\{\tilde{n}_i\tilde{n}_j\tilde{m}_k\} \\ &\quad + \widetilde{\mathbf{J}}(\tilde{n}_i\tilde{n}_j\tilde{n}_k - \sum_{(i,j,k)}\{\tilde{m}_i\tilde{m}_j\tilde{n}_k\}),\end{aligned} \tag{1.60}$$

where we set

$$(a)\ \ \widetilde{\mathbf{H}} = \widetilde{C}(\tilde{m},\tilde{m},\tilde{m}); \quad (b)\ \ \widetilde{\mathbf{I}} = \widetilde{C}(\tilde{n},\tilde{n},\tilde{m}); \quad (c)\ \ \widetilde{\mathbf{J}} = \widetilde{C}(\tilde{n},\tilde{n},\tilde{n}). \tag{1.61}$$

We call $\widetilde{\mathbf{H}}, \widetilde{\mathbf{I}}, \widetilde{\mathbf{J}}$ the **main scalar fields** of $\widetilde{\mathbb{F}}^3$ (cf., Matsumoto [6], p. 194).

Similarly, by using (1.56), (1.7.52b), (1.7.53c), and (1.7.61) we derive

$$\begin{aligned}\widetilde{S}^*_{ijkh} &= \tilde{g}\left(\widetilde{S}^*(\frac{\partial}{\partial y^h}, \frac{\partial}{\partial y^k})\frac{\partial}{\partial y^j}, \frac{\partial}{\partial y^i}\right) \\ &= \tilde{g}(\widetilde{S}^*(\tilde{m}, \tilde{n})\tilde{m}, \tilde{n})(\tilde{m}_i\tilde{n}_j - \tilde{n}_i\tilde{m}_j)(\tilde{m}_k\tilde{n}_h - \tilde{m}_h\tilde{n}_k). \quad (1.62)\end{aligned}$$

The monographs of Asanov [2] and Matsumoto [6] contain more results on the geometry of $\mathbb{F}^2$ and $\widetilde{\mathbb{F}}^3$ via the Berwald frame and Moor frame, respectively. It is also important to note that the Berwald and Moor frames have been extended by Miron to Finsler manifolds of higher dimension (see Matsumoto–Miron [1] and Matsumoto [6], p. 179). It is expected that the **Miron frame** will be an important tool in studying the geometry of Finsler (pseudo–Finsler) manifolds of dimension greater than 3.

2. The Berwald–Moor Angle

Let $\mathbb{F}^2 = (M, M', F)$ be a Finsler surface in a 3–dimensional Finsler manifold $\widetilde{\mathbb{F}}^3 = (\widetilde{M}, \widetilde{M}', \widetilde{F})$. Then we consider the Berwald frame $\{\ell, m\}$ and the Moor frame $\{\tilde{\ell}, \tilde{m}, \tilde{n}\}$ on $\mathbb{F}^2$ and $\widetilde{\mathbb{F}}^3$ respectively. On the other hand, the unit normal Finsler vector field B enables us to consider the orthonormal frame $\{\ell, m, B\}$ on $\widetilde{\mathbb{F}}^3$ along $\mathbb{F}^2$. We call $\{\ell, m, B\}$ the **immersion frame** on $\widetilde{\mathbb{F}}^3$ along $\mathbb{F}^2$. Without loss of generality we may suppose that $\{\ell, m, B\}$ and $\{\tilde{\ell}, \tilde{m}, \tilde{n}\}$ have the same orientation.

We have seen in Chapter 5 that M_α, $M_{\alpha\beta}$ and μ (see (5.1.4) and (5.1.33b)) play an important role in studying the geometry of a pseudo–Finsler hypersurface. The above frames will enable us to get more information on these geometric objects.

First, we note that (see (5.2.47))

$$g_\alpha = \tilde{g}_i B^i_\alpha - M_\alpha. \quad (2.1)$$

Then contracting (2.1) by $g^{\alpha\gamma}$ and using (2.1.13) we obtain

$$g^\gamma = \tilde{g}^i \widetilde{B}^\gamma_i - M^\gamma. \quad (2.2)$$

Now we say that the immersion of $\mathbb{F}^2$ in $\widetilde{\mathbb{F}}^3$ is of **Berwald–Moor type** if we have

$$g^\gamma \frac{\partial}{\partial v^\gamma} = \tilde{g}^i \frac{\partial}{\partial y^i}. \quad (2.3)$$

According to the definitions of m and $\tilde{m}$ we see that in this case we have $m = \tilde{m}$. Then $B = \tilde{n}$ since $\{\ell, m, B\}$ and $\{\tilde{\ell}, \tilde{m}, \tilde{n}\}$ have the same orientation.

Thus the immersion frame coincides with the Moor frame. Moreover, we prove the following.

PROPOSITION 2.1 *If the immersion of* $\mathbb{F}^2$ *in* $\widetilde{\mathbb{F}}^3$ *is of Berwald–Moor type, then* $M_\alpha = 0$ *on* M', *for* $\alpha \in \{1, 2\}$.

PROOF. Contracting (2.2) by B^j_γ and taking into account (5.1.2e) we infer that

$$g^\gamma B^j_\gamma = \tilde{g}^j - (\tilde{g}^i \widetilde{B}_i) B^j - M^\gamma B^j_\gamma. \tag{2.4}$$

By (2.1.6) we see that (2.3) is equivalent to

$$g^\gamma B^j_\gamma = \tilde{g}^j. \tag{2.5}$$

Finally, since $B = \tilde{n}$ we deduce that B is orthogonal to $\tilde{m}$, that is, $\widetilde{B}_i \tilde{g}^i = 0$. Thus the assertion follows by using (2.5) in (2.4). ■

Next, we define the **Berwald–Moor angle** at $(u, v) \in M'$ as the angle $\theta \in [0, \pi]$ between m and $\tilde{m}$. Then the immersion frame and the Moor frame are related by

$$(a) \quad \begin{cases} \ell &= \tilde{\ell}, \\ m &= \cos\theta\ \tilde{m} + \sin\theta\ \tilde{n}, \\ B &= -\sin\theta\ \tilde{m} + \cos\theta\ \tilde{n}, \end{cases}$$

and

$$(b) \quad \begin{cases} \tilde{\ell} &= \ell, \\ \tilde{m} &= \cos\theta\ m - \sin\theta\ B, \\ \tilde{n} &= \sin\theta\ m + \cos\theta\ B. \end{cases} \tag{2.6}$$

By using (1.1) and (2.1.6) we obtain

$$\ell = \ell^\alpha \frac{\partial}{\partial v^\alpha} = \ell^\alpha B^i_\alpha \frac{\partial}{\partial y^i}. \tag{2.7}$$

Similarly, we have

$$m = m^\alpha \frac{\partial}{\partial v^\alpha} = m^\alpha B^i_\alpha \frac{\partial}{\partial y^i}. \tag{2.8}$$

Hence (2.6) becomes

$$(a) \quad \begin{cases} \ell^\alpha B^i_\alpha &= \tilde{\ell}^i, \\ m^\alpha B^i_\alpha &= \cos\theta\ \tilde{m}^i + \sin\theta\ \tilde{n}^i, \\ B^i &= -\sin\theta\ \tilde{m}^i + \cos\theta\ \tilde{n}^i, \end{cases}$$

and

$$(b)\quad \begin{cases} \tilde{\ell}^i &= \ell^\alpha B^i_\alpha, \\ \tilde{m}^i &= \cos\theta\, m^\alpha B^i_\alpha - \sin\theta\, B^i, \\ \tilde{n}^i &= \sin\theta\, m^\alpha B^i_\alpha + \cos\theta\, B^i. \end{cases} \tag{2.9}$$

The above transformations allow us to express $M_\alpha, M_{\alpha\beta}$ and μ in terms of θ, m_α and the main scalars of $\widetilde{\mathbb{F}}^3$ restricted to M'.

PROPOSITION 2.2 *Let $\mathbb{F}^2$ be a Finsler surface of a Finsler manifold $\widetilde{\mathbb{F}}^3$. Then we have*

$$\begin{aligned} M_\alpha &= (\sin^2\theta\cos\theta\widetilde{\mathbf{H}} + (\cos^3\theta - 2\sin^2\theta\cos\theta)\widetilde{\mathbf{I}} \\ &\quad +(3\cos^2\theta\sin\theta - \sin^3\theta)\widetilde{\mathbf{J}})m_\alpha, \end{aligned} \tag{2.10}$$

$$\begin{aligned} M_{\alpha\beta} &= (-\cos^2\theta\sin\theta\widetilde{\mathbf{H}} + (2\cos^2\theta\sin\theta - \sin^3\theta)\widetilde{\mathbf{I}} \\ &\quad +(3\sin^2\theta\cos\theta - \cos^3\theta)\widetilde{\mathbf{J}})m_\alpha m_\beta, \end{aligned} \tag{2.11}$$

and

$$\mu = -\sin^3\theta\widetilde{\mathbf{H}} - 3\cos^2\theta\sin\theta\widetilde{\mathbf{I}} + (\cos^3\theta - 3\sin^2\theta\cos\theta)\widetilde{\mathbf{J}}. \tag{2.12}$$

PROOF. By using (1.7) and the first two relations in (2.6a) we deduce that

$$\frac{\partial}{\partial v^\alpha} = \ell_\alpha\tilde{\ell} + m_\alpha\cos\theta\;\tilde{m} + m_\alpha\sin\theta\;\tilde{n}. \tag{2.13}$$

Then taking into account (5.1.4b) and using the Cartan tensor field $\widetilde{C}$ of $\widetilde{\mathbb{F}}^3$ we infer that

$$M_\alpha = \widetilde{C}\left(B, B, \frac{\partial}{\partial v^\alpha}\right). \tag{2.14}$$

Next, we substitute in (2.14), B and $\partial/\partial v^\alpha$ as given in (2.6a) and (2.13) respectively. After a long calculation, by using (1.61) we obtain (2.10). Similarly, starting with

$$(a)\;\; M_{\alpha\beta} = \widetilde{C}\left(B, \frac{\partial}{\partial v^\alpha}, \frac{\partial}{\partial v^\beta}\right) \quad\text{and}\quad (b)\;\; \mu = \widetilde{C}(B, B, B), \tag{2.15}$$

we derive (2.11) and (2.12) respectively. ∎

From Proposition 2.2 we deduce that at each point $(u, v) \in M'$ there exist at least one value and at most three values for $\theta \in [0, \pi]$ such that one of the quantities $M_\alpha, M_{\alpha\beta}$ and μ vanish at (u, v). It is important to find a value of θ for which $M_\alpha, M_{\alpha\beta}$ and μ vanish simultaneously. The following corollary deals with this question.

COROLLARY 2.1 *Let $\mathbb{F}^2$ be a Finsler surface of $\widetilde{\mathbb{F}}^3$ such that $\widetilde{\mathbf{I}} = \widetilde{\mathbf{J}} = 0$ and $\widetilde{\mathbf{H}} \neq 0$ on M'. Then μ, M_α and $M_{\alpha\beta}$ vanish simultaneously on M' if and only if the Berwald–Moor angle vanishes on M'.*

PROOF. Since $\widetilde{\mathbf{I}} = \widetilde{\mathbf{J}} = 0$, from (2.10)–(2.12) we deduce that:

$$(a)\ \ M_\alpha = \sin^2\theta\cos\theta\widetilde{\mathbf{H}}m_\alpha; \quad (b)\ \ M_{\alpha\beta} = -\cos^2\theta\sin\theta\widetilde{\mathbf{H}}m_\alpha m_\beta;$$

$$(c)\ \ \mu = -\sin^3\theta\widetilde{\mathbf{H}}. \tag{2.16}$$

Taking into account that $\widetilde{\mathbf{H}} \neq 0$ on M', from (2.16) we obtain the assertion of the corollary. ∎

The angle θ between m and $\tilde{m}$ enables us to express the main scalar field of $\mathbb{F}^2$ in terms of the main scalar fields of $\widetilde{\mathbb{F}}^3$ as follows

$$\mathbf{I} = F(\cos^3\theta\widetilde{\mathbf{H}} + 3\sin^2\theta\cos\theta\widetilde{\mathbf{I}} + (\sin^3\theta - 3\cos^2\theta\sin\theta)\widetilde{\mathbf{J}}). \tag{2.17}$$

To show this we start with

$$g_{\alpha\beta\gamma} = \widetilde{C}\left(\frac{\partial}{\partial v^\alpha}, \frac{\partial}{\partial v^\beta}, \frac{\partial}{\partial v^\gamma}\right). \tag{2.18}$$

Then by using (2.13) and performing similar calculations as in Proposition 2.2 we obtain

$$g_{\alpha\beta\gamma} = (\cos^3\theta\widetilde{\mathbf{H}} + 3\sin^2\theta\cos\theta\widetilde{\mathbf{I}} + (\sin^3\theta - 3\cos^2\theta\sin\theta)\widetilde{\mathbf{J}})m_\alpha m_\beta m_\gamma. \tag{2.19}$$

Thus (2.17) follows by comparing (1.9a) and (2.19).

From the above study we have seen that the Berwald–Moor angle appears in all the important geometric objects of the immersion. Certainly, Finsler immersions with constant θ are of special interest. To support this assertion we prove the following.

THEOREM 2.1 *Any immersion of $\mathbb{F}^2$ in a C-reducible Finsler manifold $\widetilde{\mathbb{F}}^3$ with $\theta = \pi/2$ is Riemannian.*

PROOF. By using (1.61c) and (1.8.45) we obtain

$$\widetilde{\mathbf{J}} = \frac{1}{4}\left(\tilde{a}_{ij}\tilde{g}_k + \tilde{a}_{jk}\tilde{g}_i + \tilde{a}_{ki}\tilde{g}_j\right)\tilde{n}^i\tilde{n}^j\tilde{n}^k = 0,$$

since $\tilde{g}_i\tilde{n}^i = 0$. Then for $\theta = \pi/2$ in (2.19) we deduce that $g_{\alpha\beta\gamma} = 0$, and hence the immersion is Riemannian. ∎

We should note that even in the case when the immersion is Riemannian, since the ambient manifold is Finslerian, some of the geometric objects still survive on $\mathbb{F}^2$. As an example we can easily check that $\widetilde{\mathbf{H}} \neq 0$ and $\widetilde{\mathbf{I}} \neq 0$ on $\widetilde{\mathbb{F}}^3$ from Theorem 2.1. Thus from (2.11) and (2.12) we deduce that $M_{\alpha\beta} \neq 0$ and $\mu \neq 0$ on M'. From the result stated in Theorem 2.1 we may conjecture the following. *Any immersion of $\mathbb{F}^2$ in $\widetilde{\mathbb{F}}^3$ with constant Berwald–Moor angle is Riemannian.*

3. Curvature of Finsler Surfaces in Minkowski Spaces

Let $\mathbb{F}^2 = (M, M', F)$ be a Finsler surface in the Minkowski space $\widetilde{\mathbb{F}}^3 = (\mathbb{R}^3, \widetilde{M}', \widetilde{F})$. Since the Berwald and Rund connections coincide on $\widetilde{\mathbb{F}}^3$ we have $\overset{\mathrm{b}}{H}_{\alpha\beta} = \overset{\mathrm{r}}{H}_{\alpha\beta} = H_{\alpha\beta}$. On the other hand, we have seen that the second fundamental form induced by the Cartan connection, in general, is not symmetric. As we need a symmetric second fundamental form we shall use throughout the section only $H = (H_{\alpha\beta})$.

First, according to (5.3.6a) and (5.1.14c) we deduce that

$$R_{\alpha\beta\gamma} = H_\beta H_{\alpha\gamma} - H_\gamma H_{\alpha\beta}. \tag{3.1}$$

As the fibers of the vertical bundle VM' are 2–dimensional, we have only one horizontal flag at each $(u, v) \in M'$. Then the induced horizontal flag curvature of $\mathbb{F}^2$ at (u, v) is the function (see (5.2.13))

$$K(u, v) = \frac{R_{\alpha\beta\gamma} X^\alpha v^\beta X^\gamma}{F^2 a_{\alpha\gamma} X^\alpha X^\gamma},$$

where $X = X^\alpha(\partial/\partial v^\alpha)$ is a Finsler vector such that $\{\ell, X\}$ is a basis in $VM'_{(u,v)}$. In particular, we take $X = m$ (the second vector in the Berwald frame) and get

$$K(u, v) = \frac{1}{F} R_{\alpha\beta\gamma} m^\alpha \ell^\beta m^\gamma, \tag{3.2}$$

since by (1.7.26), (1.4b) and (1.4c) we have

$$a_{\alpha\gamma} m^\alpha m^\gamma = g_{\alpha\gamma} m^\alpha m^\gamma - \ell_\alpha \ell_\gamma m^\alpha m^\gamma = 1.$$

Next, we consider the Rund connection $\widetilde{\mathbf{RFC}} = (G\widetilde{M}', \widetilde{\nabla}'')$ on $\widetilde{\mathbb{F}}^3$ and write the Gauss formula (2.1.33) in the following form

$$\widetilde{\nabla}''_{hX} QhY = \nabla_{hX} QhY + H(hX, hY)B,$$

for any $X, Y \in \Gamma(TM')$. Here the horizontal second fundamental form H is a symmetric, $\mathcal{F}(M')$–bilinear form on $\Gamma(HM')$, where HM' is the induced non–linear connection on M'. Thus the above components $H_{\alpha\beta}$ are given by

$$H_{\alpha\beta} = H\left(\frac{\delta}{\delta u^\alpha}, \frac{\delta}{\delta u^\beta}\right). \tag{3.3}$$

Now, we say that a Finsler vector $X = X^\alpha(\partial/\partial v^\alpha)$ is a **principal Finsler vector** if it is an eigenvector of the Gauss–Rund shape operator $H^\gamma_\alpha = g^{\gamma\mu}H_{\mu\alpha}$. This means that there exists a scalar k such that

$$H^\gamma_\alpha X^\alpha = kX^\gamma. \tag{3.4}$$

The corresponding eigenvalue k is called the **principal curvature**. It is easy to see that (3.4) is equivalent to

$$(H_{\alpha\beta} - kg_{\alpha\beta})X^\beta = 0. \tag{3.5}$$

Therefore, there exist principal Finsler vectors if and only if

$$\det\,[H_{\alpha\beta} - kg_{\alpha\beta}] = \begin{vmatrix} H_{11} - kg_{11} & H_{12} - kg_{12} \\ H_{21} - kg_{21} & H_{22} - kg_{22} \end{vmatrix} = 0. \tag{3.6}$$

Thus the principal curvatures are given by (3.6). Now we prove the following result.

THEOREM 3.1 *Let $\mathbb{F}^2$ be a proper Finsler surface in a Minkowski space $\widetilde{\mathbb{F}}^3$. If the Liouville vector field is a principal Finsler vector field on M' then $\mathbb{F}^2$ is totally geodesic immersed in $\widetilde{\mathbb{F}}^3$.*

PROOF. Since $L = v^\alpha(\partial/\partial v^\alpha)$ is a principal Finsler vector field, by (3.5) we have

$$H_{\alpha\beta}(u,v)v^\beta - k(u,v)g_{\alpha\beta}(u,v)v^\beta = 0. \tag{3.7}$$

Taking the vertical covariant derivative of (3.7) with respect to $I\overset{r}{F}C$ and using (5.3.12b), (2.5.5b), and (1.6.22) we obtain

$$H_{\alpha\gamma} - k_{\|\gamma}g_{\alpha\beta}v^\beta - kg_{\alpha\gamma} = 0, \tag{3.8}$$

since $v^\alpha{}_{\|\gamma} = \delta^\alpha_\gamma$. Contracting (3.8) by v^α and using (3.7) we deduce that

$$k_{\|\gamma}F^2 = 0,$$

which implies $k_{\|\gamma} = 0$. Hence (3.8) becomes

$$H_{\alpha\gamma} = kg_{\alpha\gamma},$$

that is, $\mathbb{F}^2$ is r–totally umbilical. Thus we apply Theorem 5.3.5 and obtain our assertion. ■

Next, by using the Berwald frame $\{\ell, m\}$ on $\mathbb{F}^2$ and the horizontal second fundamental form H we define

$$\mathcal{H} = H(Q\ell, Q\ell) + H(Qm, Qm), \tag{3.9}$$

and

$$\mathcal{K} = \begin{vmatrix} H(Q\ell, Q\ell) & H(Q\ell, Qm) \\ H(Q\ell, Qm) & H(Qm, Qm) \end{vmatrix}, \tag{3.10}$$

where Q is the associate almost product structure to HM' (see (1.4.8)). It is easy to check that both $\mathcal{H}$ and $\mathcal{K}$ are independent of the choice of the orthonormal basis in $\Gamma(HM')$. To derive new formulas for $\mathcal{H}$ and $\mathcal{K}$ we note that, by using (1.7), we have

$$(a)\ \ g_{\alpha\beta} = \ell_\alpha \ell_\beta + m_\alpha m_\beta \quad \text{and} \quad (b)\ \ g^{\alpha\beta} = \ell^\alpha \ell^\beta + m^\alpha m^\beta. \tag{3.11}$$

Then by using (3.3) and (3.11b) in (3.9) we infer that

$$\mathcal{H} = H_{\alpha\beta} g^{\alpha\beta}. \tag{3.12}$$

Also, by using the transformations of components of H and g corresponding to the transformations of frames $\{Q\ell, Qm\}$ and $\{\delta/\delta u^1,\ \delta/\delta u^2\}$ on HM', we obtain

$$\mathcal{K} = \frac{\det\, [H_{\alpha\beta}]}{\det\, [g_{\alpha\beta}]}. \tag{3.13}$$

Hence (3.6) is written as follows

$$k^2 - \mathcal{H}k + \mathcal{K} = 0. \tag{3.14}$$

Moreover, using (3.9) and (3.10) in (3.14) we deduce an equivalent form for (3.6) given by

$$\begin{vmatrix} H(Q\ell, Q\ell) - k & H(Q\ell, Qm) \\ H(Q\ell, Qm) & H(Qm, Qm) - k \end{vmatrix} = 0 \tag{3.15}$$

Thus from (3.15) it follows that the equation (3.6) has real roots k_1, k_2 always. Moreover, we may prove the following.

THEOREM 3.2 *Let $\mathbb{F}^2$ be a proper Finsler surface of a Minkowski space $\widetilde{\mathbb{F}}^3$. If $k_1 = k_2$, then $\mathbb{F}^2$ must be totally geodesic immersed in $\widetilde{\mathbb{F}}^3$.*

PROOF. Since $k_1 = k_2 = k$, and the matrix $[H_{\alpha\beta}]$ is symmetric we deduce that the eigenspace at $(u, v) \in M$ is just $HM'_{(u,v)}$. Thus we have

$$H_{\alpha\beta}X^{\beta} = kg_{\alpha\beta}X^{\beta}, \quad \forall X \in \Gamma(HM').$$

Hence $\mathbb{F}^2$ is r–totally umbilical. Then the assertion follows from Theorem 5.3.5. ∎

Finally, (3.14) implies that

$$(a) \;\; \mathcal{H} = k_1 + k_2 \quad \text{and} \quad (b) \;\; \mathcal{K} = k_1 k_2. \tag{3.16}$$

According to the terminology from Riemannian geometry, the above study enables us to call $\mathcal{H}$ and $\mathcal{K}$ the **mean curvature** and the **induced Gaussian curvature** of $\mathbb{F}^2$, respectively.

We now prove an important theorem which actually justifies the name we gave to $\mathcal{K}$.

THEOREM 3.3 *Let $\mathbb{F}^2$ be a Finsler surface in a Minkowski space $\widetilde{\mathbb{F}}^3$. Then the induced horizontal flag curvature coincides with the induced Gaussian curvature.*

PROOF. By using (3.1) and (5.1.10d) in (3.2) we obtain

$$\begin{aligned} K(u, v) &= \frac{1}{F}\left(H_{\beta}H_{\alpha\gamma} - H_{\gamma}H_{\alpha\beta}\right) m^{\alpha}\ell^{\beta}m^{\gamma} \\ &= H_{\beta\varepsilon}\ell^{\beta}\ell^{\varepsilon}H_{\alpha\gamma}m^{\alpha}m^{\gamma} - H_{\varepsilon\gamma}\ell^{\varepsilon}m^{\gamma}H_{\alpha\beta}m^{\alpha}\ell^{\beta}. \end{aligned}$$

Then by direct calculations in (3.10) we obtain the same expression for $\mathcal{K}$. ∎

We should remark that in the case of Riemannian immersions, Theorem 3.3 becomes the well known Gauss *Theorema Egregium* (see Spivak [1], Vol. 3, p. 78). Our theorem proves that in general the induced Gaussian curvature depends on the immersion. However, for some particular Finslerian immersions we state the following.

THEOREM 3.4 (Finslerian Theorema Egregium). *Let $\mathbb{F}^2$ be a Finsler surface in a Minkowski space $\widetilde{\mathbb{F}}^3$ such that the induced Finsler connection by the Rund connection of $\widetilde{\mathbb{F}}^3$ coincides with the intrinsic Rund connection of $\mathbb{F}^2$. Then the induced Gaussian curvature of $\mathbb{F}^2$ depends only on the fundamental function F of $\mathbb{F}^2$.*

PROOF. Since $\overset{r}{IFC} =$ **RFC**, we have $N_{\alpha}^{\beta} = G_{\alpha}^{\beta}$, which implies that $R_{\alpha\beta\gamma} = R^{*}_{\alpha\beta\gamma}$. Thus by Theorem 3.2 $\mathcal{K}$ is given by

$$\mathcal{K} = \frac{1}{F}R^{*}_{\alpha\beta\gamma}m^{\alpha}\ell^{\beta}m^{\gamma},$$

where all the geometric objects on the right hand side are defined only by means of the fundamental function F of $\mathbb{F}^2$. ■

Certainly we should ask ourselves if, appart from Riemannian immersions, we have some Finslerian immersions satisfying the conditions from Theorem 3.4. The next corollary provides us with a large class of such immersions.

COROLLARY 3.1 *Let $\mathbb{F}^2 = (M, M', F)$ be a Finsler surface in a $C-$reducible Minkowski space $\widetilde{\mathbb{F}}^3$ such that M is tangent to the structure vector field $\tilde{c}$ of $\widetilde{\mathbb{F}}^3$. Then the induced Gaussian curvature of $\mathbb{F}^2$ depends only on the fundamental function F of $\mathbb{F}^2$.*

PROOF. According to Theorem 3.3.1, in this case we have $I\overset{r}{F}C$= **RFC** on $\mathbb{F}^2$. Then we apply Theorem 3.4 and obtain the assertion of the corollary. ■

It is well known that in Riemannian geometry the vanishing of the mean curvature is a necessary and sufficient condition for the surface to be minimal. Thus it is interesting to examine the vanishing of $\mathcal{H}$ in the case of Finsler surfaces. Surprisingly, we have the following result.

THEOREM 3.5 *Let $\mathbb{F}^2 = (M, M', F)$ be a proper Finsler surface in a Minkowski space $\widetilde{\mathbb{F}}^3$. If the mean curvature of $\mathbb{F}^2$ vanishes identically on M', then $\mathbb{F}^2$ is totally geodesic immersed in $\widetilde{\mathbb{F}}^3$.*

PROOF. From (3.12) we deduce that

$$H_{\alpha\beta} g^{\alpha\beta} = 0, \tag{3.17}$$

on M'. Taking the v–convariant derivative of (3.17) with respect to $I\overset{r}{F}C$ and using (3.12b) we obtain

$$H_{\alpha\beta} g^{\alpha\beta}{}_{\|\gamma} = 0. \tag{3.18}$$

On the other hand, by using (2.5.5b) we easily obtain

$$g^{\alpha\beta}{}_{\|\gamma} = -2 g_{\mu\nu\gamma} g^{\mu\alpha} g^{\nu\beta}.$$

Thus (3.18) becomes

$$H_{\alpha\beta} g^{\alpha\mu} g^{\beta v} g_{\mu v\gamma} = 0. \tag{3.19}$$

By using (1.9a) in (3.19) and taking into account that $\mathbf{I} \neq 0$ on M', we infer that

$$H_{\alpha\beta} m^{\alpha} m^{\beta} m_{\gamma} = 0,$$

which gives $H(Qm, Qm) = 0$. As $\mathcal{H}$ vanishes on M', from (3.9) we deduce that $H(Q\ell, Q\ell) = 0$. Hence $H_0 = H_{\alpha\beta}v^{\alpha}v^{\beta} = 0$, and the assertion follows from Theorem 3.1.2. ■

Taking into account Theorem 3.5 and Proposition 3.1.3 we may state the following corollary.

COROLLARY 2.2 *Let $\mathbb{F}^2 = (M, M', F)$ be a proper Finsler surface in a Minkowski space $\widetilde{\mathbb{F}}^3$. Then the mean curvature of $\mathbb{F}^2$ vanishes identically on M' if and only if $M = \mathbb{R}^2$ or M is an open submanifold of $\mathbb{R}^2$.*

As we mentioned above, in Riemannian geometry a minimal surface is characterized by the vanishing of the mean curvature. As far as we know, this is not the case in Finslerian geometry. From several studies (see Wegener [1], [2], Davies [1], Barthel [1] and Matsumoto [5]) it follows that the existence of minimal Finsler submanifolds requires some conditions on the ambient manifold. Thus according to the theories developed by Wegener and Barthel, minimal Finsler submanifolds are characterized only by the vanishing of the mean curvature, provided that the Cartan torsion of the ambient space vanishes, i.e., in our notations, $\tilde{g}^i = 0$. Also Matsumoto, [5] p. 663, wrote that a hyperplane is minimal if and only if the ambient space satisfies the condition $\tilde{g}^i{}_{|*i} = 0$. On the other hand, so far, no examples of minimal proper Finsler submanifolds have been produced, except the totally geodesic ones. As a conclusion based on our Theorem 3.5 and the above comments, we believe that the theory of minimal Finsler submanifolds is far from being settled.

Finally, we present a short history of the curvatures of a Finsler surface. It was Finsler [1] who first introduced a mean curvature and a Gaussian curvature for a Finsler surface. His approach was similar to the one developed in Euclidean geometry, and the Landsberg angle (see Landsberg [1]–[3]) was fully used in the calculations. The same approach was used by Cartan [1] to study curves on Finsler surfaces. Next, Berwald [3], defined the mean curvature and Gaussian curvature of $\mathbb{F}^2$ by formulas similar to our (3.12) and (3.13). The difference consists in the fact that our $H_{\alpha\beta}$ are replaced by his $\Omega_{\alpha\beta}$ given by (5.1.20). The mean curvature $\mathcal{H}$ and the induced Gaussian curvature $\mathcal{K}$ we presented in this section have been first considered by Rund [3], p. 199 and Brown [2], respectively.

BASIC NOTATIONS AND TERMINOLOGY

Throughout the book we use the Einstein convention, that is, repeated indices with one upper index and one lower index denote summation over their range.

All manifolds and mappings are supposed to be smooth, that is, differentiable of class C^∞. However, most of the results presented in the book hold under differentiability conditions of class C^5.

The quotations of formulas, theorems, etc., are made as follows:

1. Formula (1.6.3), Theorem 1.6.3, Proposition 1.6.3, Corollary 1.6.3, Remark 1.6.3 or Example 1.6.3, means that these have the number 6.3 in Chapter 1. When we do not mention the first number it is understood that we refer to a formula, etc., in the chapter where the quotation is made. Thus Theorem 6.3 means the theorem with number 6.3 in the chapter where we make the quotation.
2. Section 1.6 means Section 6 of Chapter 1, while Section 6 means that we refer to the Section 6 of the chapter where we make the quotation.

We now present the basic notations and symbols which appear frequently throughout the book.

$\mathbb{R}^m$ — the space of m–tuples $(x^1, \cdots, x^m)$ of real numbers
M — an m–dimensional smooth manifold
TM — tangent bundle of M
T_xM — tangent space of M at x
T^*M — cotangent bundle of M
T^*_xM — contangent space of M at x
$\Pi : TM \to M$ — projection map
$\theta(M)$ — zero section of TM
$TM^0 = TM \backslash \theta(M)$
M' — open submanifold of TM such that $\Pi(M') = M$ and $\theta(M) \cap M' = \phi$
$M'_x = T_xM \cap M'$
$\mathbb{F}^m = (M, M', F^*)$ — pseudo–Finsler manifold
$F(x, y) = |F^*(x, y)|^{1/2}$ — the fundamental function for $\mathbb{F}^m$
VM' — vertical vector bundle of $\mathbb{F}^m$
HM' — non–linear connection on $\mathbb{F}^m$
GM' — canonical non–linear connection on $\mathbb{F}^m$

L — Liouville vector field

V^*M' — dual vector bundle of VM'

For a vector bundle E over M we denote by E_x and E_x^* the fibres of E and of its dual E^* at $x \in M$

$T_s^r(E)_x = L_{r+s}((E_x^*)^r \times (E_x)^s, \mathbb{R})$ — the vector space of $(r+s)$–linear mappings on $(E_x^*)^r \times (E_x)^s$

$\mathcal{F}(M)$ — the algebra of smooth functions on M

$\Gamma(E)$ — the $\mathcal{F}(M)$-module of smooth sections of E

For any two vector bundles E and F over M we set $L(E^r, F) = \bigcup_{x\in M} L((E_x)^r, F_x)$, where $L((E_x)^r, F_x)$ is the vector space of all F_x–valued r–linear mappings on E_x

$[X, Y]$ — Lie bracket of vector fields X and Y

∇_X — covariant derivative with respect to X

L_X — Lie derivative with respect to X

Ω — curvature form of a linear connection on a vector bundle

R — curvature tensor field of a linear connection on a manifold

T — torsion tensor field of a linear connection on a manifold

$\dfrac{\delta}{\delta x^i} = \dfrac{\partial}{\partial x^i} - N_i^j \dfrac{\partial}{\partial y^j}$; (N_i^j) being the local coefficients of a non–linear connection HM'

$\dfrac{\delta^*}{\delta^* x^i} = \dfrac{\partial}{\partial x^i} - G_i^j \dfrac{\partial}{\partial y^j}$; (G_i^j) being the local coefficients of the canonical non–linear connection GM'

$Q : \Gamma(TM') \to \Gamma(TM')$, associate almost product structure to a non–linear connection

h and v — projection morphisms of TM' on HM' and VM' respectively

$S_{|i}$ — horizontal covariant derivative of S

$S_{||i}$ — vertical covariant derivative of S

$R_{a\ ij}^{\ b}, P_{a\ ij}^{\ b}, S_{a\ ij}^{\ b}$ — the horizontal, mixed and vertical curvatures of a vectorial Finsler connection

$R_{i\ jk}^{\ h}, P_{i\ jh}^{\ h}, S_{i\ jk}^{\ h}$ — the h–curvature, hv–curvature and v–curvature of a Finsler connection

$\sum_{(i,j,k)}$ — cyclic sum with respect to (i, j, k)

$\mathcal{A}_{(ij)}$ — the interchange of indices i, j and subtraction, as follows:

$$\mathcal{A}_{(ij)}\{S_{ik}T_j\} = S_{ik}T_j - S_{jk}T_i$$

BFC, **FC***, and **RFC** — Berwald connection, Cartan connection, and Rund connection respectively

$g_{i\ j}^{\ k}$ — the local components of the Cartan tensor field

$S_{|*i}$ and $S_{||*i}$ — the horizontal and vertical covariant derivatives of S with respect to **FC***

$S_{|_{\mathbf{b}}i}$ and $S_{||_{\mathbf{b}}i}$ — the horizontal and vertical covariant derivatives of S with respect to **BFC**
$S_{|_{\mathbf{r}}i}$ and $S_{||_{\mathbf{r}}i}$ — the horizontal and vertical covariant derivatives of S with respect to **RFC**
g — pseudo–Finsler metric of $\mathbb{F}^m$, that is, pseudo–Riemannian metric on VM' induced by F^*
a — angular metric of $\mathbb{F}^m$
G — Sasaki–Finsler metric on M'
$VM'^{\perp}$ — pseudo–Finsler normal bundle
IFC — induced Finsler connection on a pseudo–Finsler submanifold
$\mathbb{F}^m = (M, M', F^*)$ of $\widetilde{\mathbb{F}}^{m+n} = (\widetilde{M}, \widetilde{M}', \widetilde{F}^*)$ means $\mathbb{F}^m$ is a pseudo–Finsler submanifold of $\widetilde{\mathbb{F}}^{m+n}$
$\widetilde{FC}$ — vectorial Finsler connection on $V\widetilde{M}'_{|M'}$
NFC — normal Finsler connection on $\mathbb{F}^m$
B — second fundamental form of $\mathbb{F}^m$
H — h–second fundamental form of $\mathbb{F}^m$ -
V — v–second fundamental form of $\mathbb{F}^m$
A_W — shape operator (Weingarten operator)
A^h_W — h–shape operator
A^v_W — v–shape operator
$\overset{\mathrm{c}}{IFC}$, $\overset{\mathrm{b}}{IFC}$ and $\overset{\mathrm{r}}{IFC}$ — Finsler connections induced on $\mathbb{F}^m$ by $\widetilde{\mathbf{FC}^*}$, $\widetilde{\mathbf{BFC}}$ and $\widetilde{\mathbf{RFC}}$ of $\widetilde{\mathbb{F}}^{m+n}$ respectively
$\mathbf{n}^a$ — normal curvatures of a pseudo–Finsler submanifold

REFERENCES

Abate, M. and Patrizio, G.

1. *Finsler Metrics – A Global Approach*, Lecture Notes in Math., 1591, Springer–Verlag, Berlin, 1994.

Akbar–Zadeh, H.

1. Les espaces de Finsler et certaines de leurs généralisations, *Ann. Sci. École Norm. Sup.*, (3), 80, (1963), 1–79.
2. Sur les espaces de Finsler à courbures sectionelles constantes, *Bull. Acad. Roy. Bel. cl. Sci.*, (5), 74, (1988), 281–322.

Antonelli, P.L., Ingarden, R.S. and Matsumoto, M.

1. *The Theory of Sprays and Finsler Spaces with Applications in Physics and Biology*, Kluwer Acad. Publish., Dordrecht, 1993.

Asanov, G.S.

1. *Finsler Geometry, Relativity and Gauge Theories*, D. Reidel, Dordrecht, 1985.
2. *Two–Dimensional Finsler Spaces*, Univ. Athens, Sem. P. Zevros, Memo. Vol. A. Kawaguchi, Athens, 1990.

Bao, D., Chern, S.S. and Shen, Z.

1. On the Gauss–Bonnet integrand for 4–dimensional Landsberg spaces, *Contemporary Math.*, 196, (1996), 15–26.
2. Finsler geometry over the reals, *Contemporary Math.*, 196, (1996), 3–13.

Barthel, W.

1. Über die Minimalflächen in gefaserten Finslerräumen, *Ann. di Mat.*, 36, (1954), 159–190.
2. Nichtlineare Zusammenhänge und deren Holonomiegruppen, *J. Reine Angew. Math.*, 212, (1963), 120–149.

Beem, J.K.

1. Indefinite Finsler spaces and timelike spaces, *Canad. J. Math.*, 22, (1970), 1035–1039.
2. Motions in two dimensional indefinite Finsler spaces, *Indiana Univ. Math. J.*, 21, (1971/1972), 551–555.

3. On the indicatrix and isotropy group in Finsler spaces with Lorentz signature, *Atti Accad. Naz. Lincei, Rend. Cl. Sci. Fis. Mat. Natur.*, (8), 54, (1973), 385–392.
4. Characterizing Finsler spaces which are pseudo–Riemannian of constant curvature, *Pacific J. Math.*, 64, (1976), 67–77.

Bejancu, A.
1. Vectorial Finsler connections and theory of Finsler subspaces, *Sem. Geometry and Topology, Timisoara*, 1985, 38p.
2. Geometry of Finsler subspaces (II), *Bul. Inst. Politehnic Iasi, S.* 1, 35, (1985), 67–73.
3. Geometry of Finsler subspaces (I), *An. St. Univ. "Al. I. Cuza" Iasi*, 32, (1986), 69–83.
4. Geometry of vertical vector bundle and its applications, *Math. Rep. Toyama Univ.*, 10, (1987), 133–168.
5. Structure Equations for Riemann–Finsler subspaces, *C.R. Acad. Bulg. Sci.*, 40, (1987), 37–40.
6. A new viewpoint in geometry of Finsler subspaces, *Bul. Inst. Politehnic Iasi, S.* 1, 33, (1987), 13–19.
7. Special immersions of Finsler spaces, *Stud. Cercet. Mat.*, 39, (1987), 463–487.
8. *Finsler Geometry and Applications*, Ellis Horwood, New York, 1990.
9. Null hypersurfaces of Finsler spaces, *Houston J. Math.*, 22, (1996), 547–558.

Bejancu, A. and Deshmukh, S.
1. The transversal vector bundle of a lightlike Finsler submanifold, *Arab J. Math. Sci.*, 3, (1997), 37–51.
2. On the geometry of curves in Finsler manifolds, *Publ. Math. Debrecen*, 53, (1998), 293–307.

Bejancu, A. and Farran, H.R.
1. On the vertical bundle of a pseudo–Finsler manifold, *Internat. J. Math. and Math. Sci.*, 22, No. 3, (1999), 637–642.
2. A comparison between the induced and the intrinsic Finsler connections on a Finsler submanifold, *Algebras, Groups and Geometries*, 16, (1999), 11–22.
3. A geometric characterization of Finsler manifolds of constant curvature $K = 1$, *Internat. J.Math. and Math. Sci.*, 23, No 7, (2000), 1–9.
4. Finsler manifolds of constant curvature and generalized Landsberg manifolds, *to appear.*
5. Generalized Landsberg manifolds of scalar curvature, *to appear in Bull. Korean Math. Soc.*

6. Finsler submanifolds of C–reducible Finsler manifolds, *to appear.*

Berwald, L.

1. Untersuchung der Krümmung allgemeiner metrischer Räume auf Grund des in ihnen herrschenden Parallelismus, *Math. Z.*, 25, (1926), 40–73.
2. Über zweidimensionale allgemeine metrische Räume, *I, II, J. Reine Angew. Math.*, 156, (1927), 191–210, 211–222.
3. Über die Hauptkrümmungen einer Fläche im dreidimensionalen Finslerschen Raum, *Monatsh. Math. Phys.*, 43, (1936), 1–14.
4. On Finsler and Cartan Geometries III; Two dimensional Finsler spaces with rectilinear extremals, *Ann. of Math.*, 42, (1941), 84–112.
5. Über Finslersche und Cartansche Geometrie IV. Projectivkrümmung allgemeiner affiner Räume und Finslersche Räume skalarer Krümmung, *Ann. of Math.*, (2), 48, (1947), 755–781.

Blair, D.

1. *Contact Manifolds in Riemannian Geometry*, Lecture Notes in Math., 509, Springer–Verlag, Berlin, 1976.

Brickell, F.

1. A theorem on homogeneous functions, *J. London Math. Soc.*, 42, 1967, 325–329.

Brown, G.M.

1. A study of tensors which characterize a hypersurface of a Finsler space, *Canad. J. Math.*, 20, (1968), 1025–1036.
2. Gaussian curvature of a subspace in a Finsler space, *Tensor, N.S.*, 19, (1968), 195–202.

Carmo, M. do.

1. *Riemannian Geometry*, Birkhäuser, Boston, 1992.

Cartan, E.

1. *Les Espaces de Finsler*, Actualités 79, Hermann, Paris, 1934.

Chen, B.Y.

1. *Geometry of Submanifolds*, Marcel Dekker, New York, 1973.

Chevalley, C.

1. *Theory of Lie Groups*, Princeton Univ. Press, 1946.

Comic, I.

1. The induced curvature tensors of a subspace in a Finsler space, *Tensor, N.S.*, 23, (1972), 21–34.
2. The intrinsic curvature tensors of a subspace in a Finsler space, *Tensor, N.S.*, 24, (1972), 19–28.
3. Relations between induced and intrinsic curvature tensors of a subspace in a Finsler space, *Mat. Vesnik*, 29, (1977), 65–72.

Dajczer, M., Antonucci, M., Oliveira, G., Lima–Filho, P. and Tajeiro, R.

1. *Submanifolds and Isometric Immersions*, Publish or Perish, Houston, Texas, 1990.

Davies, E.T.

1. Subspaces of a Finsler space, *Proc. London Math. Soc.*, 49, (1945), 19–39.

Dhawan, M. and Prakash, N.

1. Generalizations of Gauss–Codazzi equations in a subspace imbedded in a Finsler manifold, *Tensor, N.S.*, 15, (1964), 159–164.

Dragomir, S.

1. Submanifolds of Finsler spaces, *Conf. Sem. Mat. Univ. Bari*, 217, (1986), 15p.
2. Cauchy–Riemann submanifolds of Kaehlerian Finsler spaces, *Collect. Math.*, 40, (1989), 225–240.

Eliopoulos, H.A.

1. Subspaces of a generalized metric space, *Canad. J. Math.*, 11, (1959), 235–255.
2. A generalized metric space for electromagnetic theory, *Acad. Roy. Belg. Bull. Cl. Sci.*, (5), 51, (1965), 986–995.

Finsler, P.

1. *Über Kurven and Flächen in Allgemeinen Räumen*, Dissertation, Göttingen, 1918, Verlag Birkhäuser, Basel, 1951.

Haimovici, M.

1. Formules fondamentales dans la théorie des hypersurfaces d'un espace de Finsler, *C.R. Acad. Sci. Paris*, 198, (1934), 426–427.
2. Les formules fondamentales dans la théorie des hypersurfaces d'un espace général, *Ann. Sci. Univ. Jassy*, 20, (1935), 39–58.

3. Sulle superficie totalmente geodetiche negli spazi di Finsler, *Rend. Acad. Naz. Lincei*, 27, (1938), 633–641.
4. Variétés totalement extrémales et variétés totalement géodésiques dans les espaces de Finsler, *Ann. Sci. Univ. Jassy*, 25, (1939), 559–644.

Hashiguchi, M., Hōjō, S. and Matsumoto, M.
1. On Landsberg spaces of two dimensions with (α, β)–metric, *J. Korean Math. Soc.*, 10, (1973), 17–26.

Hassan, B.T.
1. Hypersurfaces of a Minkowski space, *Tensor, N.S.*, 41, (1984), 1–9.

Hombu, H.
1. Die Krümmungs theorie in Finslerschen Raume, *J. Fac. Sci. Hokkaido Univ.*, 5, (1936), 67–94.

Horváth, J.I.
1. Contribution to Stephenson–Kilmister's unified theory of gravitation and electromagnetism, *Nuovo Cimento*, (10), 4, (1956), 571–576.
2. New geometrical methods of the theory of physical fields, *Nuovo Cimento*, (10), 9, (1958), Suppl., 444–496.

Ingarden, R.S.
1. On the geometrically absolute optical representation in the electron microscope, *Trav. Soc. Sci. Lettr. Wroclaw B*, 45, (1957), 60 p.

Ingarden R.S. and Tamássy, L.
1. The point Finsler spaces and their physical applications to electron optics and thermodynamics, *Math. Comput. Modelling*, 20, No. 4/5, (1994), 93–107.

Kawaguchi, A.
1. On the theory of non–linear connections I. Introduction to the theory of general non–linear connections, *Tensor, N.S.*, 2, (1952), 123–142.
2. On the theory of non–linear connections II. Theory of Minkowski spaces and of non–linear connections in a Finsler space, *Tensor, N.S.*, 6, (1956), 165–199.

Kikuchi, S.
1. On the theory of subspace in a Finsler space, *Tensor, N.S.*, 2, (1952), 67–79.
2. On some special Finsler spaces, *Tensor, N.S.*, 19, (1968), 238–240.

Kobayashi, S. and Nomizu, K.

1. *Foundations of Differential Geometry*, Vol. I, Interscience, New York, 1963.

Kropina, V.K.

1. On projective Finsler spaces with a metric of some special form, *(Russian), Naucn. Dokl. Vyss. Skoly. Fiz.-Mat. Nauki*, 2, (1959), 38–42.

Landsberg, G.

1. Über die Totalkrümmung, *Jber. Deutsch. Math. Verein.*, 16, (1907), 36–46.
2. Krümmungstheorie und Variationsrechnung, *Jber. Deutsch. Math. Verein*, 16, (1907), 547–551.
3. Über die Krümmung in der Variationsrechnung, *Math. Ann.*, 65, (1908), 313–349.

Matsumoto, M.

1. On C–reducible Finsler spaces, *Tensor, N.S.*, 24, (1972), 29–37.
2. On Finsler spaces with Randers metrics and special forms of important tensors, *J. Math. Kyoto Univ.*, 14, (1974), 477–498.
3. Differential–geometric properties of indicatrix bundle over Finsler space, *Publ. Math. Debrecen*, 28, (1981), 281–291.
4. The induced and intrinsic Finsler connections of a hypersurface and Finslerian projective geometry, *J. Math. Kyoto Univ.*, 25, (1985), 107–144.
5. Theory of Y–extremal and minimal hypersurfaces in a Finsler space, *J. Math. Kyoto Univ.*, 26, (1986), 647–665.
6. *Foundations of Finsler Geometry and Special Finsler Spaces*, Kaiseisha Press, Saikawa, Ōtsu, 1986.
7. Theory of Finsler spaces with (α, β)–metric, *Reports in Math. Physics*, 31, (1992), 43–83.
8. Remarks on Berwald and Landsberg spaces, *Contemporary Math.*, 196, (1996), 79–81.

Matsumoto, M. and Hōjō, S.

1. A conclusive theorem on C–reducible Finsler spaces, *Tensor, N.S.*, 32, (1978), 225–230.

Matsumoto, M. and Miron, R.

1. On an invariant theory of the Finsler spaces, *Periodica Math. Hungar.*, 8, (1977), 73–82.

Miron, R.

1. Sur les connexions pseudo–euclidiennes des espaces de Finsler à métrique indéfinie *(Romanian), Acad. R.P. Romane Fil. Iasi, St. Mat.*, 12, (1961), 125–134.
2. A non–standard theory of hypersurfaces in Finsler spaces, *An. St. Univ. "Al. I. Cuza " Iasi*, 30, (1984), 35–53.
3. *The Geometry of Higher–Order Finsler Spaces*, Hadronic Press, Inc., 1998.

Miron, R. and Anastasiei, M.

1. *The Geometry of Lagrange Spaces: Theory and Applications*, Kluwer Acad. Publish., Dordrecht, 1994.

Miron, R. and Bejancu, A.

1. A new method in geometry of Finsler subspaces, *An. St. Univ. "Al. I. Cuza" Iasi*, 30, (1984), 55–59.
2. A non–standard theory of Finsler subspaces, *Colloq. Math. Soc. J. Bolyai, Vol.* 46, (1984), 815–851.

Moor, A.

1. Über die Torsions und Krümmungsinvarianton der dreidimensionalen Finslerschen Räume, *Math. Nachr.*, 16, (1957), 85–99.

Numata, S.

1. On Landsberg spaces of scalar curvature, *J. Korean Math. Soc.*, 12, (1975), 97–100.

O'Neill, B.

1. *Semi–Riemannian Geometry with Applications to Relativity*, Academic Press, New York, 1983.

Oproiu, V.

1. Some properties of the tangent bundle related to Finsler geometry, *Proc. Nat. Sem. Finsler Spaces, Brasov*, 1980, 195–207.

Prakash, N. and Behari, R.

1. Generalizations of Codazzi's equations in a subspace imbedded in a Finsler manifold, *Proc. Nat. Inst. Sci. India*, 26, (1960), 532–540.

Randers, G.

1. On an asymmetric metric in the four–space of general relativity, *Phys. Rev.*, (2), 59, (1941), 195–199.

Rapcsák, A.

1. Eine neue Characterisierung Finslerscher Räume skalarer und Konstanter Krümmung und Projectiv–ebene Räume, *Acta Math. Acad. Sci. Hungar.*, 8, (1957), 1–18.

Rastogi, S.C.

1. Submanifolds of a Finsler manifold, *Tensor, N.S.*, 30, (1976), 140–144.

Rund, H.

1. The theory of subspaces of a Finsler space, Part I, *Math. Z.*, 56, (1952), 363–375.
2. The theory of subspaces of a Finsler space, Part II, *Math. Z.*, 57, (1953), 193–210.
3. *The Differential Geometry of Finsler Spaces*, Grundlehr. Math. Wiss., 101, Springer, Berlin, 1959.
4. Curvature properties of hypersurfaces of Finsler and Minkowskian spaces, *Tensor, N.S.*, 14, (1963), 226–244.
5. The intrinsic and induced curvature theories of subspaces of a Finsler space, *Tensor, N.S.*, 16, (1965), 294–312.

Sakaguchi, T.

1. Subspaces in Finsler space, *Mem. Nat. Def. Acad.*, 28, (1988), 1–37.
2. Subspaces in Finsler space, II, *Mem. Nat. Def. Acad.*, 29, (1989), 1–9.

Shen, Z.

1. On Finsler geometry of submanifolds, *Math. Ann.*, 311, (1998), 549–576.
2. *Differential Geometry of Spray and Finsler Spaces*, book to appear.

Shibata, C., Shimada, H., Azuma, M. and Yasuda, H.

1. On Finsler spaces with Randers metric, *Tensor, N.S.*, 31, (1977), 219–226.

Shibata, C., Singh, U.P. and Singh, A.K.

1. On induced and intrinsic theories of hypersurfaces of Kropina spaces, *J. Hokkaido Univ. Ed.*, 34, (1983), 1–11.

Sinha, B.B.

1. A generalization of Gauss and Codazzi's equations for intrinsic curvature tensor, *Indian J. Pure Appl. Math.*, 2, (1971), 270–274.

Spivak, M.

1. *A Comprehensive Introduction to Differential Geometry*, Vol. 1–4, Publish or Perish, Inc., Houston, Texas, 1979.

Stephenson, G.

1. Affine field structure of gravitation and electromagnetism, *Nuovo Cimento*, (9), 10, (1953), 354–355.
2. La géométrie de Finsler et les théories du champ unifié, *Ann. Inst. Poincaré*, 15, (1956), 205–215.

Stephenson, G. and Kilmister, G.W.

1. A unified field theory of gravitation and electromagnetism, *Nuovo Cimento*, (9), 10, (1953), 230–235.

Sternberg, S.

1. *Lectures on Differential Geometry*, Chelsea Publ. Comp., New York, 1983.

Taylor, J.H.

1. A generalization of Levi–Civita's parallelism and the Frenet formulas, *Trans. Amer. Math. Soc.*, 27, (1925), 246–264.

Teodorescu, N.

1. Sur les géodésiques de longeur nulle de certain éléments linéaires finslériens, *Bull. École Polytech. Bucharest*, 12, (1941), 9-16.

Vaisman, I.

1. Sur quelques formules du calcul de Ricci global, *Comm. Math. Helvetici*, 41, (1966), 73–87.

Varga, O.

1. Zur Differentialgeometrie de Hyperflächen in Finslerschen Räumen, *Deutsch. Math.*, 6, (1941), 192–212.
2. Über den inneren und induzierten Zusammenhang für Hyperflächen in Finslerschen Räumen, *Publ. Math. Debrecen*, 8, (1961), 208–217.
3. Über Hyperflächen konstanter Normalkrümmung in Minkowskischen Räumen, *Tensor, N.S.*, 13, (1963), 246–250.
4. Hyperflächen mit Minkowskischer Massbestimmung in Finslerräumen, *Publ. Math. Debrecen*, 11, (1964), 301–309.
5. Hyperflächen konstanter Normalkrümmung in Finslerschen Räumen, *Math. Nachr.*, 38, (1968), 47–52.

Verma, M.

1. The intrinsic and induced curvature theories of subspaces of a generalized Finsler space, *Math. Nachr.*, 51, (1971), 35–41.

Wegener, J.M.

1. Untersuchungen über Finslerschen Räume, *Lotos Prag*, 84, (1936), 4–7.
2. Hyperflächen in Finslerschen Räumen als Transversalflächen einer Schar von Extremalen, *Monatsh. Math. Phys.*, 44, (1936), 115–130.

Wolf, J.A.

1. *Spaces of Constant Curvature*, McGraw–Hill, New York, 1967.

Yano, K. and Kon, M.

1. *Structures on Manifolds*, World Scientific, Singapore, 1984.

Yasuda, H.

1. On TM–connections of a Finsler space and the induced TM–connections on its hypersurfaces, *Ann. Rep. Asahikawa Med. Coll.*, 6, (1984), 1–38.
2. A theory of subspaces in a Finsler space, *Ann. Rep. Asahikawa Med. Coll.*, 8, (1987), 1–43.
3. Special subspaces in a Finsler space, *Ann. Rep. Asahikawa Med. Coll.*, 9, (1988), 1–14.

Yasuda, H. and Shimada, H.

1. On Randers spaces of scalar curvature, *Rep. Math. Phys.*, 11, (1977), 347–360.

SUBJECT INDEX